GCSE
Physics
for You

Keith Johnson

HUTCHINSON
London Melbourne Sydney Auckland Johannesburg

Hutchinson Education

An imprint of Century Hutchinson Ltd
62–65 Chandos Place, London WC2N 4NW

Century Hutchinson Australia Pty Ltd
PO Box 496, 16–22 Church Street, Hawthorn,
Victoria 3122, Australia

Century Hutchinson New Zealand Ltd
PO Box 40–086, Glenfield, Auckland 10, New Zealand

Century Hutchinson South Africa (Pty) Ltd
PO Box 337, Bergvlei, 2012 South Africa

First published 1980
Reprinted 1980, 1981, 1982, 1983 (twice), 1984, 1985, 1986
Revised edition 1986, 1987

© Keith and Ann Johnson 1978, 1980, 1986

British Library Cataloguing in Publication Data

Johnson, Keith
    GCSE physics for you.
    1. Physics — Experiments
    I. Title
    530′.0724   QC33
     ISBN 0–09–167491–3

Printed in Great Britain by
Butler & Tanner Ltd, Frome and London

'What is the use of a book,' thought Alice, 'without pictures or conversations?'

Lewis Carroll, *Alice in Wonderland*

I do not know what I may appear to the world, but to myself I seem to have been only a boy playing on the seashore, and diverting myself in now and then finding a smoother pebble or a prettier shell than ordinary, while the great ocean of truth lay all undiscovered before me.

Sir Isaac Newton

Introduction

Physics For You is designed to introduce you to the basic ideas of physics and show you how these ideas can help to explain the world in which we live.

This book is based on the successful GCE book of the same name, but many extra pages and questions have been added to cover the requirements of the GCSE Examination.

The book is carefully laid out so that each new idea is introduced and developed on a single page or on two facing pages. Words have been kept to a minimum and as straightforward as possible. Pages with a red band in the top corner are the more difficult pages and may be left out at first.

Throughout the books there are many simple experiments for you to do. Plenty of guidance is given on the results of these experiments, in case you find the work difficult.

Each important fact or formula is printed in **heavy type** or is in a box. There is a summary of important facts at the end of each chapter.

At the back of the book there is advice on practical work, careers, revision and examination techniques as well as some help with mathematics.

Questions at the end of a chapter range from simple fill-in-a-missing-word sentences (useful for writing notes in your notebook) to more difficult problems that will need more thought. In calculations, simple numbers have been used to keep the arithmetic as straightforward as possible.

At the end of each main topic you will find a section of further questions taken from actual examination papers.

Multiple choice questions are included to give you practice in this kind of question. SI units are used throughout, often written in full.

Throughout the book, cartoons and rhymes are used to explain ideas and ask questions for you to answer. In many of the cartoons, Professor Messer makes a mistake because he does not understand Physics very well. Professor Messer does not think very clearly, but I expect you will be able to see his mistakes and explain where he has gone wrong. Here I would like to thank my wife, Ann, for her constant encouragement and help with the many diagrams and cartoons.

I hope you will find Physics interesting as well as useful. Above all, I hope you will enjoy **Physics For You**.

Keith Johnson

Professor Messer gets in messes,
Things go wrong when he makes guesses.
As you will see, he's not too bright,
It's up to you to put him right.

Contents

Waves, Light and Sound

Electricity and Magnetism

Nuclear Physics

Extra Sections

chapter 1

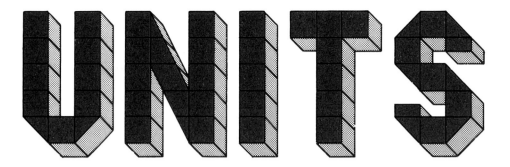

Length

Length is measured in a unit called the *metre* (often shortened to m).
A door knob is usually about 1 metre from the ground; doorways are about 2m high.

For shorter distances we often use centimetres (100cm = 1 metre) or millimetres (1000mm = 1 metre).

Experiment 1.1

a) Look at a metre rule. Which marks are centimetre and which are millimetre?

b) It is useful to know the length of your handspan. Mine is 22cm (0.22m); what is yours?

c) Use the metre rule to measure
   – the length of your foot
   – the length of a long stride
   – your height.
   In each case write down the answer in millimetres and also in metres.

When measuring in Physics we try to do it as accurately as we can.
Professor Messer is trying to measure the length of a block of wood with a metre rule but he has made at least six mistakes.

How many mistakes can you find?

Experiment 1.2
Measure the length of a block of wood taking care not to make any of the Professor's mistakes.

Mass

If you buy a bag of sugar in a shop, you will find the *mass* of sugar marked on the bag. It is written in *grams* (g) or in *kilograms* (kg). 'Kilo' always means a thousand, so 1 kilogram = 1000 grams.

The mass of this book is about $\frac{3}{4}$ kg. People often get confused between mass and weight, but they are *not* the same (see pages 85 and 88).

Experiment 1.3
Look at and feel some masses labelled 1 kg, 2 kg, 5 kg and 1 g.

In Maths and in Physics, a 'k'
Means a thousand of whatever
                              you say
For grams and for metres
And even, for teachers,
The size of their annual pay.

Time

In Physics, time is always measured in *seconds* (sometimes shortened to s).

You can count seconds very roughly, without a watch, by saying at a steady rate: ONE (thousand) TWO (thousand) THREE (thousand) FOUR .....

Experiment 1.4
Use a stopclock or stopwatch to measure the time for a complete swing of a pendulum or the beating of your heart.
What is the time taken for 100 of your heartbeats?
What is the time for one heartbeat?
By how much does it change if you run upstairs?

All the other units you will meet in this book are based on the metre, the kilogram and the second. They are called *SI units*.

Very large and small numbers

For very large or very small numbers, we sometimes use a shorthand way of writing them, by counting the number of zeros (see also page 403).

For example:

a) 1 million = 1 000 000 (6 zeros) = 10^6

b) 0·000 001 = $\frac{1}{1\ 000\ 000}$ (1 millionth) = 10^{-6}

In this shorthand way, write down:
one thousand, one thousandth, 10 million, one hundredth.

Approximate length of time in seconds	Events
10^{18}	Expected lifetime of the Sun
10^{17}	Age of the Earth
10^{15}	Time since the dinosaurs lived
10^{13}	Time since the earliest man
10^{10}	Time since Isaac Newton lived
10^9	Average human life span
10^7	A school term
10^5	One day
10^0	One second
10^{-2}	Time for sound to cross a room
10^{-7}	Time for electron to travel down a T.V. tube
10^{-8}	Time for light to cross a room
10^{-11}	Time for light to pass through spectacles
10^{-22}	Time for some events inside atoms

chapter 2

I wish I could find some use for all that energy!

Energy can exist in different forms, as you can see in the cartoon.

People get their energy from the *chemical energy* in food. Cars run on the chemical energy in petrol. The firework in the cartoon has chemical energy which it converts into *thermal* energy (*heat*) and *light* and *sound* energy when it explodes.

The moving pellets from the catapult and the moving people all have movement energy, called *kinetic energy*.

The bucket over the door and the stretched elastic in the catapult both have stored energy called *potential energy*, which they are going to convert to kinetic energy. Kinetic energy and potential energy are both called mechanical energy.

The television set is taking in *electrical energy* and converting it to thermal energy and light and sound energy.

Another form of energy is *nuclear energy*, which is used in nuclear power stations.

These forms of energy are shown in the diagram on the opposite page (see also page 128).

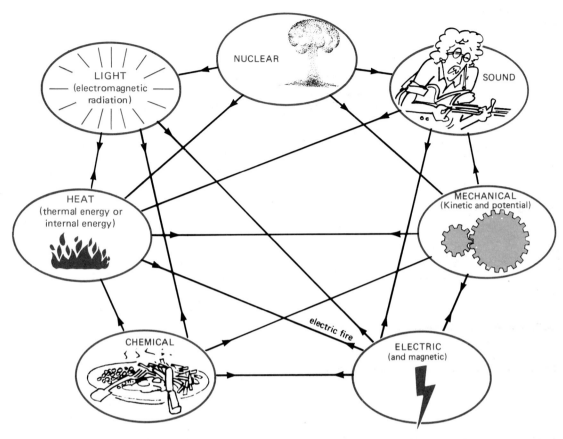

The connecting lines on the diagram show the different ways
that energy can be converted from one form to another.

See if you can decide the energy changes in the following objects.

For example:

A firework converts *chemical* energy to *thermal* energy, *light and
sound* energy. Copy and complete these sentences.

1. A T.V. set converts .... energy to .... energy.
2. A match converts .... energy to .... energy.
3. A light bulb converts .... energy to .... energy.
4. A catapult converts .... energy to .... energy.
5. A falling bucket converts ..... energy to .... energy.
6. An electric fire converts ..... energy to .... energy.
7. A human body converts ..... energy to .... energy.
8. A microphone converts ...... energy to .... energy.
9. An atomic bomb converts .... energy to .... energy.
10. A car engine converts .... energy to .... energy.

You can learn more about energy changes in Chapter 17.

Energy changes

In the diagram on the previous page (page 9), one energy change has been labelled 'electric fire'. Copy out the diagram into your book and then add the correct label to each arrow. Use words from the following list: coal fire, electric motor, steam engine, loudspeaker, car engine, atom bomb (on four arrows), battery, dynamo, very hot object, friction, vibrations, microphone, fluorescent lamp, thermocouple, plants, glow-worm, photographer's light-meter, sound-absorbing material, light-absorbing material, electroplating.

Saving money

Energy costs money and we use large amounts of energy/money each day.

There are several ways of saving energy/money in your home (and making your home more comfortable at the same time). The table shows the time taken before they have paid for themselves and start to show a 'profit'.

How well is your home insulated?
How well is your school insulated? Write a list of recommendations.

Energy is measured in units called *joules*.
The joule is a small unit. To lift this book through a height of 10 cm needs about 1 joule.
When you walk upstairs you use over 1000 joule.
The diagram below shows the energy (in joule) involved in different events.
(Remember: 10^5 = number with 5 zeros = 100 000 and $10^{-5} = \frac{1}{100\,000}$)

Method	'Profit' times*
lagging the hot water cylinder	less than a month
draught excluders	a few weeks
lagging the loft (see page 51)	about 3 years
wall cavity insulation	4-7 years
double glazing in windows	about 10 years

* all these times are shorter if you get a government grant

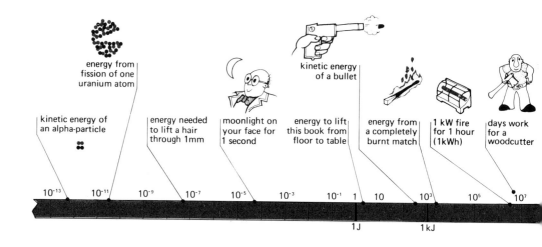

The energy crisis

We are only just beginning to realise fully that our planet Earth is a spaceship with *limited* food and fuel (and we are taking on more passengers each year as the population increases).

Our supplies of energy cannot last for ever.
Oil and *natural gas* will be the first to disappear. If the whole world used oil at the rate it is used in America and Europe, our oil supplies would end in about 4 years!
As it is, the world's oil supplies might last for about 30 years.

How old will you be then? In what ways will your life be different without oil (and therefore without petrol and plastics)?

Coal will last longer — perhaps 300 years with careful mining. We might see (improved) steam engines on the railways again!

Nuclear energy might help for a while — but it causes problems due to the very dangerous radioactive waste that is produced. Also, each power station lasts only about 30 years and cannot be dismantled because of the radioactivity.

We *waste* huge amounts of energy. It takes over 5 million joules of energy to make one fizzy-drink-can and we throw away 700 million of them each year! Making paper and steel uses particularly large amounts of energy, but very little is re-cycled.

We *must* find new ways of obtaining energy. The Sun's energy is free but it is not easy to capture it. Scientists are experimenting with new sources of energy (see next page). They will probably not give enough energy for the future. Our main hope is that the H-bomb *fusion* process (page 381) will eventually be controlled.

Power stations are wasteful. The overall efficiency (from power station to your home) is about 25%. ¾ of the energy is entirely wasted! Conversion to community steam-heating could raise the efficiency to 70%.

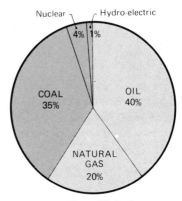

Energy supplies of Britain at present. What will happen when the oil and natural gas run out?

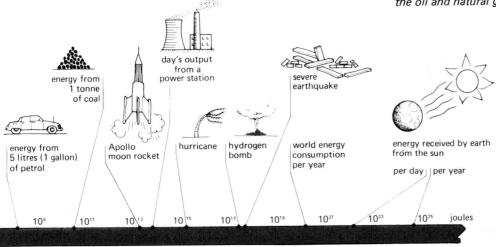

New sources of energy

Some sources of energy are *renewable*. They are not used up like coal or oil.
Physicists and engineers are working hard to develop machines to use these sources of energy.

Source of energy	Original source
solar	sun
biomass	sun
wind	sun
waves	sun
hydro-electric	sun
tides	moon
geothermal	earth

Solar energy
The earth receives an enormous amount of energy directly from the sun each day, but we use very little of it. Some homes have solar panels on the roof (see page 58). In hot countries solar ovens can be used for cooking (see page 56).

Space ships and satellites use *solar cells* to convert sunlight into electricity.
Covering part of the Sahara Desert with solar cells would produce energy but it would be very expensive. To equal one modern power station you would need 40 square kilometres of solar cells.

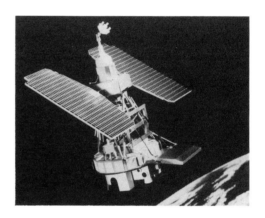

Biomass
Some of the sunlight shining on the earth is trapped by plants, as they grow. We use this *biomass* when we eat plants or when we burn wood.

In Brazil they grow sugar cane and then use the sugar to make alcohol. The alcohol is then used in cars, instead of petrol.

When plants rot, they can produce a gas called methane which is the same as the 'natural gas' we use for cooking. If the plants rot in a closed tank, called a *digester*, the gas can be piped away and used as fuel for cooking.
This is often used in China and India.

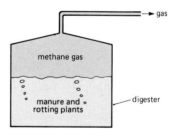

Wind energy
This energy also comes from the sun, because winds are caused by the sun heating different parts of the earth unequally.
Modern windmills are very efficient but we would need about 2000 very large windmills to provide as much power as one modern power station.

Wave energy

Waves are caused by the winds blowing across the sea.
They contain a lot of free energy.
One method of getting this energy might be to use
large floats which move up and down with the waves.
The movement energy could be converted to electricity
However, we would need about 20 kilometres of floats
to produce as much energy as one power station.

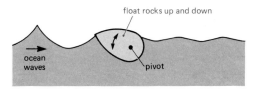

Side view of a wave energy machine

Hydro-electric energy

Dams can be used to store rain-water, and then the
falling water can be used to make electricity (see
experiment 17.6 on page 131).
This is a very useful and clean source of energy for
mountainous countries like Norway and China.

The same idea can be used to store energy from power
stations that cannot easily shut down. At night, when
demand is low, spare electricity can be used to pump
water up to a high lake. During the day, the water
can be allowed to fall back down to produce
electricity when it is needed.

North Wales pumped storage scheme from the air

Tidal energy

As the moon goes round the earth it pulls on the seas
so that the height of the tide varies.

If a dam is built across an estuary, it can have
gates which trap the water at high tide. Then at low
tide, the water can be allowed to fall back through
the dam and make electricity (see experiment 17.6).

Geothermal energy

The inside of the earth is hot (due to radioactivity,
see chapter 42). In some parts of the world (like New
Zealand) hot water comes to the surface naturally. In
other countries cold water is pumped down very deep
holes and hot water comes back to the surface.

A tidal power station in France

Summary

Energy exists in several different forms:
chemical, electric, magnetic, kinetic, potential,
sound, nuclear, electromagnetic radiation
(including light) and heat (also called internal
energy or thermal energy)
Energy can be changed from one form to another.
It is measured in joules.
Some sources of energy are non-renewable and will
be used up: coal, oil, gas, nuclear.
Other sources are renewable: solar, biomass,
wind, wave, hydro-electric, tidal and geothermal.
See also chapter 17.

chapter 3

MOLECULES

Do you know that as you are reading this sentence you are punched on the nose more than 100 billion billion times ! ! ?

This is because the air is made up of billions and billions of tiny particles called *molecules* — rather like the title at the top of this page is made up of tiny dots. But the molecules are so small that we cannot see them even through a powerful microscope. Since each molecule is very small and light, it does not hurt as much as when a big fist hits your nose.

High speed gas

The molecules hitting your face are moving very fast. Their average speed is about 1000 m.p.h. (1600 km/hr), but with some moving much faster and some slower.

The billions and billions of air molecules around you are bouncing and colliding with each other at this very high speed. It's just as though the room is full of tiny balls bouncing everywhere and filling the whole room.

If the air is cooled down, the molecules have less energy and move more slowly.
If the air is heated up, the molecules have more energy and so move faster (and hit your nose harder).

This idea about molecules is called the *kinetic theory*.

Three states of matter

Every substance can exist in three states: *gas, liquid* and *solid.*

The air in this room is a gas. But if it is cooled down it can become liquid or even solid.

In the same way, water can be a solid (ice), a liquid, or a gas (steam):

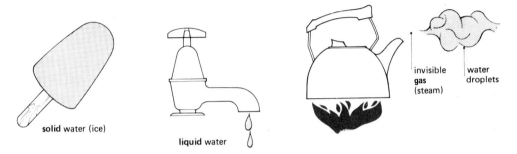

solid water (ice)

liquid water

invisible
gas
(steam)

water
droplets

In a **gas**, the molecules are moving very quickly.
In a **liquid**, the molecules are moving more slowly and are held together more closely by forces between the molecules. This is why a large volume of gas *condenses* into a smaller volume of liquid.

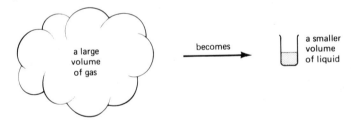

a large
volume
of gas

becomes

a smaller
volume
of liquid

These forces between molecules make the liquid have a definite size, although it can still change its shape and flow.

In a **solid**, the forces between molecules are so strong that the solid has a definite shape as well as a definite size.
There are two kinds of forces between molecules:
— *attractive* forces if molecules try to move apart
— *repulsive* forces if they try to move closer.
Normally these forces balance.
If you try to stretch a solid (and so pull the molecules apart) you can feel the attractive forces.
If you squeeze a solid (and so push its molecules closer together) you can feel the repulsive forces.
These forces are *electric* forces (see page 267).

Molecular forces are strong

15

Experiments

You will need some marbles (to act as 'molecules') and a flat box or tray with verticles sides.

Experiment 3.1   A solid

Place the marbles in the tray. Tilt the tray slightly so that the 'molecules' collect at one end.

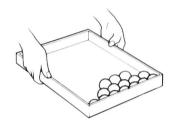

You can see that the group of 'molecules' has a definite size and shape. This is imitating a *solid* (a very cold solid). Sketch the pattern of the molecules (a regular *crystalline* pattern).
The molecules in a solid are continually vibrating. If the solid gets hotter, its molecules vibrate more.

Experiment 3.2   A liquid

Now rock the tray gently from side to side.
You can see that the group of 'molecules' now has more energy and does not have a definite shape. This is imitating a *liquid*.

Experiment 3.3   A gas

Next, with the tray flat on the table, shake the tray to and fro and from side to side to give the 'molecules' a lot more energy. You can see that the 'molecules' fill all the available space in the tray. This is imitating a *gas*.
Heat up the gas by shaking the tray faster.

Experiment 3.4   Brownian Movement

No-one has ever seen an air molecule, but here is an experiment to help us believe that they exist.

A microscope is used to look into a small glass box which contains some smoke as well as air molecules.

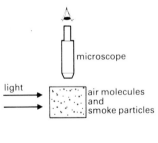

In a bright light, the smoke particles show up as bright specks which move in a continuous and jerky random movement. This is called *Brownian Movement*.
It happens because the smoke particles (like your nose) are being punched by molecules. Because the smoke particles are so light, they twist and jerk as you can see in the microscope.

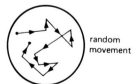

Experiment 3.5

While shaking your marbles tray to imitate a gas, drop into it a *larger* marble (to act as a particle of smoke).
Watch the random movement of the 'smoke particle' as it is punched by the 'molecules'. This is Brownian Movement.

Diffusion

When food is being cooked in the kitchen, it can be smelt in other rooms in the house, even when there are no draughts to move the air.
This is because molecules from the food are moving around at high speed and after billions of collisions eventually reach all parts of the house.
This is called *diffusion*.

Experiment 3.6   Diffusion
While shaking your marbles tray to imitate a gas, drop into one corner a coloured marble (acting as a 'smell molecule'). Watch it while you keep shaking. You will see that eventually the 'smell molecule' will diffuse to any part of the tray.

Draw a rough sketch of the random movement of the 'molecule'.

Experiment 3.7   Diffusion in a liquid
Fill a tall beaker with water and leave it for a while so that it becomes still and at room temperature.
Carefully drop in a single crystal of potassium permanganate. When this dissolves it makes the water purple.
Leave the beaker where it will not be touched or knocked for several weeks and observe it.

It takes a long time for liquids to diffuse. Why is this?

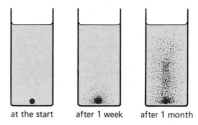

at the start   after 1 week   after 1 month

Experiment 3.8   Diffusion in a gas
Your teacher may be able to show you the diffusion of some bromine. This is a gas that you can see (it is brown), but it is very poisonous so special apparatus and care must be used: When the tap is opened, the bromine runs down and starts diffusing.

Why is the diffusion in this experiment faster than in the previous experiment?

Experiment 3.9   Diffusion into a vacuum
Your teacher may be able to repeat the bromine experiment but with the main tube completely empty of air molecules (by using a vacuum pump to take out all the molecules).
What happens this time as the tap is opened to let in the bromine? (Don't blink or you might miss it.)

Because there are no air molecules to get in the way, this shows you the speed of bromine molecules (it is about 500 m.p.h. at room temperature!).

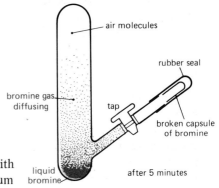

Molecular Forces in Liquids

Experiment 3.10
Look at the shape of water drops which are
dripping from a tap when it is almost turned off.

Experiment 3.11
Blow some soap bubbles and look at their shape.

Experiment 3.12
Tie a cotton thread across a wire frame as shown:
Dip it into soap solution to get a *film* of soap
solution across it.
What happens when you break the soap film on one
side of the thread?

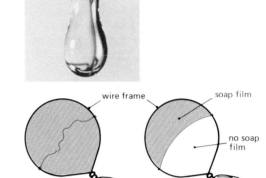

wire frame — soap film — no soap film

thread is slack   thread is pulled tight

These experiments all show that there is a surface
force or *surface tension* on the surface of
liquids. This is due to the forces attracting each
surface molecule to its neighbouring molecules.

There is a tension because the molecules on the surface are slightly
farther apart (like the molecules in a stretched elastic band).
The tension tries to reduce the surface area. This is why bubbles are
spherical (as this shape has the smallest area for a given volume).

This force of attraction between molecules of the *same* substance is
called *cohesion*. This is the force that holds a raindrop together.

The force of attraction between molecules of *different* substances is
called *adhesion*. This is the force that holds glue or paint to a wall.

Experiment 3.13
Carefully float a dry razor blade or sewing needle
on water.
The needle has not broken the surface of the water
but has stretched it like an elastic 'skin'.

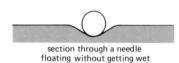

section through a needle
floating without getting wet

Some insects can walk on water because of this.
The photograph shows a 'pond-skater'.

Can you see where the water surface is pushed down
by its legs?

18

Experiment 3.14 Capillary rise

You need two *narrow* 'capillary' tubes (of
different sizes).
Place the clean glass tubes so that they dip into
a beaker of coloured water.
Watch the water rise up the narrow tubes. This is
called *capillary rise*.

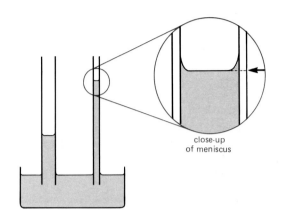

close-up
of meniscus

In which tube does it rise higher?
The water rises because the adhesion (between
water and glass molecules) is greater than the
cohesion (between water molecules).

This also explains why the *meniscus* shape at the
top of the water is curved as shown.

If you repeated this experiment with mercury, you
would find there is a capillary *fall*. This is
because the adhesion (between mercury and glass
molecules) is *less* than the cohesion (between
mercury molecules).
Which way does a mercury meniscus curve? (See the
diagram on page 101).

Using capillary rise

What happens when you use a towel to dry your face?

The water soaks up into the tiny tubes in the cloth
of the towel.
The same thing happens with nappies, paper tissues,
blotting paper and kitchen cloths.

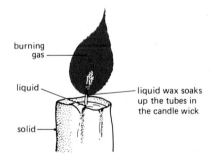

burning
gas

liquid

solid

liquid wax soaks
up the tubes in
the candle wick

The wick of a candle or a paraffin heater also uses
capillary action.
Rain water soaks through soil in the same way.

Can you invent a device to feed your house plants
from a tank of water so they are kept watered when
you are on holiday?

Preventing capillary rise

Water will soak up through the floor of a house
unless we prevent it with a sheet of plastic.

Bricks are *porous* and will allow 'rising damp'
unless a 'damp course' is included just above
ground level.

Laying a damp-proof course.

19

The Size of Molecules

A molecule is obviously very small – but how small?

Experiment 3.15   Estimating the size of a molecule
Fill a clean tray with clean water and sprinkle the surface with talcum powder. (Or use a sink, first letting water run out of the overflow for a while.)

Dip a V-shaped piece of thin wire into some olive oil to collect a small drop of oil. Keep trying until the drop is $\frac{1}{2}$ mm in diameter (d). Use a ruler and a magnifying glass to check this.

Dip the wire into the middle of the water so that the oil molecules spread out. They spread out until the oil patch is just *one molecule* thick (rather like a bag of marbles being poured into a tray).
The powder shows you the size of the oil patch.
Measure its diameter (D).

Since the volume of the oil stays constant while its shape changes, we can work out the thickness of the oil patch.
To get an *estimate* of the size, we simplify the drop to a *cube* and the patch to a *square*, as shown:

Example
If d = $\frac{1}{2}$ mm = 5×10^{-4} m and D = 25 cm = 0·25 m

Then $h = \dfrac{d^3}{D^2} = \dfrac{(5 \times 10^{-4})^3}{0\cdot25^2} = 2 \times 10^{-9}$ m

$\qquad\qquad\qquad\qquad = 2$ nanometre (nm)

This means that the length of a molecule is about one-millionth of a millimetre!
Atoms are about ten times smaller than this (10^{-10} m). The thickness of this page is equal to about one million atoms.

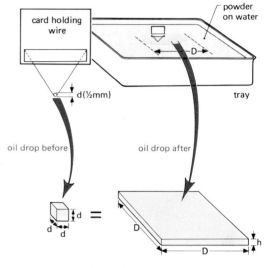

volume before = volume after

$$\therefore d^3 = D^2 h$$

$$\therefore \text{length of molecule } h = \frac{d^3}{D^2}$$

Summary

All substances can exist in 3 states:– *solid, liquid* and *gas.*
All substances are made of *molecules.*
If the substance is heated, the molecules have more energy and move faster.

In a *gas*, the molecules are moving freely at high speed.
In a *liquid*, the molecules are loosely held together by forces.

A *solid* has a definite shape because its molecular forces are very strong.

Molecules can *diffuse* through other molecules.
Brownian movement of smoke particles helps us to believe in molecules.

In liquids, molecular forces cause a 'skin' or *surface tension* and cause liquids to rise up narrow tubes.

Questions

1. Copy out and fill in the missing words:
   a) The three states of matter are . . . . . , ... and . . . . . All matter consists of tiny . . . . . travelling at . . . . . speed.
   b) A solid has a definite size and a definite . . . . . because the forces between its . . . . . are very . . . . . A liquid has weaker forces between its . . . . . and so does not have a definite . . . . .
   c) If a substance is cooled down, its molecules move at a . . . . . speed. When some molecules pass through some other molecules, we call it . . . . . The twisting and jerking of smoke particles when they are hit by air . . . . ., is called . . . . .

2. Copy out and fill in the missing words:
   The forces between the . . . . . in liquids cause a kind of 'skin' or . . . . The forces between . . . . . also cause water to . . . . . inside narrow tubes. This is called . . . . The curved surface at the top of the water is called a . . . . . This occurs because the force of adhesion is . . . . . than the force of . . . . .

3. Describe, with the aid of diagrams, how molecules, behave in
   a) solids   b) liquids   c) gases.

4. What can you say about the molecules of air in a refrigerator compared with the molecules of air in an oven?

5. Explain why
   a) solids have a definite shape but liquids flow
   b) solids and liquids have a fixed size but gases fill whatever container they are in.

6. Explain why perfume can be smelled some distance away from the person wearing it.

7. a) Explain what is meant by Brownian movement and how it helps us to believe in molecules.
   b) What difference would you expect to see in the movement if the air was cooled down?

8. Complete the rhyme and explain why you do not see this happening:
   > A rather small student called Brown,
   > Was asked why he danced up and down,
   > He said "Look, you fools,
   > It's the air . . . . .,
   > They constantly knock me around."

9. A tin can containing air is sealed. If it is then heated, what can you say about:
   a) the average speed of the molecules,
   b) how often the molecules hit the walls of the can,
   c) how hard the molecules hit the walls of the can,
   d) the air pressure inside the can?

10. Explain the following:
    a) Raindrops are almost spherical.
    b) A needle can be floated on water.
    c) The needle sinks if detergent is added.
    d) Hairs (on your head or on a paint brush) cling together when they are wet but not when they are dry.
    e) Small insects can stand on the surface of a pond without sinking.
    f) You can dry your hands with a towel but not with a sheet of polythene.
    g) Paint sticks to a wall.

11. Copy the diagram and mark the water levels in the tubes.

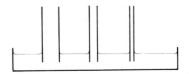

12. Bricks are full of tiny holes. Why are houses built with
    a) two walls separated by a cavity,
    b) a damp-proof layer about two bricks above ground level?

Further questions on page 78

21

chapter 4

Expansion

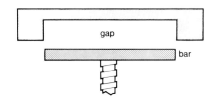

Experiment 4.1

This metal bar will *just* fit into the gap when they are both cold. Heat the bar with a Bunsen burner.

Does it fit the gap now?
What happens when it cools down?

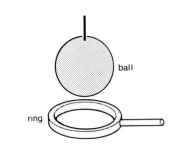

Experiment 4.2

This metal ball will *just* slip through the metal ring when they are both cold.

Does it go through when the ball is heated?

When objects get hotter they grow bigger. We say they *expand.*
When objects cool down they get smaller. We say they *contract.*
It is difficult to see this change in size because it is so small.

Experiment 4.3

This thick bar is heated with a Bunsen burner and then the big nut is tightened.

What happens to the cast-iron pin when the bar cools down? (Do not stand too near).

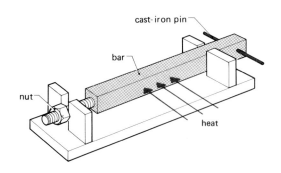

This shows that:—
— when bars *contract* you get big *pulling* forces.
— when bars *expand* you get big *pushing* forces.

Calculating expansion

The amount of expansion depends on three things:—

1—the length of the object at the beginning (called the *original length*)
2—a number which depends on the substance. (It is a very small number and is called the *linear expansivity*)
   Brass has got a bigger number than iron because it expands more.
3—how much hotter it gets (called the *rise in temperature*)

linear expansivity (per °C)	
Aluminium	0.000 026
Brass	0.000 019
Iron	0.000 012
Steel	0.000 011
Invar	almost zero
Glass	0.000 009
Pyrex glass	0.000 003

To calculate the expansion we multiply these three things together:

$$\text{Expansion} = \frac{\text{original}}{\text{length}} \times \frac{\text{linear}}{\text{expansivity}} \times \frac{\text{rise in}}{\text{temperature}}$$

Example

The Eiffel Tower is 300 metre high and is made of steel. How much taller does it get when the temperature rises by 20°C? (linear expansivity of steel = 0·000 011 per °C)

formula
first:- Expansion $= \dfrac{\text{original}}{\text{length}} \times \dfrac{\text{linear}}{\text{expansivity}} \times \dfrac{\text{rise in}}{\text{temperature}}$

then numbers:- $= 300 \text{ m} \times 0\cdot000\ 011 \times 20$

$= 0\cdot066$ metre

$= \underline{6\cdot6 \text{ cm}}$

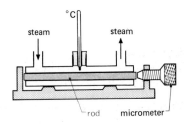

Experiment 4.4 To find the linear expansivity of a substance

Changing round the formula shown above, we get:

$$\frac{\text{linear}}{\text{expansivity}} = \frac{\text{expansion}}{\text{original length} \times \text{rise in temperature}}$$

To calculate the linear expansivity we must measure the 3 things on the right hand side of the equation.

One way is to use the apparatus shown:

a) First measure the original length (about 0·5 m) of a rod of the substance, using a metre rule.

b) Place the rod in the apparatus as shown and take the initial temperature of the thermometer and the initial reading on the micrometer.

c) Unscrew the micrometer and pass steam to heat up the rod. After a few minutes, use the micrometer again and again until it gives a constant reading (to show the rod is fully expanded).

d) Take the micrometer reading and calculate the expansion (in m). Take the final temperature and calculate the rise in temperature.

e) Use the formula to calculate the linear expansivity.

Expansion can cause trouble

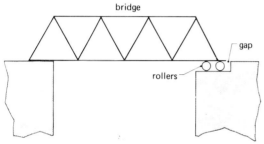

Bridges are made of big steel bars. When they get hotter, they get longer.
The Forth Railway Bridge is more than 1 metre longer in summer than in winter.

Bridges are usually put on rollers so that they can expand and contract without causing damage.

There must be an *expansion gap* in the road at the end of a bridge.
Have you seen one?

What would happen if there was no gap?

Roads are often made of large concrete slabs. They also expand. There are *expansion gaps* between the slabs, filled with a soft substance which can be easily squeezed in hot weather. If you look carefully, you may find similar expansion gaps in the floor of your school.

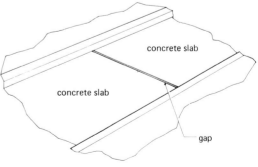

Telephone wires look like this in summer.
What will they look like in winter?

What would happen if the wires were made tight
in summer and then it went cold?

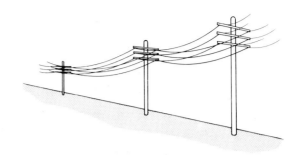

Thick glass beakers and milk-bottles sometimes
crack if boiling water is put in them. This is
because the inside tries to expand while the outside stays the same size.

To wash out a bottle I tried,
By pouring hot water inside,
But the glass was too thick,
(The expansion too quick)
The result was a crack very wide.

'Pyrex' glass beakers do not crack because 'pyrex'
does not expand as much as ordinary glass.

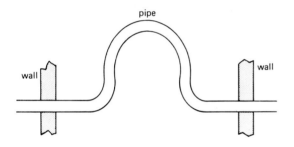

Steam and water pipes often have a large bend
in them. The bend lets the pipe expand without
breaking anything.

Steel railway lines need expansion gaps. Sometimes the gaps are at the end of each length of
rail (you can hear the clicks as the wheels go over
them). Nowadays many rails are welded together
and the gaps are much farther apart.

A railway mechanic was sent
To lay down some track across Kent.
He forgot to design
Some gaps in the line,
And when it got hot, it went bent.

25

Expansion and contraction can be useful

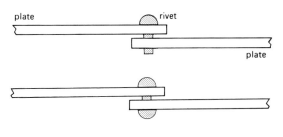

Rivets are sometimes used to hold steel plates together very tightly. A very hot rivet is pushed through the two plates.

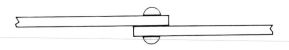

Then its end is hammered over.

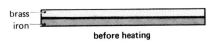

When the rivet cools down it pulls the two plates together very tightly (remember the pin-breaker experiment).

The same idea is used to fix steel tyres on to railway wheels.

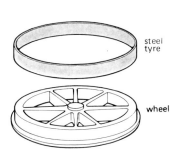

The steel tyre is just too small when it is cold. It is heated up and then slipped on the wheel. When it cools down it grips the wheel very tightly.

Experiment 4.5

Look at a *bi-metallic strip*. It is made of two metal strips (often Brass and Iron) which are placed side by side and then riveted or welded together.

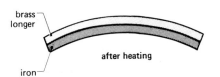

Heat it gently with a Bunsen burner.
What happens?

When it is heated, brass expands more than iron and so the strip bends with brass on the outside because it is longer.

Using bi-metallic strips

Experiment 4.6   A fire alarm

The diagram shows how you can make a fire-alarm.

Hold the bi-metallic strip in a clamp, with the brass on top.

Heat the bi-metallic strip gently. What happens?

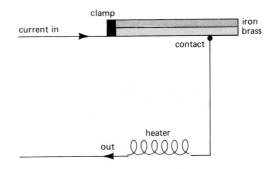

Electric thermostat

A *thermostat* is used to keep something at the same temperature, without getting too hot or too cold.
In the diagram, the electric current is flowing through the bi-metallic strip and the contact, to the heater.

What happens if it gets too hot?
What happens later when it cools?

This is used in electric irons.

Gas thermostat for a gas oven

This uses a rod made of a special metal called 'Invar' which does not expand. When the oven gets hot, the brass tube expands but the Invar does not—and so the valve is pulled to the left. This closes the gap so that the gas cannot get through.

What happens when the oven cools down?

For safety, there should be a small 'by-pass' to let some gas through even when the valve is closed. What would happen if there was no 'bypass'?

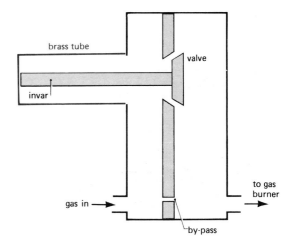

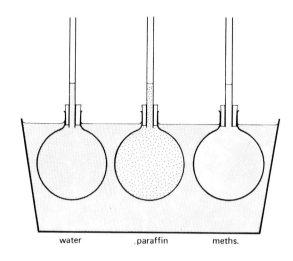

water   .paraffin   meths.

Expansion of liquids

Experiment 4.7
We need three glass bottles or flasks which are all the same size and have narrow tubes coming out of the top.

Fill them with three different liquids up to the same level as in the diagram. Put them in a dish and then fill the dish with hot water.

What happens?
Which liquid expands the most?
Do liquids expand more or less than solids?

Liquids expand more than solids

Most liquids expand when they get hotter. But *water is different* from other liquids:−

If you heat water from freezing point (0°C) up to 4°C, it gets *smaller* even though you are making it hotter. Above 4°C, it behaves like other liquids and expands as it gets hotter. This is shown on the graph:−
You can see that water is smallest at 4°C. This means that water is *densest* at 4°C.

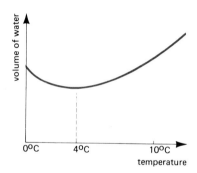

Now think what this means to a fish in a pond in winter. Because the densest water falls to the bottom of the pond, the bottom stays at 4°C even when the top freezes over.

So the bottom is warmer than the top. The fish can still swim around and stay alive underneath the ice.

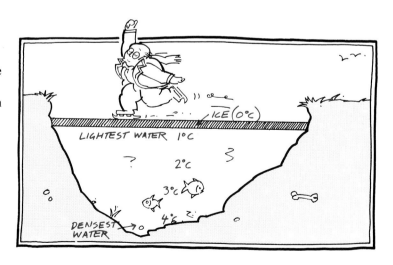

28

Expansion of gases

Experiment 4.8
Get a glass flask with a rubber bung and a tube as in the diagram.
Dip the end of the tube in a beaker of water and then warm the
flask with your hands.

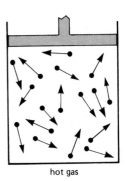

What do you see?
What is expanding?
Do gases expand more or less than liquids?

Now take your hands off and wait for the flask to cool down.

What happens?
Why?

Gases expand more than liquids.
We will see how this is used in rockets and other engines when we
come to Chapter 10 (page 70).

Explaining expansion

If a group of people are just standing still then
they can stay close together and they don't take
up much space.
But if the people start to dance then they take up
more space—the group expands.

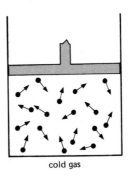

In the same way, if a substance is heated, the
molecules start moving more and so they take up
more space—the substance expands.

Notice that the molecules themselves do not get
any bigger — but as a group they take up more
space.

cold gas   hot gas

Summary

When objects are heated, they expand and push with large forces. Expansion gaps are needed in bridges, roads etc.

A bi-metallic strip is made of two different metals fixed together. When heated, it bends (with the metal with the bigger expansion on the outside). It is used in thermostats and fire alarms.

$$\text{Expansion} = \text{original length} \times \frac{\text{linear}}{\text{expansivity}} \times \frac{\text{rise in}}{\text{temperature}}$$

Liquids expand more than solids.
Water is different from other liquids—it contracts as it is heated from $0°C$ up to $4°C$ and then it expands above $4°C$.

Gases expand more than liquids.

Substances expand because the molecules are moving more and take up more space.

Can you explain this cartoon?

Questions

1. Copy out and fill in the missing words:
   a) When an object is heated it . . . . and when it cools it . . . . If we try to prevent it from contracting, it exerts a very large . . . . If a bi-metallic strip made of brass and iron is heated, it . . . . because . . . . . expands more than . . . . . . and the . . . . . . goes on the outside of the bend.
   Liquids expand more than . . . . but less than . . . . .
   b) When water is heated from $0°C$ up to $4°C$ it . . . . When it is heated above $4°C$ it . . . . . . When substances are heated, they get bigger because the . . . . . . are moving more.

2. Explain how expansion affects the design of
   a) bridges
   b) railway lines and
   c) steam pipes.

3. Which contracts the most when it is cooled – iron, water or air?

4. The diagram shows a bi-metallic strip:—
   Draw diagrams to show what it looks like
   a) if it is heated up
   b) if it is cooled down.

brass
iron

5. a) Draw a labelled diagram of a fire-alarm and explain how it works.
    b) How could you change it to a frost alarm?

6. a) Draw a labelled diagram of a thermostat for an electric iron and explain how it works.
    b) How could you change it to make a thermostat for a refrigerator?

7. Draw a labelled diagram of a thermostat for a gas oven and explain how it works.

8. Why are the doors and shelves of ovens made loosely fitting?

9. a) What happens to your ruler when the room gets warmer?
    b) What does this mean about the lines you measure?
    c) What are the best rulers made of? (see the list on page 23).

10. What will happen to the pendulum of a clock when the room gets warmer? What effect will this have on the clock's timekeeping?

11. In a motor-car engine:—
    a) why are the cylinder liners cooled before they are pushed into the cylinder block?
    b) why are piston rings needed and why do they have a gap in them?

12. Look at the list on page 23. Which two substances would be best for making a bi-metallic strip?

13. A steel bridge is 100m long at 0°C. How much longer is it at 20°C? (linear expansivity of steel = 0·000 011 per °C)

14. The steel railway track from London to Edinburgh is 600 kilometres long. By how much would it expand if the temperature rises by 30°C?

15. If a 100cm copper rod changes length by 0·17cm when it is heated up through 100°C, what is the linear expansivity of copper?

16. What is the temperature of the water in the pond
    a) at A?
    b) at B?
    Explain this.

17. a) If a bottle was filled full with a liquid, tightly sealed and then heated, what might happen?
    b) Why does a bottle of lemonade always have a space between the top of the liquid and the cap?

18. A damaged table-tennis ball can often be mended by warming it in hot water. Explain this.

19. Why must you never put sealed bottles or 'aerosol' spray cans on a fire?

20. Can you explain this cartoon?

Further questions on page 79.

chapter 5

thermometers

Thermometers are used to measure *temperature*.
Temperature is not the same thing as heat or thermal energy.

To understand this, let's compare a white-hot
'sparkler' firework with a bath-full of warm water:

The tiny 'sparkles' are at a very high temperature but contain little energy because they are very small.

The water is at a lower temperature but it contains more internal energy because it has many more molecules.

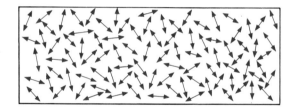

Each sparkle contains a few molecules vibrating at very high temperature but with little total energy.

The bath has many molecules, each vibrating at low temperature but with more total energy.

You can see that the bath has more *thermal energy*.
Its molecules have more *internal energy*.
To raise the temperature of an object you must
give it more energy.

A laboratory thermometer.

This uses the expansion of mercury to measure
temperatures accurately. Look at the large diagram
opposite. If the mercury gets warmer, it expands
along the tube, where there is a scale of numbers
marked °C or *degrees Celsius* (or degrees centi-
grade). There is a vacuum (no air) above the
mercury so it can easily move along the tube.

To make the thermometer *sensitive*, with a long
scale, it should have a narrow bore tube and a
large bulb.

To make the thermometer *quick-acting*, it should
have a bulb made of thin glass so that the heat
can get through easily.

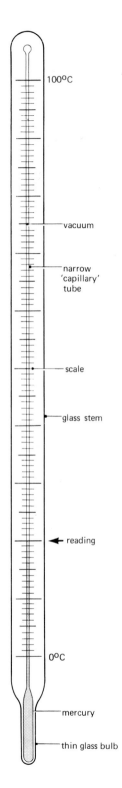

100°C

vacuum

narrow
'capillary'
tube

scale

glass stem

reading

0°C

mercury

thin glass bulb

What is the reading on the thermometer in the diagram? (This is the temperature of a warm room).

Experiment 5.1
Practise reading the scale of a thermometer by measuring the temperature of the room; of the cold water supply; of the hot water supply; of your armpit (take care — thermometers break easily).

Calibration
How are the marks put on the thermometer in the correct places?
Two *fixed points* are marked first of all:
The *upper fixed point*, marked 100°C, is the temperature of the steam over pure boiling water at standard atmospheric pressure (see page 101).
The *lower fixed point*, marked 0°C, is the temperature of pure melting ice at standard atmospheric pressure.
When the two fixed points are marked, the scale between them is divided into 100 equal divisions.

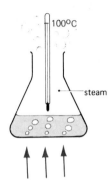

100°C

steam

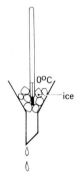

0°C

ice

Experiment 5.2
Check the fixed points on a thermometer (or 'calibrate' an unmarked thermometer) by using the apparatus shown.

Choice of liquids

Mercury is used because (a) it expands evenly as the temperature rises and (b) it is a good conductor of heat. Unfortunately it freezes solid if it is used in very cold places.

Alcohol is sometimes used instead, because (a) it can be used at very low temperatures and (b) its expansion is six times greater than mercury. Unfortunately it cannot be used in hot places because it boils if placed in very hot water.

Clinical thermometer

A doctor needs a thermometer that will continue to show its maximum temperature after it is taken out of your mouth.

A clinical thermometer has a very narrow constriction in the tube just above the bulb. When the mercury expands it pushes past the constriction, but when it is taken out of your mouth it does not go back to the bulb – and so the doctor can read your temperature.

The mercury can be shaken back into the bulb after it has been read.

Doctor Shoddy, old and frail,
Can't see the numbers on this scale,
Take the reading – tell old Shoddy,
This is the temperature of your body!

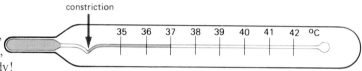

Another maximum thermometer

This contains a little iron *index* which is pushed along by the mercury.

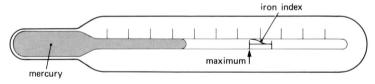

How could you use a magnet to reset this thermometer?

A minimum thermometer

This can be used to find the coldest temperature during the night. It contains an iron index placed inside the liquid which is alcohol. As it cools down, the surface meniscus of the alcohol pulls the index back.

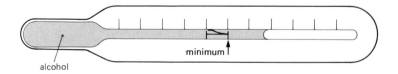

How would you reset this thermometer?

Absolute zero

As substances get colder, gases condense into liquids and liquids freeze into solids. If the temperature continues to get colder and colder, the molecules vibrate less and less until eventually they have their lowest possible energy. This happens at the coldest possible temperature, at −273°C, called *absolute zero*. (See also page 39.)

Scientists often measure temperatures on the *Kelvin scale*, which begins at absolute zero but increases just like the Celsius scale. This means 0°C becomes 273 kelvin (written 273K) and 100°C = 373K.

On a hot day, the temperature is 27°C – what is this in kelvin?

Other thermometers

1. Since gases expand by a greater amount and more evenly than liquids, *gas thermometers* are more accurate than liquid thermometers and can work at higher and lower temperatures. (See also page 40).

2. A *resistance thermometer* uses the fact that electricity does not flow so easily through a wire when the wire gets hot (see also page 290).

3. A *thermistor thermometer* uses the fact that electricity flows through it more easily when it is hot (see also page 351).

4. A *thermocouple thermometer* consists of two wires made of different metals and connected to a sensitive ammeter. Different temperatures produce different electric currents. 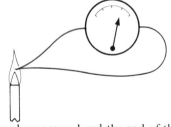 High and low temperatures can be measured and the end of the thermometer can be made very small.

5. The temperature of a very hot object like a furnace or the sun can be measured by a *pyrometer* that measures the amount of radiation given off by the object.

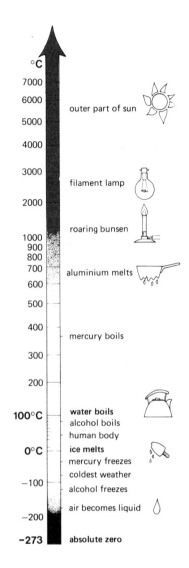

°C
7000
6000 — outer part of sun
5000
4000
3000 — filament lamp
2000
1000 — roaring bunsen
900
800
700 — aluminium melts
600
500
400 — mercury boils
300
200
100°C — water boils / alcohol boils / human body
0°C — ice melts / mercury freezes / coldest weather
−100 — alcohol freezes
−200 — air becomes liquid
−273 — absolute zero

Summary

The temperature of an object depends upon how much its molecules are vibrating.
Internal energy (or thermal energy) is the total amount of energy in an object.
A clinical thermometer has a constriction to prevent the mercury from returning to the bulb (until it is shaken).

The upper fixed point, 100°C, is the temperature of the steam over pure boiling water at standard atmospheric pressure.
The lower fixed point, 0°C, is the temperature of pure melting ice at standard atmospheric pressure.
Absolute zero is the coldest possible temperature at −273°C. Temperature in kelvin = °C + 273.

Questions

1. Copy out and fill in the missing words:
   a) Thermometers are used to measure . . . . . which depends on how much the . . . . . are vibrating. Internal energy is the amount of . . . . . in an object.
   b) The Upper Fixed Point of the Celsius scale is marked . . . . . °C and is the temperature of . . . . . over pure . . . . . water at normal atmospheric . . . . . The Lower Fixed Point of the Celsius scale is marked . . . °C and is the . . . . . of pure . . . . . ice at . . . . ' atmospheric . . . . . The scale between the two Fixed Points is marked into . . . . . equal divisions.
   c) The coldest possible temperature is . . . . . °C, called . . . .

2. Copy out and complete the rhyme:
   Thermometers are often made
   With marks described as c . . . . . .
   But recently there's been a fuss
   To change the name to C . . . .

3. If you were travelling to the South Pole, would you take a mercury thermometer or an alcohol thermometer? Explain your answer.

4. Convert (a) 10°C to kelvin (b) 300K to °C (c) What is your body temperature in kelvin?

5) a) Explain how a clinical thermometer works.
   b) Explain how it is possible to return the mercury to the bulb after using a clinical thermometer.

6. Describe and explain the features of a thermometer which will make it a) sensitive, b) quick-acting.

7. Explain the difference between heat and temperature.

8. Copy out and place ticks in the table to show which liquid is better in each case:

	mercury	alcohol
expands more evenly expands more a better conductor of heat useful at higher temperatures useful at lower temperatures		

9. Professor Messer blows his top, Tell him why he's such a flop:

Further questions on page 79.

chapter 6 THE GAS LAWS

When investigating the behaviour of gases, we must consider *three* varying quantities: **pressure**
  volume
  temperature.

In the following experiments we always keep one of these variables constant and investigate the other two variables.

Experiment 6.1 Keeping the temperature constant – **Boyle's Law**
Use the apparatus in the diagram to investigate how the **volume** of air depends on the **pressure** applied (with the temperature kept constant).

Use the vertical scale to find the **volume** of the trapped air.
What is the volume of air in the diagram?
What is the volume of air in your apparatus?

The **pressure** on the trapped air is measured using a Bourdon pressure gauge (see also page 103).
What is the pressure on the gauge in the diagram?
What is the pressure in your apparatus?
Why is the pressure not zero? (See page 100.)

Now use the pump to exert more pressure and squeeze the air. This will make the trapped air slightly warmer, so wait a minute to let it cool back to room temperature.

Now measure the new pressure and volume and put your results in a table.
Use the pump again, to get more results.

Than calculate pV (pressure x volume) for each set of results.
What do you notice?

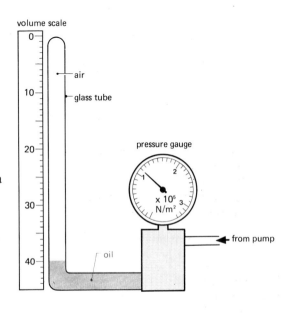

Some possible results:

pressure p	volume V	$p \times V$
1·1	40	44
1·7	26	44
2·2	20	
2·6	17	

This is called **Boyle's Law** after its discoverer, Robert Boyle:

> **For a fixed mass of gas, at constant temperature, pV = constant**

Did you notice that if p is doubled, V is halved?
If p increases to 3 times as much, V decreases to $\frac{1}{3}$rd. This means:

> **volume is *inversely proportional* to pressure, or $V \propto \frac{1}{p}$**

Charles' Law

Experiment 6.2   Keeping the pressure constant –
Charles' Law

In this experiment you will investigate how the
volume of air changes as the **temperature** changes
(with the pressure kept constant).

In the usual apparatus, the air is held inside a glass
capillary tube by a short length of concentrated
sulphuric acid (mercury can be used but the acid
dries the air to give better results).

The length *y* of the trapped air is a measure of the
volume of the air.

Hold the glass tube on to a ruler with rubber
bands. Move the tube until the bottom of the
trapped air is opposite the zero mark of the ruler.

An alternative is to use a plastic syringe which is
lubricated with silicone oil and sealed by a rubber
tube and a clip so that it is half-full of air.

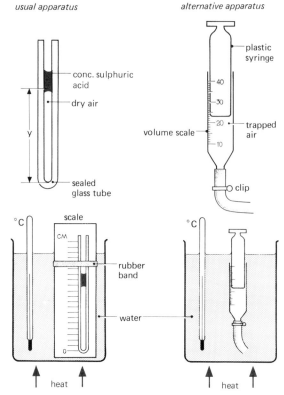

Put your apparatus in a tall beaker of cold water with a thermometer.
Wait until the trapped air is the same temperature as the water and
then measure the **volume** and the **temperature**.

Then heat the water until it is about 20°C hotter. Again, wait before
taking the readings of volume and temperature.
Repeat this at other temperatures until the water boils.
Put your results in a table.

Then plot a graph to show how the **volume** varies with **temperature**
(at constant pressure).

You used air in your experiment but other gases give the same result.

At constant pressure

Volume	temperature(°C)

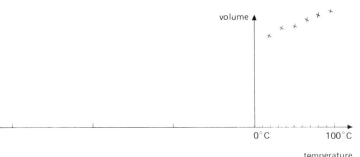

38

Through the points on your graph, draw the *best* straight line (it helps to hold the paper near your eye and look along the crosses to see the best line).

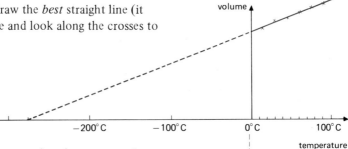

Because the graph is a *straight* line it means that the gas expands *uniformly* with temperature (when the pressure is constant).

What happens to the volume when the gas is cooled? From your graph find the temperature at which it seems the volume would become zero. Mark it on your graph.

Although the gas would liquefy before cooling to this temperature, this experiment suggests that there is a limit to how cold objects can be. This is called the **absolute zero** of temperature (see also page 35). Careful experiments show that absolute zero = **–273°C**.

In science, we often measure temperature on the **Kelvin scale**. This begins at absolute zero and increases just like the Celsius scale. This means that 0°C becomes 273 kelvin (written 273 K) and 100°C becomes 373 K.
Your body temperature is 37°C – what is this in kelvin?

Redraw your graph (or re-label the temperature axis) so that it shows the absolute temperature, in kelvin.

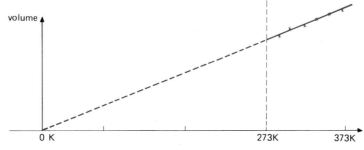

Since the graph is a *straight line* passing *through the origin* it means that:

> For a fixed mass of gas, at constant pressure,
> Volume (*V*) is directly proportional to the absolute temperature (*T*)

or, in symbols: $V \propto T$

∴ $V = \text{constant} \times T$

∴ $\boxed{\dfrac{V}{T} = \textbf{constant}}$ (if the pressure is constant)

This is called **Charles' Law**.
T must be the absolute temperature, measured in kelvin.

39

Pressure Law

Experiment 6.3   Keeping the volume constant – the Pressure Law

Use the apparatus shown in the diagram to investigate how the **pressure** of air changes as the **temperature** changes (with the volume kept constant).

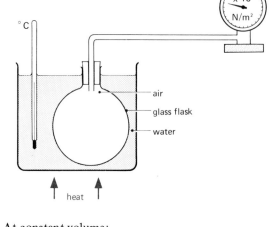

The **pressure** is measured using a Bourdon pressure gauge (see also page 103). What is the pressure on the gauge in the diagram?

What is the **pressure** in your apparatus?
What is the **temperature** of your apparatus?

Now heat the water by about 20°C. Then wait for several minutes to allow the air in the flask to reach the temperature of the water.
Take the readings of pressure and temperature.
Repeat this at several temperatures until the water boils. Put your results in a table.

Plot a graph of pressure : temperature in °C.
Compare it with the graph on the previous page.
According to your graph, at what temperature would the pressure become zero? How does this compare with the accepted value?
What is this temperature called?

What happens to the molecules of a gas as it cools? What could you say about the molecules if the gas cooled to absolute zero?

Redraw your graph (or re-label the temperature axis) so that it shows
pressure: *absolute* temperature.

At constant volume:

pressure $\times 10^5$ N/m^2	temperature °C	absolute temperature K

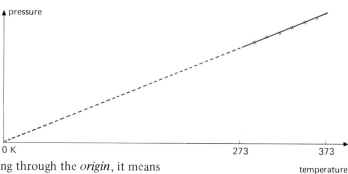

Since this graph is a *straight* line passing through the *origin*, it means that:

> **For a fixed mass of gas, at constant volume,**
> **pressure (p) is directly proportional to the absolute temperature (T)**

or, in symbols: $p \propto T$

$\therefore \ p = \text{constant} \times T$

$\therefore \ \boxed{\dfrac{p}{T} = \text{constant}}$ (if the volume is constant)

This is called the Pressure Law.
As before, T must be in kelvin.

This apparatus can be used as a thermometer: *a constant volume gas thermometer* (see also p. 35).
The dial of the pressure gauge can be marked out in °C.

40

The Gas Equation

From the last three experiments, we have 3 equations:

$pV = \text{constant (Boyle's Law)}$ $\qquad \dfrac{V}{T} = \text{constant (Charles' Law)}$ $\qquad \dfrac{p}{T} = \text{constant (Pressure Law)}$

These 3 equations can be combined into one equation, called the **ideal gas equation**:

$$\boxed{\dfrac{pV}{T} = \text{constant}}$$

If a fixed mass of gas has values p_1, V_1, and T_1 and then some time
later has values p_2, V_2 and T_2, then the equation becomes:

$$\boxed{\dfrac{p_1 V_1}{T_1} = \dfrac{p_2 V_2}{T_2}}$$

Example

A bicycle pump contains 70 cm^3 of air at a pressure of 1·0
atmosphere and a temperature of 7°C.
When the air is compressed to 30 cm^3 at a temperature of 27°C,
what is the pressure?

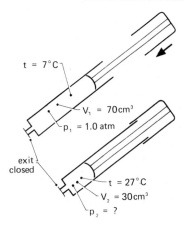

what do we know?

$p_1 = 1\cdot0$ atmosphere $\qquad p_2 = ?$
$V_1 = 70$ cm^3 $\qquad\qquad V_2 = 30$ cm^3
$T_1 = 273 + 7 = 280$ K $\quad T_2 = 273 + 27 = 300$ K

then the formula: $\qquad \dfrac{p_1 V_1}{T_1} = \dfrac{p_2 V_2}{T_2}$

put in the
numbers:

$$\therefore \dfrac{1 \times 70}{280} = \dfrac{p_2 \times 30}{300}$$

$$\therefore p_2 = \dfrac{70 \times 300}{30 \times 280} = \underline{2\cdot5 \text{ atmosphere}}$$

Note 1. p_2 has the same unit as p_1. Similarly with V_2 and V_1.
$\qquad\quad$ T_1 and T_2 must be in kelvin.
Note 2. If questions refer to **s.t.p.** (standard temperature and
$\qquad\quad$ pressure) this means $T = 273$ K (0°C) and $p = 760$ mm Hg (see page 101)

The Gas Laws and the Kinetic Theory

We saw on page 14 that air pressure is caused by
the bombardment of billions of molecules.

Boyle's Law

If the volume of a fixed mass of gas is made
smaller then the molecules will be closer together.
If the volume is halved, then the number of
molecules per cm^3 will be doubled and so the
pressure will be doubled (Boyle's Law).

Absolute Zero

As the temperature falls, the molecules have less
energy. Absolute zero is the temperature at which
molecules have their lowest possible energy.

The Pressure Law

If the temperature rises, then the molecules have
more energy and move faster.
If the volume of the container is kept constant
then the molecules hit the walls harder and more
often and so the pressure rises (the Pressure Law).

Charles' Law

If the pressure is to be kept constant even though
the molecules begin to move faster, then the
volume must increase. This lets the molecules
move farther apart so that the pressure can stay
the same (Charles' Law).

Summary

For a fixed mass of gas in each case:

Boyle's Law. At constant temperature: $p \propto \dfrac{1}{V}$ or pV = constant or $p_1 V_1 = p_2 V_2$

Charles' Law. At constant pressure: $V \propto T$ or $\dfrac{V}{T}$ = constant or $\dfrac{V_1}{T_1} = \dfrac{V_2}{T_2}$

Pressure Law. At constant volume: $p \propto T$ or $\dfrac{p}{T}$ = constant or $\dfrac{p_1}{T_1} = \dfrac{p_2}{T_2}$

T must be in kelvin (= 273 + temperature in °C)

Gas Equation $\dfrac{pV}{T}$ = constant or $\dfrac{p_1 V_1}{T_1} = \dfrac{p_2 V_2}{T_2}$

Questions

1. Copy and complete these sentences:
   a) Boyle's Law: for a . . . . . mass of gas, at constant . . . . . , ... × . . . . . is constant. Pressure is . . . . . proportional to . . . . .
   b) Charles' Law: for a . . . . . mass of gas, at constant . . . . . , ... divided by . . . . is constant. Volume is . . . . . proportional to . . . . . temperature. A graph of volume against temperature is a . . . . . line which intercepts the temperature axis at . . . . .
   c) Pressure Law: for a . . . . . mass of gas at constant . . . . . , ... divided by . . . . is constant. Pressure is . . . . . proportional to . . . . . temperature. A graph of pressure against temperature is a . . . . . line which intercepts the temperature axis at . . . . .
   d) The temperature of absolute zero is . . . °C or . . . kelvin. At this temperature the molecules have their . . . . . possible . . . . .
   e) S.t.p. means standard . . . . . and . . . . . (that is . . . kelvin and . . . mm Hg)

2. a) Convert these to kelvin: 27°C, −3°C, 150°C, −90°C.
   b) Convert these to °C: 373 K, 200 K, 1000 K

3. A bubble of air of volume 1 cm^3 is released by a deep-sea diver at a depth where the pressure is 4·0.atmospheres. Assuming its temperature remains constant (that is, T_1 =

T_2), what is its volume just before it reaches the surface where the pressure is 1·0 atmosphere?

4. To what temperature must 1 m^3 of air at 27°C be heated, at constant pressure, in order to increase its volume to 3 m^3?

5. The pressure of air in a car tyre is 2·7 atmospheres when the temperature is −3°C. If the temperature rises to 27°C, what is the pressure then? (Assume the volume remains constant.)

6. A mass of gas has a volume of 200 cm^3 at a temperature of −73°C and a pressure of 1·0 atmosphere. What is its volume at a pressure of 3·0 atmosphere and a temperature of 27°C?

7. A mass of gas has a volume of 4 m^3 at a temperature of 127°C and a pressure of 800 mm Hg. What is its pressure when the volume is 5 m^3 and the temperature is −23°C?

8. A mass of gas has a volume of 380 cm^3 at a pressure of 560 mm Hg and a temperature of 7°C. What is its volume at s.t.p?

9. The cylinder of a diesel engine (see page 73) contains 300 cm^3 at 27°C and 75 cmHg. The air is rapidly compressed to 20 cm^3 at a pressure of 3000 cmHg. What is the temperature then?

10. A metal tank in a garage normally contained compressed air at 4 atmospheres and 7°C. In a fire the tank exploded, although it was known to be safe up to a pressure of 14 atmospheres. If you were the detective of the insurance company, what value would you calculate for the minimum temperature of the fire near the tank?

Further questions on page 80.

chapter 7

measuring HEAT

We have seen that the moving molecules of an object have *internal energy*. When an object is heated, its internal energy increases. Like all other forms of energy, internal energy is measured in a unit called the *joule* (usually written J).

A joule is a small unit of energy – if a match burns up completely, it produces about 2000J. This can also be written as 2kJ where

    1 kJ = 1 kilojoule = 1000 joule.

For larger amounts of energy we use MJ where

    1 MJ = 1 megajoule = 1 million joule (1 000 000 J).

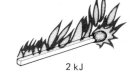

2 kJ

Your body needs energy – energy to keep warm (at 37°C) and energy to work and play. You get this energy from the fuel that you eat.

Can you work out the rhyme?
> Little Jack Horner,
> Sat in a corner,
> Feeling so chilly and cool.
> He said, "I should eat,
> And so produce . . . . . ,
> The unit of which is a . . . . ."

Your body needs a total of about 10MJ per day. Lying asleep, your body only uses about 4 kJ every minute, but for making your bed or dancing you use about 20 kJ every minute.

The table below shows some energy values, (called *calorific values*) measured in megajoule per kilogram of substance.

Too many joules, too few joules

Butter	32 MJ/kg	Eggs	7 MJ/kg	Natural Gas	55 MJ/kg
Sugar	16	Potatoes	4	Petrol	47
Meat	12	Fish	3	Coal	30
Bread	10	Fruit	2	Wood	15
Ice cream	9	Carrots	2	Dynamite	6

If you want to lose weight, which foods should you avoid?
Which food would you take on an expedition to the Arctic?

Did you hear about the cat burglar who stole the family joules?
He used them for (h)eating !

Heating water

Experiment 7.1

Measure out 1 kg of cold water into a large beaker.
Take the initial temperature of the water accurately.
Put an electric heater in the water and connect
it to a 12-volt supply through a *joulemeter*, which
will measure how much energy we give the water.

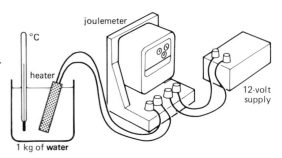

Note the joulemeter reading and switch on.
When you have given the water 10 000 J, switch
off, stir gently and note the highest temperature
that is reached.
What is the rise in temperature of the water?

(If you have not got a joulemeter, you can use a
50 watt heater switched on for 200 seconds).

Experiment 7.2

If you gave twice as much energy (20 000 J) to the same 1 kg of water,
what would be the temperature rise?
Guess and then try it.
Is it exactly what you expected?
Would you get a better result if you wrapped cotton wool round
the beaker?

Do other substances need as much energy as water?

Experiment 7.3

Put your thermometer and heater into a 1 kg block
of aluminium. Give it the same amount of energy
(10 000 J) as in experiment 7.1, and find the
temperature rise.
Is it the same as when you had 1 kg of water?

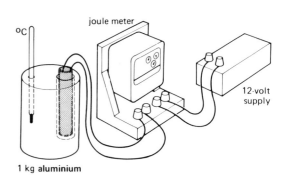

If the temperature rise is 5 times as much, it means
that aluminium needs only $\frac{1}{5}$th as much energy as
water (to give the same temperature rise to the
same mass).

Experiment 7.4

If you have time, repeat this experiment with other solids and liquids
and compare the results.

We find that different substances have different appetites or
capacities for energy.
Water is very greedy, it needs five times as much energy as the same
mass of aluminium to produce the same temperature rise.

Specific heat capacity

If you calculate how much energy 1 kg of water needs to become 1°C hotter, you will find it needs about 4200 J. This number is called the *specific heat capacity* of water.

The specific heat capacity of a substance is the amount of energy (in joule) that is needed to raise the temperature of 1 kg of the substance by 1°C.
Its unit is written J/kg°C or J/kg K.

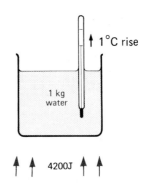

Here are some values of this number — notice that the number for aluminium is about $\frac{1}{5}$ th of the number for water.

Water	4200 J/kg°C	Ice	2100 J/kg°C
Meths.	2500	Aluminium	880
Paraffin	2200	Sand	800
Mercury	140	Copper	380

All these numbers are for 1kg and 1°C. The energy needed for 2kg through 1°C would be twice as much and the energy needed to raise 2kg through 3°C would be three times as much again — that is, six times as much. So to calculate the energy needed, we multiply like this:

Energy needed = specific heat capacity × mass × change in temperature

 joules   J/kg°C   °C

Example 1
How much energy is needed to heat 100g of water from 10°C to 30°C?

what do
we know?

specific heat capacity of water = 4200 J/kg°C.
mass of water = 100g = 0·1 kg
temperature rise = 30°C − 10°C = 20°C

write down
the formula

$$\text{Energy needed} = \text{specific heat capacity} \times \text{mass} \times \text{change in temperature}$$

then put in
the numbers

= 4200 × 0·1 × 20   joule
= 4200 × 2   joule
= 8400 J   (= energy from 4 matches)

In a similar way, using the same formula, you can work out *your* values for specific heat capacity from the results of your experiments.

If you met a girl called Julie, would you expect her to be full of energy?

Example

A 2 kW (2000 W) electric heater supplies energy to a 0·5 kg copper kettle containing 1 kg of water. Calculate the time taken to raise the temperature by 10°C. (Use values from the table on page 45.)

2 kW = 2000 W = 2000 J/s = 2000 joules in each second (see also page 133).
∴ Energy supplied in t seconds = 2000 x t joules

Assuming no energy is lost to the surroundings:

Energy lost by hot = **Energy gained by cold**
object (heater) **objects** (water and kettle)

$$2000 \times t = \left(\text{mass} \times \frac{\text{specific heat}}{\text{capacity}} \times \frac{\text{rise in}}{\text{temperature}} \right)_{\text{water}} + \left(\text{mass} \times \frac{\text{specific heat}}{\text{capacity}} \times \frac{\text{rise in}}{\text{temperature}} \right)_{\text{copper}}$$

$2000 \times t = (1 \times 4200 \times 10)_{\text{water}} + (0{\cdot}5 \times 380 \times 10)_{\text{copper}}$

$2000 \times t = (42000) + (1900) = 43\ 900$

$\therefore t = \underline{22 \text{ seconds}}$

Effect of different specific heat capacities

You could see from the table that water has easily the highest specific heat capacity. It needs more energy to heat it up. It stores more energy when it is hot and it gives out more energy when it cools down. This is why a hot water bottle is so effective in warming a bed (an equal weight of mercury would store only $\frac{1}{30}$ th as much energy!) Water is also the best liquid to use in a central heating system.

In a similar way, water is the best for *cooling* engines in cars and factories.

Similarly, the specific heat capacity of water affects our weather. Islands, being surrounded by water, tend to stay at the same temperature while large areas of land tend to get very hot in summer and very cold in winter.

Heat capacity

Sometimes, engineers prefer to talk about the heat capacity of the *whole* object, rather than just 1kg as we have done.
For example, the Heat Capacity of a pan might be 100 J/°C. This means that the pan (empty) needs 100J to raise its temperature by 1°C.

Since specific heat capacity is for 1 kg and Heat Capacity is for the whole object, we can calculate:

Heat Capacity	=	specific heat capacity	x	mass
(J/°C)		(J/kg°C)		(kg)

Summary

Thermal energy, like all other forms of energy, is measured in joule.
1 kJ = 1000 joule. 1 MJ = 1 000 000 joule.

The specific heat capacity of a substance is the amount of energy needed to raise the temperature of 1 kg of the substance through 1°C.
Unit: J/kg°C or J/kgK.

$$\frac{\text{Energy needed}}{\text{to change temperature}} = \frac{\text{specific heat}}{\text{capacity}} \times \text{mass} \times \frac{\text{change in}}{\text{temperature}}$$

Heat capacity is the amount of energy to raise the temperature of the whole object through 1°C.
Unit: J/°C or J/K.

Questions (Use the tables on page 43, 45 if necessary.)

1. Copy out and complete:
   a) The unit of energy is the . . . . .
      1 kJ stands for one . . . . .
      1 MJ stands for one . . . .
   b) The specific heat capacity of a substance
      is the amount of . . . . . needed to raise
      one . . . . . of the substance through
      . . . . . °C
      The unit of specific heat capacity is . . . . .
      The specific heat capacity of water in
      these units is . . . . .
   c) The formula for the energy needed to
      change the temperature of a substance is:
   d) The heat capacity of an object is the
      amount of . . . . . needed to raise the
      temperature of the object by . . . . °C.
      Its unit is . . . . .
   e) Energy . . . . . by . . . . . object = Energy
      . . . . . by . . . . . object.

2. Jack Spratt could not eat fat,
   His wife could not eat lean.
   While Jack ate fruit or carrots,
   His wife would lick ice-cream.
   Two kilograms they each put in
   And one grew fat and one grew thin.
   I ask you now to calculate
   How many joules of heat each ate.

3. Calculate the amounts of energy needed to
   change the temperature of
   a) 2 kg of water by 5°C
   b) 500g of water by 4°C
   c) 100g of aluminium from 20°C to 30°C
   d) 200g of copper from 60°C to 10°C

4. Professor Messer fills his car. How many faults can you find with his idea (there are at least 4).

5. A 2kg block of iron is given 10 kJ of energy
   and its temperature rises by 10°C.
   What is the specific heat capacity of iron?

6. In an experiment, 500g of water at 20°C is
   given 22 000 J of energy and the temperature
   rises to 30°C.
   a) What value for the specific heat capacity
      of water do these results give?
   b) How does this compare with the accepted
      value? Suggest as many reasons as possible
      for the difference.

7. At the sea-side, the sun shines down equally
   on the sand and the sea. Assuming they
   both absorb the same amount of energy, why
   is the sand hotter than the sea?

8. A 1000 W heater supplies energy for 100
   seconds to a 2 kg metal can containing 1 kg
   of water. The temperature rises by 20°C.
   Calculate the specific heat capacity of the
   metal.

9. 2 kg of iron at 80°C is placed in a copper
   can, mass 0·5 kg, containing 1 kg of water at
   20°C. After stirring, the temperature of the
   mixture is 30°C. Calculate the specific heat
   capacity of iron.

   Further questions on page 81.

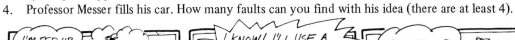

chapter 8

Conduction, Convection and Radiation

There are *three* ways in which heat can travel.

Experiment 8.1   Conduction
Get a piece of stiff copper wire about the same length as a match. Strike the match and hold the copper wire in the flame.

What happens?
Does the energy get to your hand quicker through wood or through copper?

We say that copper is a better *conductor* than wood. The heat has travelled from molecule to molecule through the copper.

hot

cold

Experiment 8.2   Convection
Hold your hand over and then under the flame of a match.

What do you notice?

The hot air expands and then rises.
We say the heat is *convected* upwards.

Radiation

The thermal energy of the sun is *radiated* to us in the same way that the light reaches us from the sun.

Here is a 'model' to help us see the difference between the three ways. Three ways of getting a book to the back of the class.

1 *Conduction:* a book can be passed from person to person — just as heat is passed from molecule to molecule.

2 *Convection:* a person can walk to the back of the class carrying the book. This is the way hot air moves in convection, taking the energy with it.

3 *Radiation:* a book can be thrown to the back of the class rather like the way energy is radiated from a hot object.

CONDUCTION

Experiment 8.3

Get some rods (all the same size) of different substances — for example, copper, iron and glass. Rest them on a tripod and fix a small nail near one end of each rod, using vaseline as 'glue'

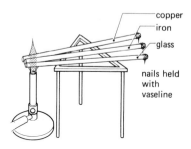

copper
iron
glass

nails held with vaseline

Heat the other ends of the rods equally with a bunsen burner.

What happens?
How many minutes does it take for the first and second nail to drop off?

All metals are good conductors of heat.
Copper is a very good conductor.
Glass is not a good conductor of heat — it is an *insulator*.

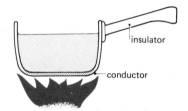

insulator

conductor

Pans for cooking are usually made with a copper or aluminium bottom and plastic handle. What would happen if the handle was made of copper?

The Prof. called his doctor by phone,
"I bought a new pan", was his groan,
"Twas badly constructed,
The handle conducted,
And my fingers are burned to the bone".

Experiment 8.4

Hold a large test tube of water at the bottom and aim a gentle bunsen flame just below the water surface until the water boils at the top.

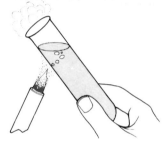

Is water a good conductor or a poor conductor?

Most liquids are poor conductors of heat.

Insulators

Is air a conductor or an insulator?

Experiment 8.5
Hold a live match about 1 cm away from a very hot bunsen flame.

Does the match get hot enough to burst into flame?

This shows that air is a very poor conductor — it is a very good insulator.

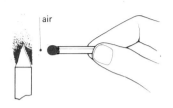

Many insulators contain tiny pockets of trapped air to stop heat being conducted away.

For example, wool feels warm because it traps a lot of air.

The air trapped in and between clothes and blankets keeps us warm.

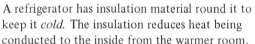

In the same way, the air trapped in fur and feathers keeps animals warm. Birds fluff up their feathers in winter to trap more air.

A refrigerator has insulation material round it to keep it *cold*. The insulation reduces heat being conducted to the inside from the warmer room.

Insulation in your home can save a lot of money

Pipes and hot water tanks should be lagged with insulation material to reduce the loss of energy.

About a third of the energy from your house will be lost through the roof unless insulating material (containing a lot of air) is put in the loft.

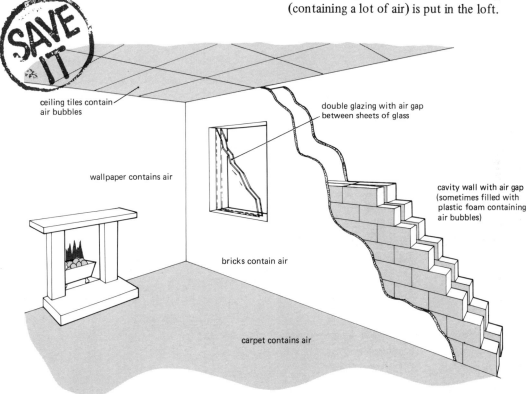

ceiling tiles contain air bubbles

double glazing with air gap between sheets of glass

wallpaper contains air

cavity wall with air gap (sometimes filled with plastic foam containing air bubbles)

bricks contain air

carpet contains air

Count up the number of ways this room is insulated. (See also page 10.)

CONVECTION

Experiment 8.6

Put a single crystal of potassium permanganate into a beaker of water, near to the side, as in the diagram.
Warm the water near the crystal with a small bunsen flame.

The crystal colours the water so that you can see the movement of the *convection currents* in the water. The warm water rises in the same way that warm air rises in a flame.
The cooler water falls to the bottom until it is heated.

Car engines are cooled by convection currents in the water pipes.

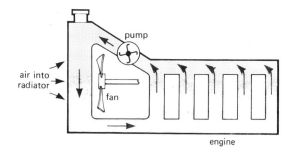

The diagram shows a simple domestic hot water system as it might be in your house.

Convection currents take energy from the boiler up to the storage tank which is connected to the hot taps.

Where would you add lagging to insulate this system?

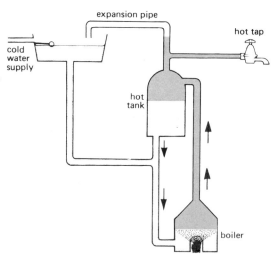

Experiment 8.7

Investigate convection currents in a model room like the one in the diagram, using smoke so that you can see how the air moves.

Hold the smouldering paper near the window before and then after lighting the fire.

What do you notice?
This shows that fires and chimneys help to ventilate rooms.

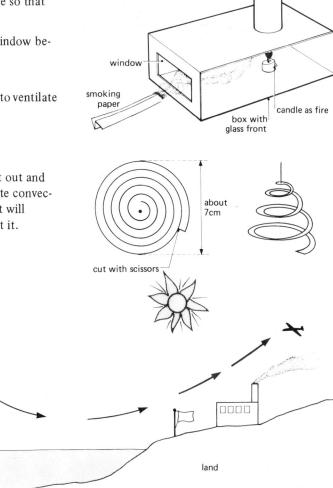

Experiment 8.8

Draw a spiral on paper as shown. Cut it out and hang it from cotton. Use it to investigate convection currents near fires and windows. It will rotate as convection currents move past it.

The sun can cause very large convection currents which we call winds.

In the day-time the land warms up more than the sea. The warm air rises over the land and cool air falls over the sea. So we feel a sea breeze.

Rising convection currents over the land are used by glider pilots to keep their planes in the air.

Can you draw a diagram of what happens at night when the land cools quicker than the sea?

A rather small boy called Maguire,
Sat next to a roaring hot fire,
Life is full of surprises.
He found hot air rises,
And he floated up higher and higher!

Hot air balloons rise in the air in the same way as convection currents.

53

RADIATION

Energy from the sun reaches us after travelling through space at the speed of light. When this energy hits an object, some of it is taken in or *absorbed*. This makes the molecules vibrate more – and so the object is hotter.

Objects take in and give out energy (as radiation) all the time. Different objects give out different amounts of radiation, depending on their *temperature* and their *surface*.

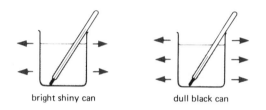

bright shiny can   dull black can

Which can cools down more quickly?
Which can is losing energy more quickly?

A dull black surface loses energy more quickly – it is a good radiator.
A bright shiny surface is a poor radiator.

Brightly polished kettles and teapots do not lose much energy by radiation – they keep warm.

Experiment 8.9
Get two cans of equal size, one which is polished and shiny on the outside and one which is black and dull.
Put a thermometer in each and pour in *equal* amounts of hot water. Let them cool, side by side. Stir and take their temperatures every minute.

The cooling fins on the back of a refrigerator, in a car radiator and on a motor bike engine are usually dull black so that they will radiate more energy away.

Central heating 'radiators' are not usually painted dull black because they lose most of their energy by convection and should be called 'convectors'.

If different surfaces give out different amounts of energy, do different surfaces take in different amounts of energy?

Experiment 8.10

Use the same two cans as in experiment 8.9, but this time pour in equal amounts of cold water. Place the two cans in full sunlight or place them equal distances from an electric fire. Stir and take the temperatures every minute.

sun or fire

bright shiny can   dull black can

Which can heats up more quickly?
Which can is absorbing energy more quickly?

A dull black surface is a good absorber of radiation (as well as a good radiator). It takes in and gives out a lot of radiation.
A bright shiny surface is a poor absorber of radiation — it reflects the radiation away.

A fire-fighting suit is bright and shiny so that it does not take in a lot of energy and burn the fireman.

In hot countries, people wear bright white clothes and paint their houses white to reduce absorption of energy from the sun.
In the same way, petrol storage tanks (and sometimes factory roofs) are sprayed with silver paint to reflect the sun's rays.

Infra-red rays

Energy to heat us up travels from the sun at the speed of light, just like the light rays. The rays which cause the most heating are called *infra-red rays*. These are like light rays but have a *longer wavelength* than the light we use to see with (see page 225).

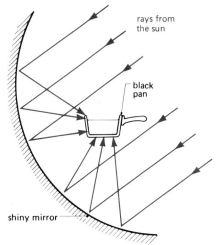

Like visible light rays, infra-red rays can be reflected by mirrors. In hot countries, a large curved (concave) mirror can be used as a 'solar furnace' to collect the rays from the sun and focus them on to a kettle or a pan for cooking.

In an electric fire, a mirror is used in the opposite way, to reflect rays out to the room.

Experiment 8.11

On a sunny day, use a concave mirror to heat a spoonful of water or to light a match.

Why does it help if you first blacken the head of the match with a pencil?

Greenhouses

The inside of a greenhouse is warmer than the outside because the rays from the very hot sun have a short wavelength which can get through the glass. These rays are absorbed by the plants which get warmer and also radiate infra-red rays. However, because the plant is not very hot, the rays have a longer wavelength and cannot get through the glass. So energy is radiated in, but cannot radiate out again.

The bowl of my goldfish, (called Egbert by me),
Was left in the sun for an hour or three.
It absorbed infra-red,
Till Egbert was dead,
So we all ate boiled Egbert for tea.

The vacuum flask

A vacuum or 'Thermos' flask will keep tea hot (or keep ice cream cold).
It does this by reducing or stopping conduction, convection and radiation.

It is a double-walled glass bottle. In the space between the two walls, both pieces of glass are coated with shiny bright 'silvering' and all the air is pumped out to form a vacuum.

A *vacuum* is used because it stops energy transfer, by stopping conduction and convection.
The *silvering* on one glass wall reduces radiation of energy and the silvering on the other glass wall reflects back any infra-red rays that may have been radiated.

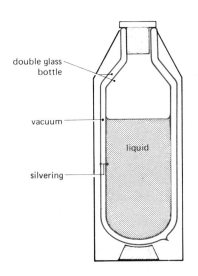

double glass bottle

vacuum

liquid

silvering

Summary

Conduction
The energy is passed from one vibrating molecule to the next.
All metals are good conductors. Water is a poor conductor. Air is a very poor conductor.

Convection
Hot liquids and gases expand and rise while the cooler liquid or gas falls. (A convection current.)

Radiation
Energy is transferred by infra-red rays which can travel through a vacuum, at the speed of light.
Dull black surfaces are good radiators and good absorbers.
Shiny bright surfaces are poor radiators and poor absorbers — they reflect the infra-red rays.

Professor Messer "RAYS-ES" the temperature!

57

Physics at work: Heat Radiation

Solar heating

Although the sun's energy is free, it is not easy to make use of it. One way is to fit solar panels to the roof of a house, in order to make hot water. The hot water can be used for washing or for central heating radiators.

The photograph shows some houses with solar panels: Which way do you think is North?

Here is a simple design for a solar panel:

1. Why is the main surface dull black?
2. Why is it covered with glass? (see page 56)
3. Why is the back insulated?
4. Should the pipes inside the box be made of copper or plastic?
5. Why is the pipe inside the box made as long as possible?
6. Why should the pipe CD be kept as short as possible?
7. Where would you add insulation?
8. Why does the water move through the pipes?
9. Why is the storage tank placed at a higher level than the solar panel?
10. Why would an electric immersion heater usually be added? (see page 304)

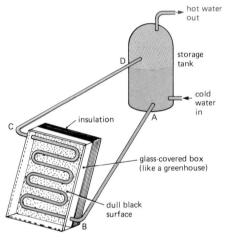

Thermography

Special photographs which are taken using infra-red rays are called thermographs. Thermographs can be used by doctors because diseased parts of the skin are often hotter (and show up lighter on the thermograph).

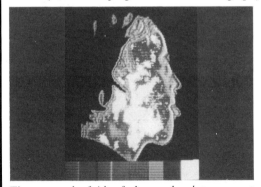

Thermograph of side of a human head, tongue out

An infra-red photograph of an airfield. The plane at the right has warm fuel in its wings. The plane in the centre has its engines running. At the left, a 'heat shadow' shows where a plane has left the airfield.

Physics at work: Keeping Warm

When string vests are worn under clothes, they trap pockets of air and keep you warm. Why are string vests no good by themselves?

Why is this eskimo wearing her coat with the fur side inwards?
Why does it help to have fur at the neck, wrists and bottom?

New-born babies are sometimes wrapped in a plastic sheet coated with shiny aluminium. This reflects the infra-red rays and keeps the baby warm.

Why do mountain-rescue teams also carry some of this shiny plastic?

Physicists have found ways to show the energy we lose from our bodies.

Here the *Schlieren* method shows us the convection currents round a person's body:

Here is a photograph taken using infra-red rays:
This thief was walking away from the camera in total darkness!
You can see that more energy was lost from his warm bare hands and where his clothes were tight (at the shoulders and trousers).
Why does it help to have loose clothes?

The latest padding for anoraks uses Physics to reduce conduction, convection *and* radiation:

cloth
cloth
Shiny plastic strips.
These trap air *and* reflect back the infra-red rays.

Things to do:
1. Devise an experiment to compare the insulation of two anoraks. You can have a beaker of hot water placed in the open sleeve of an anorak. What else would you need? List all the things you would do to ensure this was a *fair* test.
2. A competition for the class: using only a yoghurt pot and anything that will fit *inside* it, who can keep an ice-cube longest? Can you beat 3 hours?

Questions

1. Copy out and fill in the missing words:
   a) Thermal energy travels through the bottom of a pan by . . . . . The energy is passed from one vibrating molecule to the next.
   All metals are good . . . . .
   Plastics, water and air are poor . . . . .
   (good . . . . .)
   b) . . . . . currents can form when liquids and gases are heated. The hot fluid . . . . . and the cold fluid . . . . .
   c) Energy can travel through empty space by . . . . . rays, which can be . . . . . by mirrors like light rays. Dull black surfaces are . . . . . radiators and . . . . . absorbers. Shiny, bright surfaces are . . . . . radiators and . . . . . absorbers.
   d) A vacuum flask uses silvering to cut down heat transfer by . . . . . and uses a vacuum to cut down heat transfer by . . . . . and . . . . .

2. Explain the following:
   a) Copper or aluminium pans are better than iron ones.
   b) A fur coat is warmer if worn inside out.
   c) Sheets of newspaper can be used to keep ice cream cold and keep fish and chips hot.
   d) In winter fat people keep warmer than thin people.
   e) A carpet feels warmer to bare feet than lino.
   f) Two thin blankets are usually warmer than one thick one.
   g) Eskimos build their igloos from snow not ice.

3. Explain seven ways by which energy loss from a house is reduced.

4. a) Why do flames go upwards?
   b) Why do a chimney and a fire help to ventilate a room?
   c) Why would you crawl close to the floor in a smoke-filled room?
   d) Why is the element at the bottom in an electric kettle?
   e) Why is the freezer compartment at the top of a refrigerator?
   f) Why is there often a seabreeze during the day and a land-breeze at night?
   g) How can a 500kg glider stay up in the air?

5. Explain with a diagram, how a domestic hot-water system works.

6. a) Why does a shiny teapot stay hotter than a dull brown teapot?
   b) Why are fire-fighting suits made of shiny material?
   c) What would probably happen if an astronaut's spacesuit was dull black?
   d) Why is it more likely to be frosty on a clear night than on a cloudy night?
   e) Why is the back of a refrigerator painted black?
   f) If pieces of black paper and white paper are laid on snow in sunshine, what is likely to happen?
   g) What might be the effect of sprinkling black soot over the ice in the arctic or the antarctic?

7. Explain why plants are warmer in a greenhouse than outside.

8. Explain, with a diagram, how a vacuum flask can a) keep tea hot and (b) keep ice-cream cold.

Further questions on page 82.

chapter 9
Changing State

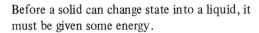 **SOLID** to **LIQUID**

Before a solid can change state into a liquid, it must be given some energy.

In fact to melt 1kg of ice at 0°C into 1kg of water at 0°C, we must give it 340 000 joule of energy.

This energy is called 'hidden heat' or *latent heat* because it does not make it any warmer − it is still at 0°C.

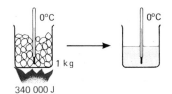

In an opposite way: to make 1kg of ice at 0°C, a refrigerator would have to take 340 000 J *away from* 1kg of water at 0°C.

This number − 340 000 J/kg of water − is called the *specific latent heat of fusion of water.* (Fusion means melting). Other substances have different values.

The specific latent heat of fusion of a substance is the amount of energy (in joule) needed to melt 1kg of the solid to liquid without changing the temperature. Its unit is J/kg.

Molecules

The latent heat energy is given to the molecules in a solid, so that they can move more freely as in a liquid. In ice there are strong forces between the molecules, so it needs a lot of energy to melt it.

Calculating how much energy is needed.

If 1kg of ice at 0°C needs 340 000 J to melt, 2kg would need twice as much and 3kg would need three times as much. We multiply like this:

> **Energy needed to melt ice at 0°C = 340 000 × mass**
>
> (J/kg)
>
> joules

Example

How much energy is needed just to melt 100g of ice at 0°C?

What do we know?	Mass of ice = 100g = 0·1kg
	Specific latent heat = 340 000 J/kg

formula:
put in
the numbers:

Energy needed = 340 000 × Mass
= 340 000 × 0·1 joule
= 34 000 J (=energy from about 20 matches)

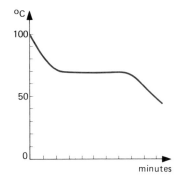

joule meter

12-volt supply

0·1kg ice

Experiment 9.1

Put 100g of ice in a beaker with a heater connected to a joulemeter.

Find how many joule are needed to just melt the ice without making it hotter.

When you have found the number of joule for 0.1kg of ice, calculate the number for 1kg of ice and see it agrees with the accepted value of 340 000 J/kg. If it is different, suggest reasons.

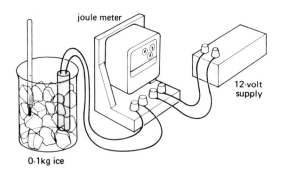

Experiment 9.2 To find the melting point of a substance

Put some of the substance (octadecanoic acid, or hexadecanol) in a test tube and melt it by putting the test tube in a beaker of boiling water.

Then take out the test tube and let it cool down, using a thermometer to take the temperature every ½-minute.

Plot a graph of your results.

From your graph, work out the melting point of the substance. Using the idea of latent heat, explain what is happening during the level part of the graph.

Changing the melting point in two ways

1 Impurities

Experiment 9.3

Put some small pieces of ice in a beaker and sprinkle some salt on the ice. Stir until the ice melts and then take its temperature.

What do you find?
This shows that when water is not pure, it melts at a temperature colder than 0°C.

Impurities lower the melting point of water.

Why is salt put on icy roads by the Council?
Why do motorists add another liquid to the water in their car?

Mrs Messer: How do you make anti-freeze?
Professor: Hide her coat!

2 Pressure

Experiment 9.4

Press two ice cubes together as hard as you can and then release the pressure.

What has happened?
Squeezing causes the ice to melt, but when you release the pressure, it refreezes and 'glues' the ice cubes together.
We make snowballs in just the same way. This shows that when the pressure is increased, ice melts at a temperature colder than 0°C.

Increased pressure lowers the melting point of water.

Experiment 9.5

How to cut a block of ice in half without breaking it!

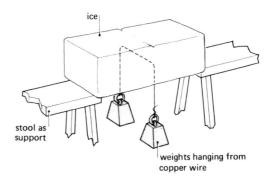

The high pressure of the thin copper wire lowers the melting point, so the ice melts and the wire moves down. When the water moves above the wire, it is not squeezed, so it re-freezes.
This is called *regelation*.

In a similar way, icy glaciers flow very slowly down mountains.

63

Changing State

When water is heated by a bunsen burner, it reaches 100°C and then stays at that temperature as the water changes into steam.

The energy from the bunsen is needed as latent heat – the energy is needed to overcome the forces between the molecules (and also to push back the surrounding air molecules).

In fact, to boil 1kg of water at 100°C into steam at the same temperature, we must give it 2 300 000 joule of energy.

In an opposite way, if 1kg of steam at 100°C condenses to hot water at 100°C, it will give up 2 300 000 J of energy.

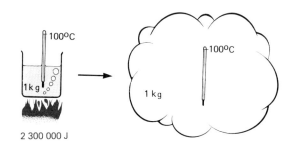

This number – 2 300 000 J/kg of water – is called the *specific latent heat of vaporisation of water.* Other substances have different values. **The specific latent heat of vaporisation of a substance is the amount of energy (in joule) needed to boil 1kg of the liquid to gas without changing the temperature. Its unit is J/kg.**

Experiment 9.6
You can do an experiment to find the number for water after reading question 6 at the end of this chapter.

To calculate the energy needed to boil a different mass of water, we use a formula like the last one:

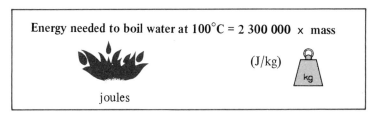

Energy needed to boil water at 100°C = 2 300 000 × mass

(J/kg)

joules

How much energy would be needed to boil 2kg of water at 100°C?

Changing the boiling point in two ways

1 Impurities

Experiment 9.7
Put some salt or other impurity into a beaker of water and heat it until it boils.
Measure the boiling point with a thermometer.

What do you find?

Impurities raise the boiling point of water.

Why do potatoes cook faster if you put salt in the water?

2 Pressure

Experiment 9.8
Put some hot (but not boiling) water in a strong flask and use a pump to reduce the air pressure.

What happens?

What is the boiling point of the water now?

Reducing the pressure lowers the boiling point of water.

This is because the molecules find it easier to escape from the liquid when the pressure is less.

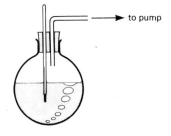

As you go higher in the air, the air pressure decreases. This means that it is difficult to make a good cup of tea on Mount Everest because the water boils at only 70°C.

What would happen to the blood in an astronaut's body if he went into space without wearing his spacesuit?

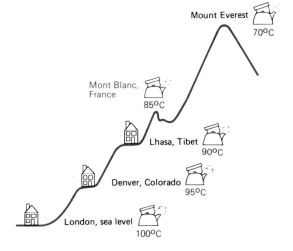

65

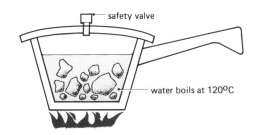

safety valve

water boils at 120°C

The pressure cooker

In an opposite way, increasing the pressure raises the boiling point of water. This is used in the *pressure cooker.*

Because the pressure is about twice normal air pressure, the water boils at about 120°C and the food cooks more quickly.

Evaporation

Experment 9.9
Put some methylated spirit or petrol on the back of your hand.
What do you feel as it evaporates away?

How can you tell the wind direction by holding up a wet finger?
Why do you often feel cold after getting out of the swimming baths?

In each case the liquid takes latent heat from your body so that the liquid can change state by evaporating. The faster (hotter) molecules leave and the slower (colder) molecules stay behind until your body warms them.

Your body sweats more in hot weather. This is so that you keep cool as the sweat takes in latent heat from your body and evaporates.

See if you can decide the four factors that affect evaporation by thinking about drying clothes on wash-day:
1.   Would you choose a dry day or a wet day?
2.   Would you choose a cold day or a hot day?
3.   Would you choose a still day or a windy day?
4.   Would you fold the clothes or spread them out to show a large area?

Freezing by evaporation

Experiment 9.10

Use a pump to suck air bubbles through a small tank of ether resting in a saucer of water.

The ether evaporates quickly, taking its latent heat from the tank and from the water.

What happens to the water after a few minutes?

A similar idea is used in a kitchen refrigerator:

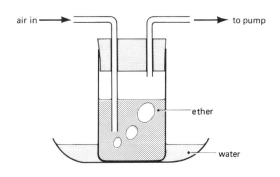

The refrigerator

An electric motor is used to pump a liquid called 'Freon' from a low pressure pipe to a high pressure pipe, from where it squirts through a narrow hole back into the low pressure pipe and round again.

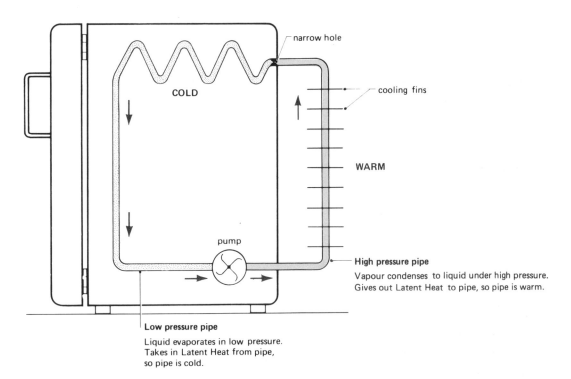

High pressure pipe
Vapour condenses to liquid under high pressure. Gives out Latent Heat to pipe, so pipe is warm.

Low pressure pipe
Liquid evaporates in low pressure. Takes in Latent Heat from pipe, so pipe is cold.

Why is the cold pipe put at the *top* of the box?
What colour would you paint the cooling fins?
Where would you put the insulation?

How to get water from the ground — even in the desert

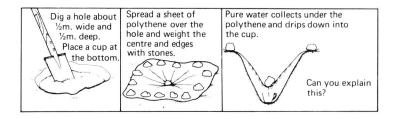

Dig a hole about ½m. wide and ½m. deep. Place a cup at the bottom.

Spread a sheet of polythene over the hole and weight the centre and edges with stones.

Pure water collects under the polythene and drips down into the cup.

Can you explain this?

Summary

Substances must be given latent heat before they can melt or boil. This energy is needed to overcome forces between the molecules.

The specific latent heat of fusion (melting) of a substance is the amount of energy (in joule) needed to melt 1 kg of the solid to the liquid, without changing the temperature. Its unit is J/kg.

The specific latent heat of vaporisation (boiling) of a substance is the amount of energy (in joule) needed to boil 1 kg of the liquid to the gas, without changing the temperature. Its unit is J/kg.

Energy needed to = specific latent × mass
  change state   heat

  joules   J/kg   kg

Impurities or a change of pressure affect the melting point and the boiling point of water as the diagram shows:

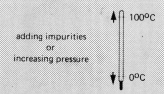

adding impurities
or
increasing pressure

100°C

0°C

Evaporation depends on the temperature, the dryness of the air, the wind and the surface area. When liquids evaporate, they cool down. This is used in the refrigerator.

Professor Messer's boiling mad!
He feels quite sure that he's been had.
The mercury just won't go high,
Please work it out and tell me why:

Questions

1. Copy out and complete:
   a) A solid must be given . . . . . heat before it can melt into a liquid and a liquid must be given . . . . . heat before it can boil into a gas. The energy is used to overcome the forces between the . . . . . so that they can move more freely.
   b) The specific latent heat of fusion of a substance is the amount of . . . . . (measured in . . . . .) needed to melt . . . kg of a solid into a . . . . . without changing the . . . . . Its unit is . . . . . / . . . . .
   c) The formula is:
   $$\frac{\text{Energy needed to}}{\text{change state}} = \text{specific} \ldots \text{heat} \times \ldots$$
   d) Impurities . . . . . the melting point of water. Increased pressure . . . . . the melting point of water.
   e) Impurities . . . . . the boiling point of water. Increased pressure . . . . . the boiling point of water.
   f) Evaporation causes cooling because only the . . . . . (hotter) molecules escape, leaving the slower (. . . . .) molecules in the liquid.

2. Explain the following, using the idea of molecules where possible.
   a) Lakes are more likely to freeze than the sea.
   b) Salt and sand are sprinkled on roads in winter.
   c) Motorists put 'anti-freeze' in the cooling system in winter.
   d) Snowballs cannot be made in very cold weather.
   e) A steam burn is worse than a hot water burn.
   f) A pressure cooker cooks food faster.
   g) A car cooling system is usually pressurised.
   h) An astronaut needs a pressurised spacesuit all round his body.
   i) It is difficult to make tea on Mount Everest.
   j) Wet clothes feel cold.
   k) Tea is cooled more rapidly by blowing on it.
   l) Athletes should put on a tracksuit soon after finishing a race.
   m) A swimming pool is colder on a windy day.

3. What factors affect the rate of evaporation of a liquid?

4. Explain, with a labelled diagram, how a refrigerator works.

5. If the specific latent heat of ice is 340 000 J/kg and the specific latent heat of steam is 2 300 000 J/kg, calculate
   a) the energy needed to melt 2kg of ice at 0°C
   b) the energy needed to melt 500g of ice at 0°C
   c) the energy needed to boil 3kg of water at 100°C
   d) the energy needed to boil 100g of water at 100°C

6. An electric kettle (marked 3kW) produces 3000 joule of heat every second. It is filled with water, weighed and switched on with the lid left off.
   After coming to the boil, it is left on for a further 80 seconds and then switched off. It is found to be 100g lighter.
   Calculate the specific latent heat of steam. Suggest reasons why your answer does not agree with the accepted value.

7. If the specific latent heat of ice is 340 000 J/kg, the specific latent heat of steam is 2 300 000 J/kg and the specific heat capacity of water is 4200 J/kg K, calculate the heat needed to change 2 kg of *ice* at 0°C to *steam* at 100°C.

8. 0·1 kg of ice at 0°C is added to 1·0 kg of water at 30°C. The mixture is stirred and becomes all water, at 20°C. If the specific heat capacity of water is 4200 J/kg K, calculate the specific latent heat of ice.

Further questions on page 83

chapter 10

HEAT ENGINES

All these engines convert fuel energy (chemical energy) into movement energy (kinetic energy).

They do this by burning the fuel at a high temperature so that the molecules have a lot of movement energy and this energy is used to push the engine forwards.

These engines are not very efficient because a lot of energy is wasted (as heat energy).

Note — these three experiments could be dangerous. Do *not* attempt them yourself unless your teacher tells you to.

hot air

Experiment 10.1
Get and empty tincan with a press-on lid. Push on the lid but not too tightly.
Place the tin over a bunsen burner and stand well back.
What happens?

As the temperature rises, the molecules move faster and faster and beat against the lid until they force it off violently.

Experiment 10.2
Repeat the experiment with a few drops of water inside the tin.

Does the lid come off sooner?
This is because the molecules of steam are hitting the lid as well as the molecules of hot air.

The high pressure of steam was used in steam engines but they were only about 8% efficient — that is, for every 100 joules of energy in the fuel (coal), only 8 joules were used to push the engine and the other 92J were wasted.

Experiment 10.3
Make a small hole in the lid of your tin and a larger hole in the bottom so that it just fits over a bunsen burner.
Use the bunsen burner to fill the tin with gas and then immediately put the tin on a tripod, light the gas at the small hole in the lid and stand well back.

gas + air

The flame burns lower and lower, using up gas, until with the correct mixture of air and gas there is a violent explosion. This is called *internal combustion*.

In the *internal combustion engine* of a motor car, the explosion is made even more violent by
1.   Using petrol vapour instead of gas
2.   Squeezing the petrol vapour to a high pressure before exploding it.

The 4-stroke petrol engine—suck, squeeze, bang and blow

1 Suck

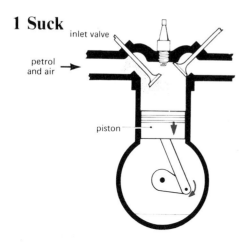

inlet valve

petrol and air →

piston

(*Induction* stroke)
The inlet valve is open and the piston is moving down. A mixture of petrol vapour and air is sucked in from the carburettor.

2 Squeeze

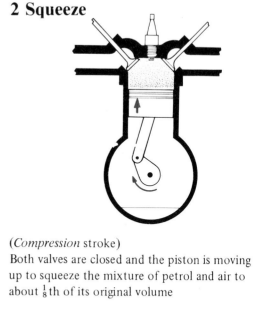

(*Compression* stroke)
Both valves are closed and the piston is moving up to squeeze the mixture of petrol and air to about $\frac{1}{8}$ th of its original volume

3 Bang

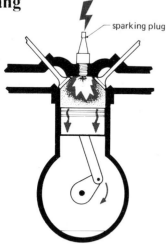

sparking plug

(*Power* stroke)
An electric spark from the sparking plug ignites the mixture which burns rapidly and forces the piston down.

4 Blow

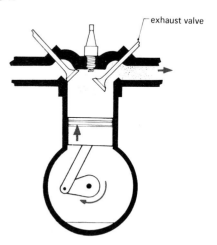

exhaust valve

(*Exhaust* stroke)
The exhaust valve is open and the piston is moving up, to push out the waste gases. The cycle then begins again.

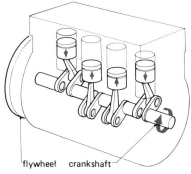

flywheel   crankshaft

Motor cars usually have four pistons connected by a crankshaft to a flywheel. The flywheel smooths out the jerks and keeps the engine turning over between explosions.

Motor cars are usually about 25% efficient — three quarters of the chemical energy of petrol is wasted.

Diesel engines

Diesels are built almost the same as petrol engines, but there are five differences.

1. They use diesel oil instead of petrol.

2. There is no carburettor – only air is sucked in. Then the fuel is squirted in through a valve when the piston is at the top of its compression stroke.

3. There is no sparking plug.

4. The engine is designed to squeeze the air more than in a petrol engine. It squeezes the air to about one-sixteenth of its volume and this makes it so hot that the fuel catches fire and explodes as soon as it is squirted in.

5. The diesel engine is about 40% efficient.

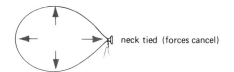

neck tied (forces cancel)

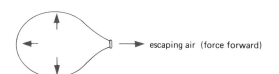

escaping air (force forward)

Jets and rockets

Experiment 10.4

Blow up a balloon and then release it.

What happens?

The balloon moves forward because the escaping air is rushing backwards.

In the same way, jet engines and rockets move forward because of the hot gases rushing out of the back.

Which way would Professor Messer's machine-gun boat move?

These are all examples of Newton's third law (see page 114).

Jet engine—suck, squeeze, burn and blow

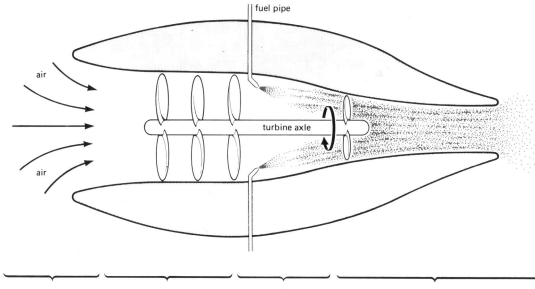

fuel pipe

air

air

turbine axle

1. SUCK	2. SQUEEZE	3. BURN	4. BLOW

1. Air sucked in.

2. Air squeezed by compressor fans.

3. Fuel squirted in, burns continuously.

4. Burnt gases blow out of the back like a blowlamp and push the engine forward (they also turn a turbine fan which keeps the compressor fan turning).

Rocket engine

A rocket engine uses the same idea but carries its own compressed air or oxygen with it, so that it can go outside the Earth's atmosphere.

fuel
(liquid hydrogen)

liquid oxygen

Questions

1. a) In a heat engine, ... energy is converted to . . . . . energy, because at high temperature, the . . . . . move more violently and exert a larger . . . . .
    b) The 4-stroke cycle in a petrol engine or a diesel engine is 1. Suck (induction), 2. . . . . . (. . . . .), 3. . . . . . (. . . . .), 4. . . . . . (. . . . .). A petrol engine is only about . . . % efficient but a diesel engine is about . . . % efficient because a diesel engine squeezes the air . . . . . than a petrol engine.
    c) A jet engine burns fuel continuously. The four stages are suck, ... , ... , ...
    d) A rocket engine carried its own . . . . . and so can work . . . . . the Earth's atmosphere.

2. Describe, with diagrams, the cylinder and piston of a 4-stroke petrol engine and explain what happens during each part of the cycle.

3. Why do motor cars have a) a carburettor b) sparking plugs c) a crankshaft d) gears e) a clutch f) four or six small cylinders rather than one large cylinder?

4. What is meant by compression ratio?
    Explain, with diagrams, the ways in which a diesel engine is different from a petrol engine.

5. Describe, with a large diagram, how a jet engine works. In what way is a rocket engine different?

6. Copy out and complete the rhymes, using the words in the correct order.

    Down the road, fast or slow,
    (suck, ... , ... , ...)
    Car or lorry, van or truck,
    (Squeeze, ... , ... , ...)
    Don't forget to oil and grease,
    (Bang, ... , ... , ...)
    Petrol burns and pistons clang,
    (Blow, ... , ... , ...)
    Efficiency is awful low!
    (Suck, ... , ... , ...).

Further questions on page 83.

Multiple choice questions on thermal energy

Most examination papers include questions for which you are given five answers (labelled **A, B, C, D, E**) and you have to choose the *ONE* correct answer.

Although the answers are given, the questions are not always as easy as they look. Take care to read all the answers before you choose the one that you think is correct.

Example

The metre is a unit of

- **A** speed
- **B** time
- **C** force
- **D** length
- **E** mass

The correct answer is **D**.

Now write down the correct letter for each of the following questions (but do not mark these pages in any way).

1. When a match burns, the main change of energy is from
   - **A** thermal energy to chemical energy.
   - **B** chemical energy to thermal energy.
   - **C** light energy to thermal energy.
   - **D** sound energy to light energy.
   - **E** electrical energy to kinetic energy.

2. Which of the following is an example of the *useful* application of expansion?
   - **A** Putting a bridge on rollers.
   - **B** Putting large bends in steam pipes.
   - **C** Cracking a glass bottle when boiling water is poured in.
   - **D** Loosening a bottletop with hot water.
   - **E** Leaving an airspace above the liquid in a sealed bottle.

3. The bulb of a liquid-in-glass thermometer is made of *thin* glass. This is so that
   - **A** the liquid can be seen more clearly.
   - **B** it acts as a magnifying glass.
   - **C** mercury can be used instead of alcohol.
   - **D** the liquid expands further up the tube.
   - **E** the liquid expands more quickly.

4. A bucket is filled with equal amounts of hot and cold water. The hot water is at $70^\circ C$ and the cold water is at $10^\circ C$. The temperature of the mixture will be approximately
   - **A** $20^\circ C$
   - **B** $30^\circ C$
   - **C** $40^\circ C$
   - **D** $50^\circ C$
   - **E** $60^\circ C$

5. 2000 joules of energy are needed to heat 1kg of paraffin through $1^\circ C$. How much energy is needed to heat 10kg of paraffin through $2^\circ C$?
   - **A** 4000 joules
   - **B** 10 000 joules
   - **C** 20 000 joules
   - **D** 24 000 joules
   - **E** 40 000 joules

6. Chris and Melanie each have a metal can into which they put some water. Using bunsen burners they each begin to heat their cans at the same time, but Chris's water boils first. Which one of the following statements could explain this?
   - **A** Chris's can is more polished than Melanie's.
   - **B** Chris's can is heavier than Melanie's.
   - **C** Chris's can is in a colder place.
   - **D** Chris put colder water in his can.
   - **E** Chris put more water in his can.

7. When an electric fire heats a room, the energy reaches other parts of the room by
   A conduction only.
   B convection only.
   C radiation only.
   D conduction and radiation only.
   E conduction, convection and radiation.

8. Which of the following statements about a vacuum ('thermos') flask is NOT correct?
   A The vacuum reduces transfer of heat by conduction.
   B The vacuum reduces transfer of heat by convection.
   C The vacuum reduces transfer of heat by radiation.
   D The silvering reduces transfer of heat by radiation.
   E The cork stopper reduces transfer of heat by convection.

9. Which of the following describes how the melting point and boiling point of water are affected by an *increase* of pressure?

	melting point	boiling point
A	lowered	lowered
B	lowered	raised
C	unchanged	unchanged
D	raised	lowered
E	raised	raised

10. Which of the following can NOT be explained by the movement of molecules?
    A Expansion
    B Conduction
    C Convection
    D Radiation
    E Evaporation

When you have finished,
a) check all your answers carefully
b) discuss your answers (and why you chose them) with your teacher.

More questions can be found in *Multiple Choice Physics for You.*

Molecules

1. This is a question about the kinetic theory
    of matter.
    a) Solids keep their shape at room tempera-
    ture. Draw a labelled diagram of a model
    which shows how the particles are
    arranged in a solid. How would you show
    a rise in temperature with your model?
    b) In liquids and gases, particles flow about.
    Why are the particles more free in a gas
    than in a liquid?
    c) Much energy is needed to boil a liquid
    and to turn it into a gas. What do the
    particles use this energy for?
    d) A gas molecule can have a speed of 400
    metres per second, yet two gases mix very
    slowly. How is this explained?
    (SR)

2. The movement of very small particles in a
    gas or liquid, known as Brownian motion,
    can be observed by illuminating the particles
    in the medium from one side, and viewing
    the effect through a microscope.
    a) Why is good illumination necessary?
    b) Describe briefly, or illustrate by means of
    a diagram, the motion which is observed.
    c) Give the names of (i) suitable particles for
    this experiment, and (ii) the medium in
    which these particles could be suspended.
    d) Why is the motion irregular, even when
    the particles suspended in the medium are
    all the same size?
    e) When particles of different sizes are
    present, why will the motion of the larger
    particles usually be different from the
    motion of the smaller particles? In what
    way will the motions differ?
    (C)

3. a) Describe how you could obtain a value
    for the order of magnitude of molecular
    sizes by an experiment you could carry
    out in a school laboratory. Explain the
    theory of the experiment clearly and

mention the principal practical difficulties
involved.
    b) State **two** ways in which a solid differs
    from a gas. How does the behaviour of
    the molecules, in each case, account for
    the difference?
    c) Explain why energy has to be given to a
    solid to convert it to a liquid at the same
    temperature.
    (O&C)

4. a) An oil drop of volume 10^{-9} m^3 spreads
    out on water to form a film of area
    $0 \cdot 2$ m^2. Estimate the length of an oil
    molecule from this information. What
    assumption have you made in your
    calculation?
    Describe what could be done to make the
    film of oil clearly visible in this experi-
    ment. What precautions should be taken
    to achieve a reliable result?
    b) Explain how a gas exerts a pressure on
    the walls of its container.
    c) Outline very briefly **two** situations in
    which the phenomenon of surface tension
    is exhibited.
    (L)

5.

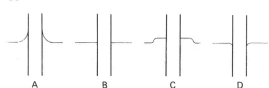

A   B   C   D

The diagrams show a clean glass rod dipping
into mercury. Which diagram shows the
mercury surface correctly?
    (WM)

6. Describe **two** simple experiments which
    demonstrate the phenomenon of surface
    tension.
    Use the kinetic theory of matter to explain
    the observations made in **one** of the
    experiments you have described.   (JMB)

Expansion

7.

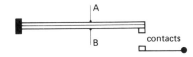

The diagram shows a bimetallic strip as used in a simple fire alarm.
a) Name suitable materials for A and B.
b) What adjustment would you make to the arrangement shown to make the alarm operate at a lower temperature? Give a reason for your answer.   (C)

8. A bimetal strip bends when heated.
   a) What fact about two different metals makes them suitable for making such a strip?
   b) With reference to the fact given in answer to (a), which of the two metals will form the 'outside' of the curved strip when it is heated?
   c) Name **two** suitable metals for making a bimetal strip.
   d) Which of the two metals you have named in answer to (c) will form the outside of the curved strip when it is heated?
                                                        (N16+)

9. Draw a diagram of a simple mercury thermometer.
   a) What properties of the liquid mercury make it useful for thermometers?
   b) Explain why a sensitive thermometer should have a fairly large bulb and a fairly narrow capillary stem.
   c) Explain why a thermometer which has to reach its readings quickly should have a fairly small cylindrical bulb made of specially thin glass.   (O)

10. Distinguish between *temperature* and *heat*. Explain why mercury is preferable to water for use as thermometric liquid.
   Why does a thick glass tumbler crack when it is suddenly filled with boiling water?
                                                        (WAEC)

11. a) The diagram shows a modern hot water cylinder. It has a volume of 200 litres and contains two immersion heaters labelled P and Q.

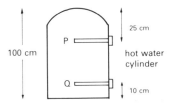

   i) Write down the approximate volume of water which element P can heat up. Show how you got your answer. [4 marks]
   ii) Electricity is available either at a cheap 'off peak' rate or a more expensive rate. Explain why:
      (A) Q is used during the off peak period
      (B) P is necessary.   [4]
  b) The temperature of the water is controlled by a device called a thermostat:

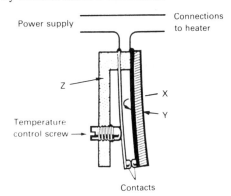

   X and Y form a bimetallic strip.
   Z is an electrical insulator.
   i) Name a suitable material for Z.
   ii) Explain why the bimetallic strip bends when it becomes hot.
   iii) Explain what effect this has on the current through the heater.
   iv) If the temperature control screw is turned so that it moves to the left, does the temperature of the water increase, decrease or stay the same? [7]
                                                        (NEA)

The gas laws

13. Explain, in terms of molecules, their properties and their motion:

a) how a fixed amount of air enclosed in a container exerts a pressure on the walls,

b) why this pressure increases if the temperature is raised,

c) why, if the temperature is kept constant, the pressure is doubled when the volume of the container is halved,

d) why smoke particles in air can be seen under the microscope to be moving about continuously with a random jerky motion.

Describe how you would verify experimentally one of the statements (b), (c) or (d). (O)

14. A tyre originally completely flat is inflated by using a pump.

a) Explain in terms of moving molecules

(i) how the atmosphere exerts a pressure on the outside of the inflated tyre,

(ii) why the air inside the tyre exerts a greater pressure, the temperature of the air in the tyre being the same as that of the air outside.

b) A tyre contains 2000 cm³ of air at atmospheric pressure P. The volume of air in the pump is 100 cm³. What will be the pressure in the tyre after one stroke of the pump, assuming the volume of the tyre and the temperature of the air do not change? (C)

15. a) Describe how you would show experimentally the relationship between the volume and pressure of a fixed mass of gas at constant temperature.

Your answer should include i) a diagram; [3] ii) a statement of the readings made; [2] iii) one precaution taken to ensure a reliable result. [1]

b) The following readings for the pressure and temperature of a fixed mass of gas were obtained in a different experiment.

Pressure/kPa	3·80	4·08	4·36	4·64	4·92	5·20
Temperature/°C	0	20	40	60	60	100

i) Plot a graph of pressure (y-axis) against temperature (x-axis). [4]

ii) From the graph read and state the temperature when the pressure is 4.50 kPa. [1]

iii) Find the pressure when the temperature is 150 K. [2]

iv) Comment on the reliability of the values obtained in parts ii) and iii). [2]

(LEAG)

16. A fixed mass of gas occupies a volume of 2·5 litres at a pressure of 5 Nm⁻². If the temperature is constant find the volume at a pressure of 8 Nm⁻². (WAEC)

17. A sample of dry air is enclosed in a capillary tube between the sealed end of the tube and a mercury index. In order to investigate the expansion of the air with increase in temperature, the length of the air column in the tube is measured at different temperatures. The following readings are obtained.

Length of air column, l(mm)	154	162	170	175	182
Temperature, θ(°C)	18·0	34·0	50·0	62·0	78·0

Plot a graph of l against θ, starting the length scale at 100 mm and the temperature scale at 0°C. Record the value l_1, of l at 0°C, and l_2, of l at 100°C. On a separate piece of graph paper plot the values l_1 and l_2 against θ. For this graph start the length scale at 0 mm and the temperature scale at −300°C, using 40 mm along the temperature axis for 100°C. Record the value of θ at which the straight line joining the two points cuts the temperature axis. Draw a diagram or sketch to show how you would arrange the apparatus to carry out this experiment. Give the reasons why

i) l may be used as a measure of the volume of the air in the tube,

ii) the pressure on the air in the tube is constant throughout the experiment.

State a conclusion which can be drawn from the result of this experiment. (C)

18. A car tyre was tested before being driven on a motorway. The reading on a tyre pressure gauge which reads the pressure above atmospheric was 120 000 N/m² and the temperature was 7°C. At the end of the journey the temperature of the tyre was found to be 35°C. What reading would you expect on the same tyre pressure gauge if the volume of air in the tyre remained constant and atmospheric pressure throughout the journey was 100 000 N/m²? (JMB)

Specific heat capacity

19. 2 kg of a substance of specific heat capacity 4 kJ/kg °C is heated from 20°C to 50°C. Assuming no heat loss, the heat required to do this is

    A   15 kJ   C   160 kJ   E   400 kJ
    B   36 kJ   D   240 kJ   (WY)

20. 8 000 J of energy are supplied to a mass of 2 kg of a metal. If the temperature rise is 10°C then the specific heat capacity of the metal, in J/kg °C, is

    A   400   C   1 600   E   16 000
    B   800   D   4 000   (NI)

21. An experiment to measure the specific heat capacity of an aluminium block uses an electrical heater to heat the block and then to heat water.

    a) What is meant by specific heat capacity?
    b) Aluminium has a low specific heat capacity. Explain why this is useful in aluminium saucepans.
    c) Water has a high specific heat capacity. Explain why this is useful in the home.
    d) Here are some results of the experiment described above.

Mass of aluminium block	= 1 kg
Starting temperature	= 15°C
Final temperature	= 39°C
Mass of water heated	= 1 kg
Starting temperature	= 15°C
Final temperature	= 20°C

    The water has a specific heat capacity of 4200 J/kg°C.
    (i) Calculate the heat transferred to the water
    (ii) The same amount of heat was supplied to the aluminium. Calculate the specific heat capacity of the aluminium block. (SR)

22. Edinburgh and Moscow are equal distances from the North Pole. In winter Edinburgh's average temperature is 15°C warmer than Moscow's, and in summer Moscow is the warmer. Explain these differences. (EA)

23. a) What is the meaning of the statement: 'The specific heat capacity of water is 4200 J/kg°C'?

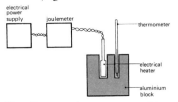

The diagram shows an experiment which can be used to measure the specific heat capacity of aluminium. The joulemeter measures the amount of energy supplied to the heater.

The results obtained were as follows:

Mass of aluminium block	1 kg
Initial temperature	15 °C
Final temperature	19 °C
Initial joulemeter reading	1500 J
Final joulemeter reading	5100 J

    b) What was the temperature rise?
    c) How much energy was supplied?
    d) Use these results to calculate the specific heat capacity of aluminium.
    e) What assumption have you made in this calculation?
    f) Suggest TWO precautions you might take if you were doing the experiment in order to make this assumption a reasonable one. (WM)

24. A 12 W electric heater, working at its stated power, is found to heat 0·01 kg of water from 20°C to 35°C in one minute. Calculate
    (i) the heat energy produced by the heater in one minute,
    (ii) the heat energy absorbed by the water in one minute.
    Account for the difference in the answers to (i) and (ii).
    [Assume that the specific heat capacity of water is 4200 J/kg K (4·2 kJ/kg K).] (JMB)

See page 366 for further questions on electrical heating.

Heat transfer

25. Explain the processes by which a drink is cooled by an ice cube floating in it.
(O&C)

26. The diagram shows a vessel which is used to prevent hot liquids from going cold. It can also be used to keep cold liquids cool in hot weather.
    a) What is the name of this vessel?
    b) (i) What is the part labelled *A* usually made from?
       (ii) Why is this material chosen for this part?
    c) The walls of this vessel are made of thin glass but its inner surfaces, B, are specially treated.
       (i) What is special about these surfaces?
       (ii) Why are these surfaces treated in this way?
    d) (i) What is contained in the space *C*?
       (ii) What is the purpose of the space *C*?
(NW)

27. This question is about solar panels, devices that are sometimes seen on the roofs of houses and are used to provide hot water. An example is shown on page 58.
    a) What is the purpose of the following:
       i) the insulation behind the absorber panel. [1 mark]
       ii) having the absorber panel painted black. [1]
       iii) having a glass cover on the top of the panel. [1]
    b) i) Name suitable materials for making the absorber panel and water-ways. (Do not use brand names). [2]
       ii) Give your reasons for the choice of such materials. [2]
    c) The pipe connecting the water outlet from the panel to the hot water storage tank should be kept short. Why is this desirable? [1]

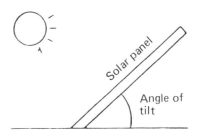

Solar panel

Angle of tilt

d) The angle of tilt of a solar panel greatly affects the amount of energy it receives at different times of the year. The diagram shows what is meant by the angle of tilt. The table of data below shows the effect of different angles of tilt for the summer months.

Maximum daily input of energy in megajoules to a 1 m² panel

Month	Angle of tilt of panel to the horizontal					
	20°	30°	40°	50°	60°	70°
Apr	23.8	24.9	24.8	24.1	22.7	20.5
May	28.4	28.8	27.4	25.2	23.0	19.8
Jun	29.2	29.2	27.4	25.2	22.3	19.1
Jul	28.8	29.2	27.4	25.6	23.0	20.2
Aug	25.6	25.9	26.3	24.8	22.7	20.5
Sept	20.5	21.6	22.3	22.7	21.6	20.5

Use the table above to answer the following question
   i)   What angle of tilt would be ideal for a solar panel in April? [1]
   ii) Is it better to have the solar panel tilted at an angle of 40° or at an angle of 50° for all the months shown in the table? Give the reasons for your answer. [3]
   iii) What is the maximum amount of energy that a 4 m² panel could receive during a day in July? [2]
(LEAG)

Latent heat

28. The following readings were taken in an experiment in which hot naphthalene was allowed to cool down to room temperature.

Reading	A	B	C	D	E	F	G	H	I	J	K
Time (minutes)	0	1	2	3	4	5	6	7	8	9	10
Temperature of naphthalene (°C)	105	95	85	81	81	81	81	80	78	76	74

Plot a graph of temperature against time.
What is happening between readings A and D?
What is happening between readings D and G?
What is happening between readings H and K?
What property of naphthalene can be deduced from the experimental results?

(AEB)

29. In a determination of the specific latent heat of ice, dried ice at 0 °C is added to water at room temperature in a thin metal calorimeter. An electrical heater of known wattage is placed in the water and switched on until all the ice has melted and the temperature has risen to room temperature.
   a) Make a list of all the readings which must be obtained in the course of this experiment.
   b) What additional information will be required in order to calculate the result?
   c) Show how the specific latent heat of ice is calculated in this experiment.
   d) Explain the advantage of starting and stopping the experiment at room temperature.
   e) Why should the ice be dried before use?
   f) Write down **two** further precautions and give a reason for each of them.

(C)

30. 200 g (0·2 kg) of water at 10°C were heated in a flask by a bunsen burner and reached 100°C after 10 minutes heating.
   a) How much heat was supplied in those 10 minutes if the specific heat capacity of water is 4 J/g °C (4000 J/kg °C)?
   The water was then boiled for a further one minute by the same bunsen and it was found that 3 g (0·003 kg) of water had boiled away in that time.
   b) Calculate the specific latent heat of vaporization of water given by these results. What assumption have you made in calculating your answer?

c) Explain carefully why latent heat has to be supplied to turn water into steam. What difference (if any) would it have made to the boiling point of the water if some salt had been dissolved in it?

(WM)

31. Explain
   a) why a kettle of water with a steady supply of heat takes a much longer time to boil dry than it does to reach its boiling point,
   b) how evaporation differs from boiling, and
   c) how the molecular theory of matter accounts for the drop in temperature which results when rapid evaporation of a volatile liquid occurs.

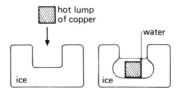

A lump of copper of mass 0·5 kg is placed in an oven for some time and then transferred quickly to a large dry block of ice at 0 °C. When the temperature of the lump of copper reaches 0 °C, 0·3 kg of the ice is found to have melted. Estimate the temperature of the oven. (Take the specific heat capacity of copper as 400 J/kg K and the specific latent heat of fusion of ice as $3·2 \times 10^5$ J/kg.)
Give **two** reasons why this estimate might be lower than the actual temperature of the oven.

(L)

Further Reading

Physics Plus: *Alternatives to Petrol, Convection Around Your Body, Centrally Heating A Home* (CRAC)
Heating and Lighting a Home – Bowers (CUP)
Multiple Choice Physics for You – Johnson (Hutchinson)
Butter Side Up – Pyke (Murray)

Parachutists have two forces acting on them.
One is the force upwards due to air resistance. What is the other?

chapter 11

Pushes and pulls are *forces*. Whenever we are pushing or pulling, lifting or bending, twisting or tearing, stretching or squeezing, we are exerting a force.

Experiment 11.1
Exert some push and pull forces on different objects including a ball, a spring, a piece of plasticine and an elastic band. What do you notice? Make a list of the things that forces can do to the objects.

Forces can
a) change the speed of an object
b) change the direction of movement of an object
c) change the size or shape of an object.

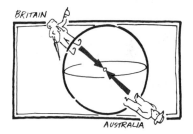

Weight is a very common force. Weight is the force of gravity due to the pull of the Earth. This is the force that is pulling you downwards now. It acts towards the centre of the Earth.

An apple fell on Newton's head,
Moved <u>downwards</u> from the tree,
And that was when Sir Isaac said:
"The force is gravity"

The weight of an object varies slightly at different places on the Earth.

If an object is taken to the Moon, it weighs only about one-sixth as much because the Moon is smaller than the Earth.
In outer space, well away from any planets, objects become weightless.

Stretching a spring

Experiment 11.2

Find out how forces change the length of a spring. Hang a spring from a retort stand and fix a pointer at the bottom of the spring so that it points to the scale of a ruler which is clamped vertically.

You will need several *equal* weights. Hang one weight on the spring and measure the *extension* of the spring.

Then add another weight so that there are two pulling down on the spring. Again measure the extension (from the original position). Increase the load by adding more weights and measure the extension each time.

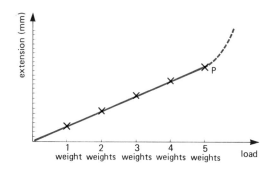

one weight

two weights

What do you notice?
If you double the load, does it double the extension?
If you put on three times the load, do you get three times the extension?

This means that:

The extension is directly proportional to the load.

This is called *Hooke's Law*. This law also applies to the stretching of metal wires and bars.

From your results, plot a graph of extension against load.

What shape is the graph?

A straight line through the origin of the graph confirms that the extension is directly proportional to the stretching force (see page 402).

What happens with very heavy loads?
Hooke's law only applies to the straight part of the graph (up to the *limit of proportionality*).
The point P is called the *elastic limit*.

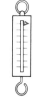

A *spring balance* uses a spring to measure the weight of an object or to measure the strength of any pulling force. The marks on the scale are equally spaced because of Hooke's Law.

Measuring forces

Forces are measured in units called *newtons* (often written N), named after Sir Isaac Newton.

An apple weighs about 1 newton (1N).

The weight of this book is about 8 newton (8 N).

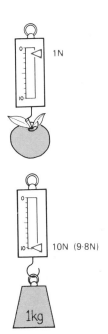

Experiment 11.3
Use a spring balance to find the weight of a 1kg mass.

A mass of 1 kg (here on Earth) weighs almost 10 newton. (To be more exact, 1 kg on Earth weighs 9·8 newton).

How much would a 1kg mass weigh on the Moon, if the pull of gravity there is about one-sixth of the pull of gravity here?

Here is a weighty question from Professor Messer: does 1 newton of lead weigh more than 1 newton of feathers?

Experiment 11.4
Use a weighing machine to find the weight of your body in newton.

Your weight changes slightly from place to place on the Earth – at the North Pole you would weigh about 3 newton heavier (like having 3 extra apples in your pocket).

How much would you weigh on the Moon?

How many press-ups could you do on the Moon?

Experiment 11.5
Use a force-meter (a very strong spring balance) to measure the strength of your arm muscles when you are pushing and when you are pulling.

Did you hear about the prisoner who stepped on some scales and got 'a-weigh'? The Police couldn't 'weight' to catch him!

Mass and inertia

If this book (which has a weight of about 8 newton here on Earth) is taken into outer space, well away from the Earth or any other object, it will become weightless. But it would still be the same book with the same *mass*.

The mass of an object is the amount of matter in it. It is measured in kilogram (kg). (See page 7.) The mass of this book is about 0·8 kg.

When an object is stationary, it needs a force to make it move. The bigger the mass, the bigger the force needed to start it moving. We say that masses have *inertia*, a reluctance to start moving.

Experiment 11.6
Hang two tins from long pieces of string. Fill one with wet sand and leave the other empty. Try pushing the cans.

Which is the harder to push?
Which has the larger inertia?
Which has the larger mass?

Since this experiment depends on the *mass* of the object, (not the weight), you would feel just the same effect on the moon or out in space – if you pushed a large mass you would feel a large amount of inertia.

Experiment 11.7
Place a piece of smooth card on top of a milk bottle and place a penny on top. Flick the card away with your finger.

What happens?
Why does the penny stay when the card moves?

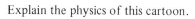

Explain the physics of this cartoon.

In a similar way, moving objects need a force to *stop* them moving.
Their inertia tends to keep them moving.

Experiment 11.8

Pull aside the cans used in experiment 11.6 and
when they swing back, stop them with your hand.
Which is the harder to stop?

Passengers in a car have a lot of inertia and so they
need seat-belts. If the car stops suddenly, the
people will tend to keep on moving (through the
windscreen), unless the seat-belts exert large forces
to stop them.

What happens when the car turns a corner?
Your body, because of its inertia, will tend to
travel straight on. You can feel your body sway as
the car turns the corner, but fortunately your seat
exerts a force on you and this pulls you round the
corner with the car. If the car is travelling round a
circle, this force acts *towards the centre* of the
circle and is called the *centripetal force*.

If you are standing on a bus, what happens when
the bus

a) starts moving
b) stops moving
c) turns a corner to the left?

Sir Isaac Newton stated all this in **Newton's first law of motion**:

> **Every mass stays at rest or moves at constant speed in a straight line
> <u>unless</u> a resultant force acts on it.**

Measuring mass

A beam balance, built like a see-saw, can be used
to compare masses.
If enough standard masses (marked in kg) are put
on to balance the object, then there must be the
same mass on each side.

What is the mass of the object in the diagram?

If it balances here on Earth, it would also balance
if it was taken to the moon. This is because the
pull of gravity decreases equally on each side.

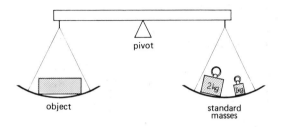

Summary

Forces can
a) change the speed of an object
b) change the direction of movement of an object
c) change the size or shape of an object

Hooke's law: The extension of a spring is proportional to the force pulling it (up to the limit of proportionality).

Weight is the force due to the pull of gravity. It can be measured in newton (N) by a spring balance and varies from place to place.

Mass is the amount of matter in an object. It can be measured in kilogram (kg) by a beam balance and does not change from one place to another. A mass of 1kg here on Earth weighs about 10N (9·8N)

A mass has a reluctance to change its motion, called inertia.

Newton's first law: Every mass stays at rest or moves at constant speed in a straight line <u>unless</u> a resultant force acts on it.

Something to make—a microbalance

Follow the numbered steps to build your own microbalance and then use it to compare the mass of 10cm of your hair with the same length of someone else's hair.

What length of sewing cotton has the same mass as 10 cm of your hair?

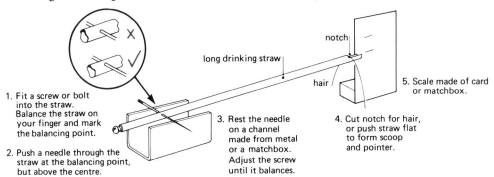

1. Fit a screw or bolt into the straw. Balance the straw on your finger and mark the balancing point.

2. Push a needle through the straw at the balancing point, but above the centre.

3. Rest the needle on a channel made from metal or a matchbox. Adjust the screw until it balances.

4. Cut notch for hair, or push straw flat to form scoop and pointer.

5. Scale made of card or matchbox.

Professor Messer puts his skates on. Explain the physics of this cartoon.

90

Questions

1. Copy out and fill in the missing words:
   a) When a spring is pulled, the . . . . . is proportional to the . . . . . (called . . . . . law).
   b) Weight is the force of . . . . . on an object due to the pull of the . . . . .
   c) Weights and other forces are measured in . . . . .
   d) Here on Earth, the pull of gravity on a mass of 1 kg is . . . . . newton.
   e) Weight is measured by a . . . . . balance and mass is measured by a . . . . . balance.
   f) Newton's first law says: every mass stays at . . . . . or moves at constant . . . . . in a . . . . . line unless a resultant . . . . . acts on it.

2. An object has a mass of 4 kg. What is its weight (in newton) here on Earth.

3. An astronaut has a mass of 60 kg.
   a) What is his weight here on Earth?
   b) What is his weight on the Moon?
   c) What is his mass on the Moon?

4. Should Professor Messer pull the tablecloth quickly or slowly? Explain the Physics of this trick, using the correct scientific words.

5. A spring is 20cm long when a load of 10N is hanging from it, and 30cm long when a load of 20N is hanging from it. Draw diagrams and work out the length of the spring when a) there is no load on it b) there is a load of 5N on it.

6. In an experiment with a spring, these results were obtained:

Load (in N)	50	100	150	200
Length of spring (in cm)	9.0	11.0	13.0	15.0

   Draw a graph of these results and from the graph find
   a) The length of the spring when unstretched
   b) The length of the spring when the load is 80N
   c) The load needed to produce an extension of 5.0cm.

7. Could the convict escape this way?

   Could he lift the ball more easily in a prison on the Moon?
   Could he jump higher in a prison on the Moon?

Further questions on page 168.

chapter 12

DENSITY

Which is heavier – iron or wood? Many people say iron – and yet an iron nail is lighter than a wooden tree!

What people mean is that iron and wood have a different *density*.

To measure density, we need to measure the mass of a definite volume of the substance.

In fact:

$$\text{Density} = \frac{\text{mass}}{\text{volume}}$$

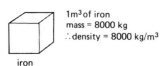

1m³ of iron
mass = 8000 kg
∴ density = 8000 kg/m³

iron

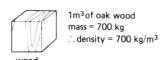

1m³ of oak wood
mass = 700 kg
∴ density = 700 kg/m³

wood

If the mass is measured in **kg** (kilogram) and the volume in **m³** (cubic metre), the density is measured in **kg/m³** (kilogram per cubic metre).

Sometimes the mass is measured in g (gram) and the volume in cm³ (cubic centimetre) so that the density is measured in g/cm³ (gram per cubic centimetre).

Example 1

A block of lead has a mass of 44 000 kg and a volume of 4m³. What is the density of lead?

formula first:	Density	=	$\dfrac{\text{mass}}{\text{volume}}$
then put in the numbers:		=	$\dfrac{44\ 000\ \text{kg}}{4\text{m}^3}$
		=	$11\ 000\ \text{kg/m}^3$

Here are the densities of several substances:

substance			density	
solid	liquid	gas	kg/m³	g/cm³
gold			19 000	19
	mercury		13 600	13·6
lead			11 000	11
iron			8 000	8
	water		1 000	1
ice			920	0·92
	petrol		800	0·80
		air	1.3	0·0013

What do you notice about the numbers?

What do you *guess* the weight of air in this room is?

Example 2
An 'empty' bedroom measures 5m x 4m x 2·5m. If the density of air is 1·3kg/m³, what is the mass of air in the room?

calculate the volume first:	Volume of room	= length x breadth x height
		= 5m x 4m x 2·5m
		= 50m³

formula: $\qquad$ Density $\quad = \dfrac{\text{mass}}{\text{volume}}$

then put in the numbers: $\qquad 1\cdot3\text{kg/m}^3 \quad = \dfrac{\text{mass}}{50\text{m}^3}$

$$\therefore \text{Mass} \quad = 1\cdot3 \ \times \ 50 \ \text{kg} = \underline{65\text{kg}}$$

This air would have a *weight* of almost 650 newton (see page 87).
Do you weigh as much as this (see page 87)?
In most rooms, the air weighs more than you do!

Measuring the volume of an irregular object

Experiment 12.1

Find the volume of a stone using a measuring cylinder.
Partly fill the measuring cylinder with water and note the reading. Remember to read the mark at the *bottom* of the meniscus (see page 19).
Gently lower the stone into the water and note the new reading. The difference gives the volume of the stone.

What is the volume of the stone in the diagram?

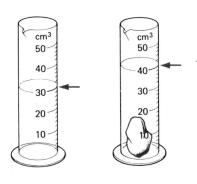

Experiment 12.2

For larger objects, an overflow can is used.
The can is filled up to the spout with water and then a measuring cylinder is placed under the spout before the object is gently lowered in.

The volume of water that overflows into the measuring cylinder must be equal to the volume of the object.

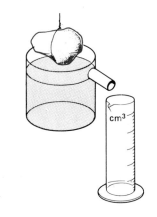

Measuring density

Experiment 12.3

Find the density of a *solid* (for example, a piece of iron), in three stages.

1. Find the mass of the solid by using a beam balance (see page 89).
2. Find the volume of the solid by using a measuring cylinder, with an overflow can if necessary (see page 93).

3. Use the formula:

$$\text{density} = \frac{\text{mass}}{\text{volume}}$$ to calculate the density.

What unit is your result calculated in?

Professor Messer is finding the density of some wood. Why does he need a needle? Is his answer correct?

Experiment 12.4

Find the density of *water* in five stages.

1. Find the mass of a dry empty beaker by using a beam balance.
2. Use a measuring cylinder to pour exactly 100cm³ of water into the beaker.
3. Now find the mass of the beaker with the water in it.
4. Work out the mass of the water from the results of steps 1 and 3.

5. Use the formula:

$$\text{density} = \frac{\text{mass}}{\text{volume}}$$ to calculate the density of water

What unit is your result measured in?
How does your result compare with the value shown on page 92?
What would your result be in units of kg/m³?

Experiment 12.5

Repeat experiment 12.4 to find the density of another liquid.

Can you find Professor Messer's mistake? What is the correct answer?

Relative density

Sometimes the density of a substance is compared with the density of water, to give a number called the *relative density*.

Relative density of a substance $=$	$\dfrac{\text{density of substance}}{\text{density of water}}$

It is just a number, with no units.

From the densities given on page 92:

$$\text{relative density of iron} = \frac{8\,000}{1\,000} = 8.$$

This means that a certain volume of iron has 8 times the mass of the same volume of water — this is the idea used in the next experiment.

Expanded polystyrene blocks have a very low density

Experiment 12.6

Find the relative density of methylated spirit, using a relative density bottle, in six stages.

1. Use a balance to find the mass of a clean, dry relative density bottle.
2. Fill the bottle with methylated spirit, dry the outside carefully and find its new mass.
3. From the results 1 and 2, work out the mass of this volume of methylated spirit.
4. Clean out the bottle and re-fill with water. Dry the outside carefully and find the mass.
5. From results 1 and 4, work out the mass of this same volume of water.

Since the volumes are the *same* in each case, you can work out the relative density from:

$$\text{relative density} = \frac{\text{mass of methylated spirit}}{\text{mass of water}}.$$

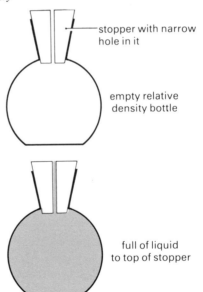

stopper with narrow hole in it

empty relative density bottle

full of liquid to top of stopper

Relative density is useful when considering whether substances will float or sink in water (see page 107).

Summary

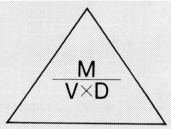

1. Density $= \dfrac{\text{mass}}{\text{volume}}$

Changing round the formula (see page 402):

2. Volume $= \dfrac{\text{mass}}{\text{density}}$

3. Mass $=$ volume x density

To find the one you want, cover up that letter in the triangle and the remaining letters show you the formula.

Density of water $= 1\,000 \text{ kg/m}^3 = 1 \text{ g/cm}^3$.

Relative density $= \dfrac{\text{density of substance}}{\text{density of water}} = \dfrac{\text{mass of substance}}{\text{mass of same volume of water}}$

Questions

1. An object has a mass of 100 gram and a volume of 20 cubic centimetre. What is its density?

2. An object has a mass of 40 000 kilogram and a volume of 5 cubic metre. What is its density?

3. An object has a volume of 3 cubic metre and a density of 6 000 kilogram per cubic metre. What is its mass?

4. Copy and complete this table.

Object	Density (kg/m³)	Mass (kg)	Volume (m³)
A		4 000	2
B	8 000		4
C	2 000	1 000	
D		2 000	4

Which object a) has the greatest mass b) has the least volume?
Which objects could be made of the same substance?
Which object would float on water?

5. A water tank measures 2m x 4m x 5m. What mass of water will it contain?

6. An object has a mass of 20 000 kilogram and a density of 4 000 kilogram per cubic metre. What is its volume?

7. The density of air is $1 \cdot 3 \text{ kg/m}^3$. What mass of air is contained in a room measuring $2 \cdot 5 \text{m x } 4 \text{m x } 10 \text{m}$?

8. A stone of mass 30 gram is placed in a measuring cylinder containing some water. The reading of the water level increases from 50cm^3 to 60cm^3. What is the density of the stone?

9. Describe, in detail, with diagrams, how you would find the density of a piece of coal.

10. Using the table on page 92, what is the relative density of a) gold b) mercury c) ice?

11. Calculate the relative density of the oil in this diagram.

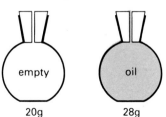

empty   oil   water

20g   28g   30g

96

Further questions on page 168.

chapter 13

PRESSURE

You can push a drawing pin into a piece of wood – but you cannot push your finger into the wood even if you exert a larger force. Why? What is the difference between a sharp knife and a blunt knife?

The difference in each case is a difference of *area* – the point of the drawing pin and the edge of the sharp knife have a small area. A force acting over a small area gives a large *pressure*.
Pressure is *force per unit area*, or

$$\text{pressure} = \frac{\textbf{force (in newton)}}{\textbf{area (in square metre)}}$$

Its unit is newton per square metre (N/m^2). This unit is also called the pascal (Pa).

Why do eskimoes wear snow-shoes?

Example 1
An elephant weighing 40 000 N stands on one foot of area 1000 cm² ($= \frac{1}{10} m^2$). What pressure is exerted on the ground?

formula first: $\text{Pressure} = \dfrac{\text{force}}{\text{area}}$

then put in
the numbers: $= \dfrac{40\,000\,N}{\frac{1}{10}\,m^2}$

$= 40\,000 \times 10\,N/m^2$

$= 400\,000\,N/m^2$

Example 2
What is the pressure exerted by a girl weighing 400N standing on one 'stiletto' heel of area 1 cm² ($= \frac{1}{10\,000}\,m^2$)?

formula first: $\text{Pressure} = \dfrac{\text{force}}{\text{area}}$

then put in
the numbers: $= \dfrac{400\,N}{\frac{1}{10\,000}\,m^2}$

$= 400 \times 10\,000\,N/m^2$

$= 4\,000\,000\,N/m^2$ (ten times bigger!)

So the elephant exerts a larger *force* (because it is heavier) but the girl's heel exerts a larger *pressure* (because of its smaller area). Her heel would sink farther into the ground.

Why do camels have large flat feet?

Sir jumps quickly to his feet
He's got the point (–he's got a scar!)
The pressure, acting on his seat,
Is force per unit area*aaaaagh!*

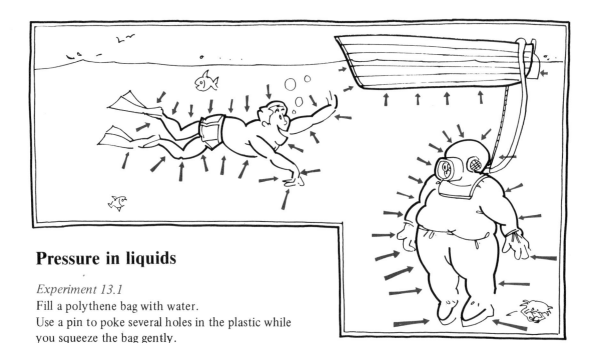

Pressure in liquids

Experiment 13.1
Fill a polythene bag with water.
Use a pin to poke several holes in the plastic while
you squeeze the bag gently.

What do you see?
Squeezing the top of the bag causes the water to
squirt out. This means that
Pressure is transmitted throughout the liquid.

Also the water squirts out in *all* directions. This
means that
Pressure acts in all directions.
In the picture above you can see that the red
arrows are acting in *all* directions on the divers —
up and down, and left and right.

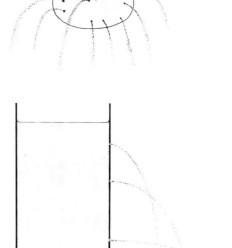

Experiment 13.2
Use a tall can with three equal holes drilled in the
side. Fill it with water.

What happens?
Where does the water squirt out farthest?
Where is the pressure greater — at the top hole or
at the bottom hole?
Pressure increases with depth.

In the picture above you can see by the arrows
that the pressure on the deep sea diver is greater
than on the swimmer.

Is the water pressure greater at a downstairs tap
or an upstairs tap? Why?

98

Experiment 13.3

Look inside a teapot (or kettle) which is full of water. What do you notice about the level of water in the spout and the level inside the pot?

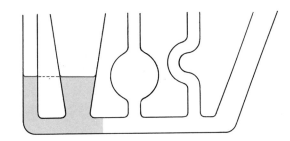

The diagram shows several containers of different shape, all connected together and filled with water. The water level is shown in two of the cans — where will it be in the other cans?

Hydraulic disc brakes

Experiment 13.1 showed that pressure is transmitted throughout a liquid. This idea is used in a car, to apply equal forces to the brakes on each side of the car.

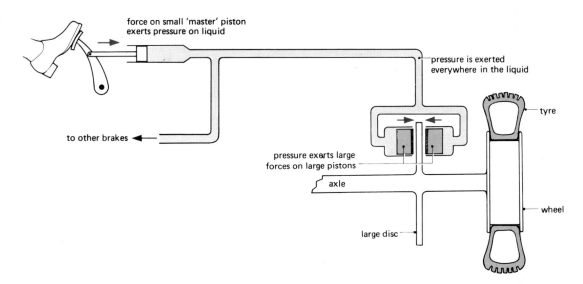

force on small 'master' piston exerts pressure on liquid

pressure is exerted everywhere in the liquid

tyre

to other brakes

pressure exerts large forces on large pistons

axle

wheel

large disc

The driver's foot pushes the piston to exert pressure on the liquid. This pressure is transmitted to pistons on each side of a large disc on the axle. The pressure makes the pistons squeeze the disc (like the brakes on a bicycle) to slow down the car. Exactly the same pressure is applied to the other brakes on the car.

If the pistons at the disc have *twice* the area of the master piston, they will each exert *twice* the force that the driver applies with his foot. The force is magnified by the increased area of the pistons.

Other hydraulic machines use the same principle — for example, a hydraulic press, a hydraulic car jack, a hydraulic garage lift.

Atmospheric pressure

We are living at the bottom of a 'sea' of air called the atmosphere, which exerts a pressure on us (just as the sea squeezes a diver).

Molecules of air are hitting us all the time to cause this pressure (see page 14). Fortunately the pressure inside our bodies is equal to this, so we are not crushed. In fact, astronauts must keep this air pressure round them by wearing spacesuits.

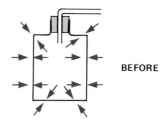

BEFORE

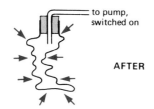

to pump, switched on

AFTER

Experiment 13.4
Remove the air from the inside of a can by connecting it to a vacuum pump. (Alternatively, the air can be removed by boiling water in the can so that the steam drives out the air before a bung is used to seal the can).

What happens?

Before the pump is switched on, molecules are hitting the outside and the inside of the can with equal pressure.
After the pump is switched on, there are almost no molecules inside the can and the pressure of the molecules outside the can crushes it!

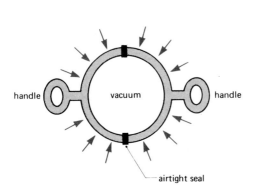

handle   vacuum   handle

airtight seal

Experiment 13.5
Put together two *Magdeburg hemispheres* and use a pump to take all the air out and so form a *vacuum*.
Hold the handles and pull.
Can you pull them apart?
Why not?
What happens if you let the air back in?

The pressure exerted on us by the atmosphere is about 100 000 newton per square metre (100 kN/m^2)!
On a dining table of area 1 square metre, the air pushes down with a force of 100 000 newton (the weight of more than 100 people).
Fortunately, the air under the table exerts an equal force upwards.

Experiment 13.6

Fill a test-tube (or a milkbottle) completely full
with water and place a piece of card on top.
Hold the card in place while you turn the test-
tube upside down. Then take your fingers off the
card.

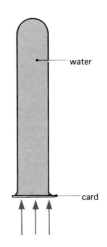

What is holding up the card against the pressure of
the water?

The pressure of the atmosphere can easily support
a few centimetres of water. If you used a longer
and longer test-tube, you would find that the
air pressure is strong enough to support about
10 metre of water! If we use a denser liquid, like
mercury, we find that the height is less.

The mercury barometer

Experiment 13.7

Your teacher may show you a mercury barometer
made by filling a long glass tube with mercury,
tapping out all the air bubbles and then turning
it upside down in a bowl of mercury.

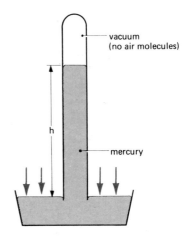

The column of mercury is held up by the air
pressure. As the air pressure varies from day to
day (depending on the weather), the height of the
mercury varies.

What is the height of the mercury today (the
distance to measure is shown in the diagram)?
A height of 760 mm is called *Standard Atmospheric
Pressure,* (written 760 mmHg.).

Experiment 13.8

Tilt the barometer tube to different angles and
measure the *vertical* height in each case.

What do you notice?

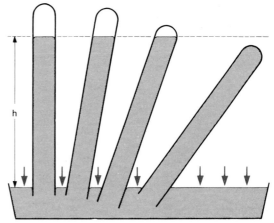

Drinking through a straw

What is forcing the liquid up the straw?

If the height of a *water* barometer is 10 metre, what is the maximum height you could possibly suck water?

Would it be easier or more difficult to drink through a straw at the top of a high mountain? Why?

Aneroid barometer

This kind of barometer uses a flexible metal can which has had air taken out of it. A strong spring stops the atmosphere from completely crushing the can.

If the air pressure increases, the top of the can is squeezed down slightly. If the air pressure decreases, the spring pulls up the top of the can. This small movement is magnified by a long pointer.

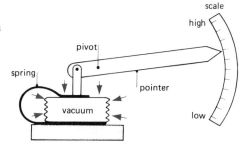

Altimeter

If a barometer is taken up a mountain, it shows a reduced pressure – because it is going nearer to the top of our 'sea' of air.
Mountaineers and aircraft pilots can measure their height by using an aneroid barometer.

It is this change in pressure that causes the boiling point of liquids to change at different heights (see page 65).

MOUNT EVEREST
(AIR PRESSURE ABOUT 30 cm Hg)

SEA LEVEL
(AIR PRESSURE ABOUT 76 cm Hg)

Pressure Gauges

1 Bourdon gauge

A bourdon gauge is often used to show the
pressure in gas cylinders and boilers (see p 37).
It works in the same way as the paper tubes you
may have blown out straight at parties.
The movement of the tube is used to turn a pointer.

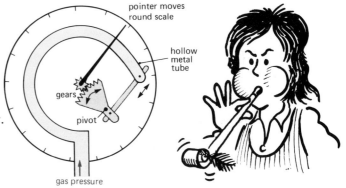

2 Manometer

A manometer is a U-tube containing some liquid,
usually water.

Experiment 13.9
Use a manometer to measure the pressure of the
gas supply.

The pressure of the gas changes the levels of the
water in the manometer. The pressure can be
found (in 'centimetres of water') by measuring the
height marked 'h'.

Experiment 13.10
How hard can you blow?
Under your teacher's supervision, use a large
manometer to measure the pressure you can exert
with your lungs.

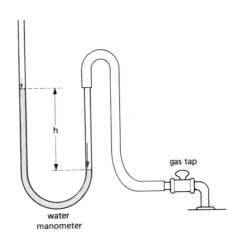

The Siphon

How can you empty a water tank using a rubber tube?

Experiment 13.11
Place one end of a rubber tube in a beaker of
water on a table. Nothing will happen until you
fill the tube with water by sucking at the lower end.

What is it that forces the water into the top of
the tube?
Once the tube is full, the water will *siphon* from
the top beaker to the bottom.

What happens if air gets into the tube?

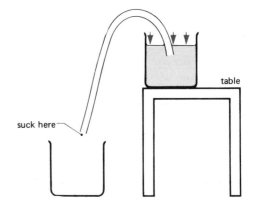

103

Pumps

1 Syringe

Experiment 13.12

Place the nozzle of a syringe into some water and pull up the piston.

What is forcing the water into the syringe?

Hold the syringe up in the air. Why does the water not fall out? What would happen if air leaked past the piston?

What happens if you push the piston down? How do water pistols work?

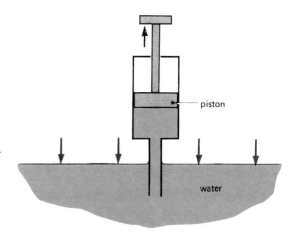

2 Force pump

A force pump is like a syringe with two valves added:
Valve A is like a little door which will let water come up, but will not let it go down again.
Valve B will allow water to be forced out to the outlet pipe, but will not let it back in.

When the piston is lifted, atmospheric pressure forces water up past valve A into the pump (air pressure also keeps valve B shut).

When the piston is pushed down, valve A is pushed shut and the water is *forced* out past valve B.

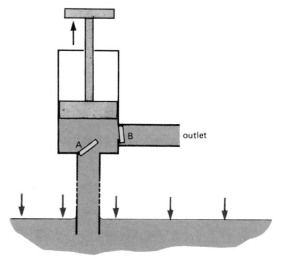

3 Lift pump

A lift pump also has two valves, but one of them is in the piston.

What happens when the piston is lifted?
What happens when the piston is pushed down?
What happens when the piston is lifted again?

If the height of a water barometer is 10 metre, what is the maximum height of valve A above the water?

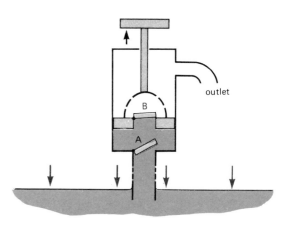

104

A formula for the pressure in a liquid.

Consider a horizontal area A at a depth h in a liquid of density d:

Volume of liquid $= h \times A$ (in m^3)

∴ Mass of liquid $=$ volume $\times$ density (see page 92 and page 402)

$\quad\quad\quad\quad\quad\quad = h \times A \times d$ (in kg)

Force on area A $=$ weight of liquid (in N)

$\quad\quad\quad\quad\quad\quad = 10 \times$ mass (see page 87)

$\quad\quad\quad\quad\quad\quad = 10 \times h \times A \times d$

∴ Pressure $= \dfrac{\text{force}}{\text{area}} = \dfrac{10 \times h \times \cancel{A} \times d}{\cancel{A}} = 10 \times h \times d$ (in N/m^2)

Pressure	$= 10 \times$ depth $\times$ density
(N/m^2)	(m) $\quad$ (kg/m^3)

What is the pressure 100 m below the surface of water (density 1000 kg/m^3)?

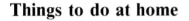
depth h — surface of liquid — liquid of density d — Area A

Things to do at home

1. An air trick

Ask your friends if they can lift up some heavy books with their breath!
The diagram shows how you can try it yourself.

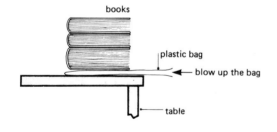

books

plastic bag

← blow up the bag

table

Can you explain this?

2. A water trick

Make some holes in the bottom of a washing up liquid bottle. Fill the bottle with water and quickly replace the top.
What is keeping the water in?
Give the bottle to a friend and ask him to take the top off and smell the liquid!
What happens?

Summary

$$\text{Pressure (in N/m}^2) = \frac{\text{force (in N)}}{\text{area (in m}^2)}$$

Pressure is transmitted throughout a fluid and acts in all directions.

Pressure in a fluid increases with depth.

Pressure = 10 x depth x density.

Atmospheric pressure is about 100 000 N/m² (or about 760 mmHg on a mercury barometer) but depends on the weather and the altitude.

Questions

1. a) Pressure = ... Its units are .....
      per square ..... (or .....).
   b) In fluids (liquids and gases), pressure acts in ..... directions and pressure ..... as the depth increases.
   c) Air pressure can be measured by a mercury ..... The ..... height of the mercury is usually about ... mm.

2. A box weighs 100N and its base has an area of 2m². What pressure does it exert on the ground?

3. If atmospheric pressure is 100 000 N/m², what force is exerted on a wall of area 10m²?

4. Explain the following:
   a) Stiletto heels are more likely to mark floors.
   b) Eskimos wear snowshoes.
   c) It is useful for camels to have large flat feet.
   d) Tractors have large tyres, bulldozers have caterpillar tracks and heavy lorries may need eight rear wheels.
   e) It hurts to hold a heavy parcel by the string.
   f) It is more comfortable to sit on a bed than on a fence.
   g) A ladder would be useful if you had to rescue someone from an icy pond.
   h) An Indian fakir can lie on a bed of nails if there is a large number of nails.

5. Explain the following:
   a) You can fill a bucket from a downstairs tap quicker than from an upstairs tap.
   b) Deep sea divers have to wear very strong diving suits.
   c) A dam is thicker at its base than at the top.
   d) A hole in a ship near the bottom is more dangerous than one nearer the surface.
   e) A giraffe must have a stronger heart than a human.
   f) A barometer will show a greater reading when taken down a coal mine.
   g) Aeroplanes are often 'pressurised'.
   h) Astronauts wear spacesuits.

6. In a hydraulic brake, a force of 500N is applied to a piston of area 5cm².
   a) What is the pressure transmitted throughout the liquid?
   b) If the other piston has an area of 20cm², what is the force exerted on it?

7. Draw a diagram of a mercury barometer. If the height of the mercury is 760 mm, how would this height be affected if
   a) The tube is tilted?
   b) The width of the tube is increased?
   c) Some air is allowed into the tube?
      Why is mercury used rather than water?

8. Explain, with diagrams, how each of the following uses atmospheric pressure:
   a) a drinking straw   b) a syringe
   c) a lift pump   d) a force pump
   e) a rubber sucker   f) a vacuum cleaner
   g) an altimeter   h) a siphon
   i) a rubber plunger used for clearing blocked sinks.

Further questions on page 169

chapter 14

Floating & Sinking

When you walk into the water at a swimming pool or at the seaside, do you feel lighter or heavier? If you walk in deep enough (up to your chin), your feet come off the ground and you start to float. This is because the water is exerting on you a force called *upthrust*.

Experiment 14.1

Hang a piece of metal on a spring balance and measure its weight.

Then lower the metal into a beaker of water. How much does it appear to weigh now?

The difference between the two readings is equal to the upthrust.
How much is the upthrust in the diagram?

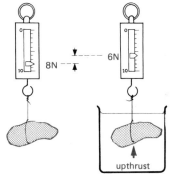

Upthrust and pressure

All liquids exert an upthrust like this because the pressure inside the liquid increases as you go deeper (see page 98). This means that the pressure on the bottom of an object is greater than on the top, and so there is a resultant force upwards.

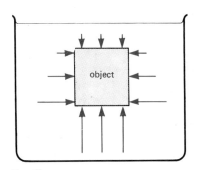

Upthrust and density

If the object has exactly the same density as the liquid, then it will stay still, neither sinking nor floating upwards, just as the liquid nearby stays still.

But if the object is denser than the liquid, then the same volume will weigh more and it will sink, because the upthrust will not be enough to support it. If the object is less dense, the upthrust will make it float up to the surface.

Look at the table on page 92 and decide which substances will sink in water. What will float on mercury? Will ice float on petrol?

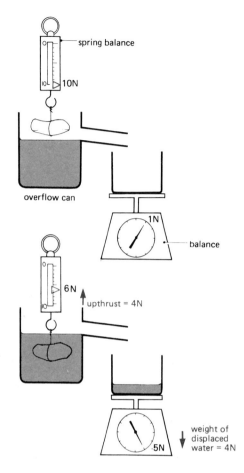

spring balance

10N

overflow can

1N

balance

6N

upthrust = 4N

.5N

weight of displaced water = 4N

Use the apparatus shown in the diagrams to investigate how the *upthrust* depends on the *weight of liquid displaced* by the object.

First weigh the object in air.
Fill an overflow can up to its spout and place a dry beaker on a balance under the spout.

Then lower the object slowly into the water, collecting all the water that is displaced and weighing it.

What do you notice about the size of the upthrust and the weight of water pushed out?
What is the upthrust in the diagram?
What is the weight of water displaced in the diagram?

It is found that the upthrust is always *equal* to the weight of liquid displaced. (Archimedes is said to have discovered this more than 2000 years ago).

This is found to be true for *all fluids* – that is all liquids *and* all gases. It applies to gases because they exert a pressure, like liquids.

SMALL UPTHRUST

LARGE UPTHRUST

EUREKA! EUREKA! (THE VERY FIRST STREAKER)

Archimedes' principle:
When an object displaces a fluid,
Upthrust = Weight of fluid displaced

Because this applies to gases and because you are displacing air at this moment, there is a small upthrust on you now (it is only about ½N).
You get a bigger upthrust when you get in your bath because water is denser than air.

Long ago Archimedes did shout,
"I have found what it is, without doubt,
The upthrust (or force),
Is equal (of course),
To the weight of the fluid pushed out."

Floating

Floating is a special case, when the upthrust is big enough to equal the weight of the object.

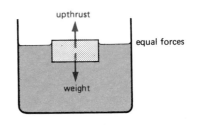

upthrust

equal forces

weight

From Archimedes' principle this should mean that the weight of fluid displaced is equal to the weight of the object.

Experiment 14.3
Put an overflow can on to a balance and fill it up to its spout. Note the reading on the balance (a).

Then float a piece of wood on the water. When the water has stopped overflowing, note the reading on the balance (b). What do you notice?

W

W

same reading

(a)   (b)

The increase in weight due to the piece of wood has been balanced out by the loss of the over-flowing water.
This is the **Principle of flotation**:

Weight of fluid **displaced**	=	**Weight of floating** **object**

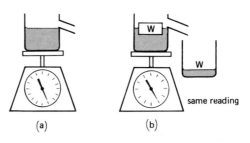

An iron ship floats because it displaces a large amount of water and the pressure of the surrounding water exerts a large upthrust on it. (see the diagrams at the top of page 98 and page 107).

If a ship weighs 100 million newton, then it must sink into the water until it displaces 100 million newton of water. If cargo weighing 1000 N is put on board then the ship will sink down until an extra 1000 N of water is displaced.

What is your weight? (see page 87). What weight of water must you displace in order to float?

A hydrogen balloon or a hot air balloon weighing 5000 newton must displace 5000 newton of air in order to float in air.

Experiment 14.4
If your teacher can supply you with a balloon filled with hydrogen, tie a long piece of string to it and then cut off bits of string until the balloon just floats in mid-air.
What happens when you take it into a colder room? Why?

Hydrometers

A hydrometer is a simple instrument used to measure the density of a liquid. It floats to different depths in different liquids, depending on their densities.

If it is put in a very dense liquid, will it float high or low? Why?

What is the density of the liquid in diagram A? What do you notice about the scale on the hydrometer?

Hydrometers are often used to test beer and milk to see if they have too much water in them.

Experiment 14.5

Make your own hydrometer by using a drinking straw and **plasticine** (see diagram B).

Mark the scale by first putting it in liquids whose densities are known. Then use it to find the densities of other liquids.

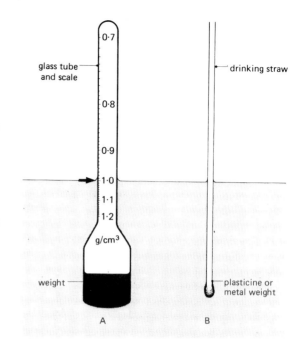

Plimsoll line

Ships act like huge hydrometers: they rise higher in salty sea-water and sink lower in freshwater.

The Plimsoll line on the side of a ship shows how deeply it may be safely loaded, depending on the density of the water it is going to sail in.

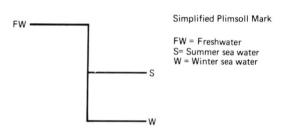

Professor Messer takes a dip, but in the water he's a drip,
He must have water on the brain — please take a look then *you* explain!

Summary

Whenever an object displaces some fluid, there is an upthrust caused by the pressure of the surrounding fluid.
Archimedes' principle:
Upthrust = Weight of fluid displaced.

For a floating object, the upthrust equals the weight of the object.
Principle of flotation:
Weight of fluid = Weight of floating
    displaced   object.
A hydrometer is used to measure the density of a liquid.

Questions

1. Copy out and complete:
   a) Whenever an object displaces a fluid, there is an . . . . . which is equal to the . . . . . of . . . . . displaced. This is called . . . . . principle.
   b) In order to sink in a fluid, the density of an object must be . . . . . than the . . . . . of the fluid.
   c) For a floating object, the . . . . . of fluid displaced is . . . . . to the . . . . . of the floating object. This is the principle of . . . . .
   d) A hydrometer is used to measure . . . . . It floats higher in a liquid whose density is . . . . .

2. A man weighs 700N. What weight of water must he displace in order to float?

3. Explain the following:
   a) Walking down a pebble beach becomes less painful as you enter the sea.
   b) A life-jacket helps you to float.
   c) A ship sinks deeper as cargo is taken aboard.
   d) An iron nail sinks but an iron ship floats.
   e) A ship rises as it sails from freshwater into seawater.
   f) It is easier to learn to swim in a calm sea than in a freshwater lake.
   g) Most of an iceberg is below water (see page 92).
   h) Hot air rises (see page 53).

4. A large balloon weighs 100N and displaces 110N of air. What happens?

5. Referring to the table of densities on page 92 which of those substances will a) sink in water b) float on water c) sink in mercury d) float on mercury?

6. A solid has a mass of 2000kg and a volume of 4m^3. Will it sink or float in water?

7. An object hanging from a spring balance weighs 40N in air. When the object is lowered into water, it is found to weigh 30N.
   a) What is the upthrust on the object?
   b) What is the weight of water displaced? (see page 108)
   c) What is the mass of water displaced? (see page 87).
   d) What is the volume of water displaced? (see page 92)
   e) What is the volume of the object?
   f) What is the mass of the object? (see page 87)
   g) What is the density of the object?

8. In a similar situation to question 7, an object weighs 60N in air and 40N when immersed in water.
   What is its density?

Professor Messer: Does stainless steel float?
Mrs Messer: In kitchens, I see stainless steel sinks!

Further questions on page 169.

chapter 15

more about

FORCES

Friction

Friction is a very common force. Whenever one object slides over another object, friction tries to stop the movement.

Friction always *opposes* the movement of an object.

Friction is often a nuisance, because it converts kinetic energy into heat and wastes it.

Experiment 15.1
Place the palms of your hands together and rub hard. What do you notice? This waste of heat energy reduces the efficiency of engines (chapter 10) and other machines (chapter 18).

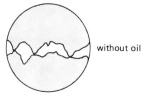

without oil

views through a microscope

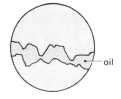

oil

Reducing friction

1. The slide in the park is polished smooth so that you can slide down easily.

    However even highly polished objects look rough through a microscope, so there is always some friction (as the lumps on one object catch and stick to the lumps on the other object).

2. A second way of reducing friction is to *lubricate* the surfaces with oil. The oil separates the surfaces so that they do not catch each other. Ice is usually slippery because it is usually covered with water, which lubricates like oil.

    If workers puff
    That rough
    Is tough,
    Reduce the toil
    By adding
    Oil.

3. Another way of reducing friction is to separate the two surfaces by air. This is how a hovercraft works.

4. A fourth way of reducing friction is to have the object *rolling* instead of sliding. This is what happens with ball bearings.

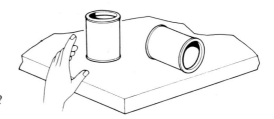

Experiment 15.2
Slide and then roll a tin can along the bench, using equal pushes. Which way does it go farther?

5. A fifth way: cars, planes and rockets are *streamlined* to reduce friction with the air. Even so, rockets can get very hot when they enter the Earth's atmosphere.

Advantages of friction

Although it is often a nuisance, friction is also very useful. Our lives depend on the friction at the brakes and tyres of cars and bicycles.
Air friction slows down the parachute of a falling man so that he can land safely.

You are able to walk only because of friction with the floor (try walking on wet ice!). Knots in string and the threads in your clothes are held together by friction. Nails and screws are held in wood by friction. What things do you think would happen if there was suddenly no friction at all in this room?

Other forces

Frictional forces only act when two objects are in contact. Other forces, like the pull of gravity, can act at a distance, when the objects are not touching each other.

Experiment 15.3
Place a bar magnet on a smooth table and then bring another magnet near it.
What happens?
Turn round one of the magnets.
What happens now?
Magnetic forces can act at a distance.

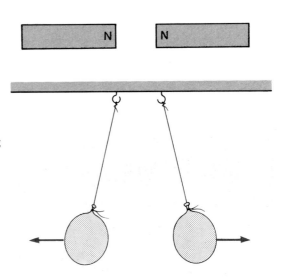

Experiment 15.4
Hang up two dry balloons side by side and touching each other. Rub each balloon on a dry cloth so that they become charged with static electricity.
What happens?
What happens when you bring the cloth near one of the balloons?
Electric forces can act at a distance.

You will learn more about magnetic and electric forces later (see pages 259, 267).

Newton's third law

The two teams are having a tug of war contest. If the rope is not moving, what can you say about the forces exerted by the two teams?

Experiment 15.5
Use two spring balances to have a tug of war.

What do you notice about the readings on the balances?

Newton noticed that forces were *always* in pairs and that the two forces were *always* equal in size but opposite in direction. He called the two forces *action* and *reaction*. **Newton's third law of motion is:**

Action and reaction are equal and opposite.

In the tug of war picture, if the rope is not moving,

the force
of team A   **=**   the force
of team B
on team B   on team A

You can see that the way we get the sentence describing the second force is by *changing round the words from the first sentence.* We use the same method in the following examples.

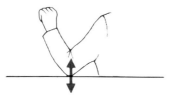

a) Rest your elbow on the table. What is the pair of forces here?

The force of your   The force of the table
elbow on the table   **=**   on your elbow
(downwards)   (upwards)

b) What are the forces when you start to walk or run?

The force of your feet   The force of the
on the Earth   **=**   Earth on your feet
(moves the Earth   (moves you forwards)
slightly backwards)

Yes! The Earth moves backwards slightly as you move forwards.

114

c) Now consider a block of wood resting on a table. There are four forces here, in two pairs:

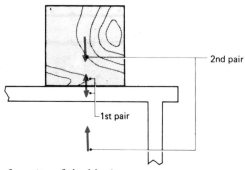

First pair

| The force of the block on the table. (downwards) | = | The force of the table on the block. (upwards) |

These forces act where the block touches the table.

Second pair

| The force of gravity of the Earth pulling down on the block (this weight in effect acts at the centre of the block – see page 122) | = | The force of gravity of the block pulling up on the Earth (this force in effect acts at the centre of the Earth). |

When the block is on the table, these four forces are equal. But if the block is allowed to fall to the floor, only the second pair of forces exists and as the block moves downwards, the Earth moves slightly upwards!

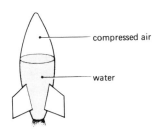

WHY IS THIS A WORLD MOVING EVENT ?

Experiment 15.6

Get a plastic water rocket and follow the instructions to prepare it. What happens when you release it and let the compressed air force out the water? Complete this equation:

| The force of the rocket on the water (downwards) | = | The force of the . . . . . on the . . . . . (. . . wards) |

This is why rockets and jet engines move forward (see page 74 and page 163).

Experiment 15.7

Blow up a balloon and release it. What happens? Why? (see page 74) Write down an equation for the pair of forces.

When a gun is fired, why does it recoil? Write down an equation for the pair of forces (see the cartoon on page 74).
Why does the bullet move farther than the gun?

115

Adding forces

If you want to move this book to another place you will have to apply a force. *Two* things about the force are important: the *size* of the force and the *direction* of the force. We show the direction of a force by an arrow and show the size of a force by the length of the arrow drawn to a chosen scale — for example, 1 cm for 1 newton.

scale: 1cm for 1N(1newton)

Because both *size* and direction are important, force is called a *vector* quantity. Some other vector quantities are: displacement, velocity, acceleration (see page 148).

Other quantities are called *scalars*. Scalars have only a size, no direction. For example: temperature, money, mass, volume and density.

Adding scalars

For scalar quantities, 2 + 2 = 4, always.

Adding vectors

For vector quantities, like forces, 2 and 2 *sometimes* equals 4!

a) Forces of 2N and 2N acting in the same direction, add up to give a resultant force of 4N.

b) Forces of 2N and 2N acting in opposite directions, cancel out to give no resultant force at all.

c) What would be the resultant force of 5N opposed by 2N? The 2N force cancels out part of the 5N force. How much is left?

d) What happens when two forces are acting at some other angle to each other, like here:

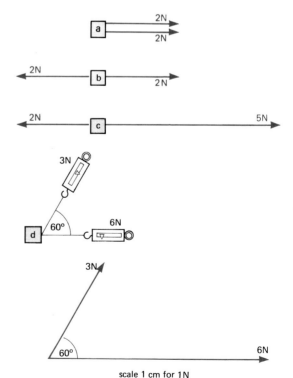

scale 1 cm for 1N

To find the answer we must draw a diagram, in four steps:

Step 1
Choose a scale, for example 1 cm for 1 N.

Step 2
Draw the two forces *to scale*, showing them by arrows at the correct angle.

Step 3

Draw two more lines to make a parallelogram (opposite sides should be parallel and equal in length).

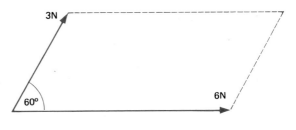

Step 4

Draw in the diagonal from where the two forces meet. The diagonal shows the direction and size of the resultant force. (Use the scale to convert the length of the diagonal to newtons.)

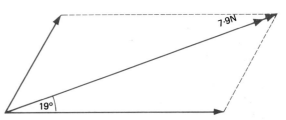

This method of adding forces is called the **Parallelogram of forces**:

> If two forces are represented in size and direction by the sides of a parallelogram (drawn to scale), the resultant force is represented in size and direction by the diagonal drawn from where the two forces act.

Experiment 15.8

You can check the parallelogram method this way.
a) Hold one end of a spring on a nail on a wooden board and pull the other end with a spring balance. Mark the position of the end A of the spring on the board and note the reading of the spring balance.

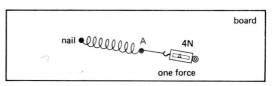

b) Now pull the end of the spring to the *same* position A using two spring balances.
Mark the positions of the strings and note the readings on the spring balances.

c) Now draw a parallelogram to add these two forces (using steps 1 to 4).
Find the resultant force.

Compare this resultant force with the size and direction of the single force in part (a) of the experiment (that had the same effect on the spring). Are they the same in size and direction? If so, it shows you have added the forces correctly.

scale: 1cm for 1N

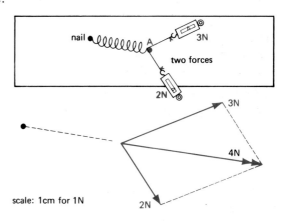

Now here's the tale of Albert Hector
Who didn't know that force is a vector
He asked for large amounts of force
(A tractor with a rope of course)
When it arrived it caused a crush

Instead of pull it gave a push
(And then too late it gave a pull
And put our Al in hospital)
Now our Albert's looking paler
He knows that force is not a scalar!

117

Adding other vectors

Other vectors are added in the same way as forces.

Example

A pilot aims his plane due east at 50 m/s while the wind is blowing to the north at 20 m/s.

The resultant speed and direction of the plane are found from the diagonal of the parallelogram.

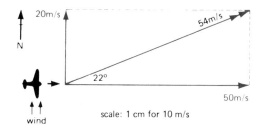

scale: 1 cm for 10 m/s

Resolving into two components

Sometimes it is useful to *reverse* the parallelogram method of adding vectors.

In the diagram above, the resultant (54 m/s) can be *resolved* into two *components* (20 m/s and 50 m/s).

A force also can be resolved into two components: In the diagram (in black) a man is pulling a log using a rope at 30° to the horizontal and exerting a force of 50 N.

Drawing a rectangle of forces (in red) shows that this 50 N can be resolved into two components:
a) a force of 43 N pulling the log horizontally *and*
b) a force of 25 N lifting the log vertically.

What would be the components if the rope was at 60° to the horizontal?

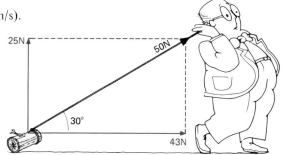

Scale : 1cm for 1N

Summary

Friction always opposes movement.
Forces are always in pairs.
Newton's third law: action and reaction are equal and opposite. Vectors have size and direction.
Scalars have size only.

Parallelogram of forces: If two forces are represented in size and direction by the sides of a parallelogram (drawn to scale) the resultant force is represented in size and direction by the diagonal drawn from where the forces meet.

Questions

1. On a bicycle, where is friction
   a) an advantage   b) a disadvantage?
2. Explain the following:
   a) Walking on wet ice is difficult.
   b) It is difficult to strike a match on a smooth surface.
   c) It is more difficult to pull a boat on the beach than in the sea.

   d) Cars are more likely to skid on loose gravel.
   e) Aborigines light fires by rubbing pieces of wood together.
   f) Spaceships get hot when they return to Earth.
   g) You can be burned by sliding down a rope quickly.

3. One night Professor Messer dreamed that all friction suddenly disappeared. Write a story about what might happen to the Professor. Why did his dream turn into a nightmare?

4. For each of the following forces, describe the reaction, giving its direction and stating what it acts on.
   a) the push (north) of a boot on a football
   b) the push (west) of a crashing car on a wall
   c) the push (east) of a ship's propeller on the water
   d) the pull of gravity (down) on a falling apple

5. Explain the following:
   a) A gun recoils when it is fired.
   b) Firemen have to brace themselves when aiming a fire hose.
   c) You, by yourself, can move our planet.
   d) An astronaut is drifting away from his spaceship. How can he return, using only an aerosol spray?

6. What is the resultant force in each of these diagrams?

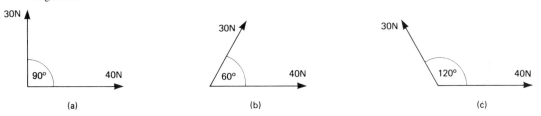

(a)   (b)   (c)

8. Explain the difference between a vector and a scalar. Divide the following into vectors and scalars: force, temperature, velocity, displacement, volume, money, acceleration, mass and weight.
   Find the sum of a) a mass of 6kg and a mass of 8kg, b) a force of 6N acting at right angles to a force of 8N.

9. Tom, Dick and Harry are all pulling on a metal ring. Tom pulls north with a force of 90 N and Dick pulls east with a force of 120 N.
   a) What is their resultant force?
   b) If the ring does not move, what force is Harry exerting?

10. A pilot aims his plane due east at 100 m/s while the wind is blowing to the north at 25 m/s. What is his actual speed and direction?

11. A boy is pulling a log using a rope at $20°$ to the horizontal and exerting a force of 100 N. Find the horizontal and vertical forces exerted by the rope on the log.

7. Explain this cartoon – how many pairs of forces can you find?

Further questions on page 170.   119

chapter 16

TURNING FORCES

If you want to undo a very tight nut on your bicycle, would you choose a short spanner or a long spanner?

Why is the handle on a door placed a long way away from the pivot? Can you close a door easily by pushing near the pivot? Who would win if a strong man pushed *at* the pivot and a small boy pushed the opposite way at the handle? Try it.

Experiment 16.1
Hold a rod horizontally and hang a weight on the rod near your hand.
Move the weight to different distances from your hand and try to keep the rod horizontal.

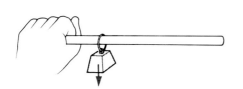

What happens?
Now try it with a heavier weight.

You can feel that the turning effect on your hand depends on the size of the force (the weight) *and* the distance from your hand. The turning effect of a force is called a *moment* or a *torque*.
It is calculated by:

Moment of a force	= force ×	perpendicular distance (from the force to the pivot)
(newton)		(metre)

The distance used is always the *shortest* (perpendicular) distance.
Moments are measured in newton-metre (often written Nm)

Example
What is the turning effect of the force in the diagram, about the point P.

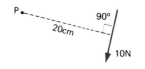

$$\text{Perpendicular distance from the force to P} = 20\text{cm} = 0.20 \text{ metre}$$

formula first: $\dfrac{\text{Moment of}}{\text{the force}}$ = force x $\dfrac{\text{perpendicular distance}}{\text{from the force to the pivot}}$

then numbers: = 10N x 0·20 m
                    = 2 Nm,   turning *clockwise*.

A car mechanic might need to apply a torque of 40Nm to a nut.

120

You know that a seesaw can be balanced even when the two people have different weights, by sitting them at different distances from the pivot.

When this seesaw is balanced and is not moving (*in equilibrium*), the moment of the girl (a clockwise moment) must *equal* the moment of the man (an anticlockwise moment).
This is called the **Principle of moments**:

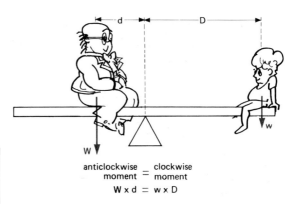

anticlockwise = clockwise
moment = moment
$$W \times d = w \times D$$

In equilibrium,		
total anti-clockwise moment	=	total clockwise moment

You can check this by experiment.

Experiment 16.2

Get a metre rule with a hole drilled at its centre; push a large pin through and clamp it to a retort stand. If the rule is not balanced, weight the higher end with plasticine until it balances.
You need weights of 1 newton (a 100g mass will do) and 2 newton (a 200g mass will do) to hang from loops of string.

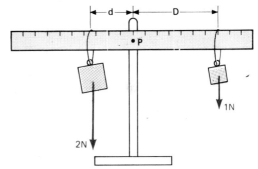

Adjust the weights until the rule balances. Measure the distances d and D and note them in the table. Repeat with these and other weights at different distances. Repeat with two weights on one side (see the example below).

What do you find about the anti-clockwise moments and the clockwise moments?

Anticlockwise			Clockwise		
Weight W (N)	distance d (cm)	Moment W×d	Weight w (N)	distance D (cm)	Moment w×D
2	20	40	1	40	40
2	15	—	1		

Example

If the ruler in the diagram is balanced, what is the weight W?

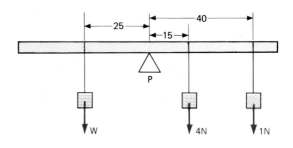

Equation first:

In equilibrium,
total anti-clockwise moments = total clockwise moments

then put in numbers
(see diagram)

$W \times 25 = (4 \times 15) + (1 \times 40)$
$W \times 25 = 60 + 40$
$W \times 25 = 100$
$W = 4$ newton

Centre of gravity (centre of mass)

What part of a ladder would you rest on your shoulder in order to carry it most comfortably?

Although the force of gravity pulls down on every little molecule, the weight of the ladder seems to act at its centre. This point on the ladder, where it balances, is called the *centre of gravity* (or *centre of mass*).

The centre of gravity is the point through which the whole weight of the object seems to act.

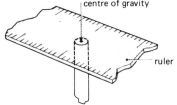

centre of gravity

ruler

Experiment 16.3

Find the centre of gravity of a ruler by moving it until it balances on the end of a pencil.

Where is the centre of gravity of a ruler?
Where is the centre of gravity of a plank?

Use a similar experiment to find the centre of gravity of a book, a retort stand or a hammer.

Experiment 16.4

Tie a small weight to a piece of string to form a *plumbline*. Let it hang down from your hand.

Why does it always hang vertically?
What can a plumbline be used for?

Experiment 16.5

Find the centre of gravity of a piece of flat card (a *lamina*) using a plumbline.
Let the card hang *freely* from a large pin held in a retort stand. Hang a plumbline from the same pin.

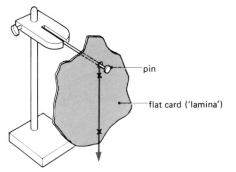

pin

flat card ('lamina')

Mark the position of the plumbline by two crosses on the card. Join the crosses with a ruler.

Just as the plumbline hangs with its centre of gravity vertically below the pivot, so also will the card. This means that the centre of gravity of the card is *somewhere* on the line you have marked. How can you find where it is on this line?

To find just where the centre of gravity is on this line, re-hang the card with the pin through *another* hole and again mark the vertical line. The only point that is on *both* lines is where they cross, so this point must be the centre of gravity. How can you check that this point really is the centre of gravity?

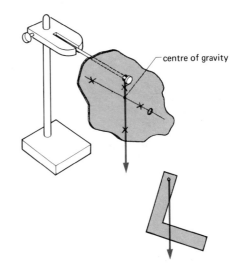
centre of gravity

Experiment 16.6

Repeat experiment 16.5 with an L-shaped piece of card. Where does the centre of gravity seem to be?

Stability

This plumbline is said to be in *stable equilibrium* because if you push it to one side, it returns to its original position. It does this because when you push it to one side its centre of gravity rises and gravity tries to pull it back to its lowest position.

If you carefully balance a ruler vertically on your finger it is in *unstable equilibrium*, because if it moves slightly, its centre of gravity falls and keeps on falling down.

A billiard ball on a perfectly level table is in *neutral equilibrium*, because if it is moved, its centre of gravity does not rise or fall.

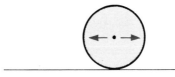

Is a tightrope walker in stable or unstable equilibrium?

A bottle is shown in equilibrium in three positions. Name the kind of equilibrium for each position.

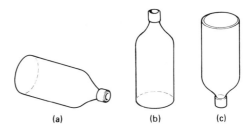

(a)   (b)   (c)

Stable and unstable objects

Most of the objects we use each day are in stable equilibrium.
How can objects be made more stable?

Experiment 16.7

Place a box or matchbox upright on a rough piece of wood (A). Slowly tilt the wood until the box topples over. Note the angle of the wood.

Raise the centre of gravity by weighting the top of the box with plasticine or sliding up the matches (B). Tilt the wood again. Is the box more stable or less stable?

Place the box in other positions and see when it topples.

To make the box more stable, should it have
a) a high or a low centre of gravity?
b) a narrow or a wide base?

Which is more stable, a racing car or a double decker bus? How could the bus be made more stable? Why are passengers not allowed to stand upstairs?

How are these objects made stable?

Why does a rowing-boat become less stable if you stand up?

bunsen burner   retort stand   wine glass

Discuss the physics of these cartoons.

Things to do at home

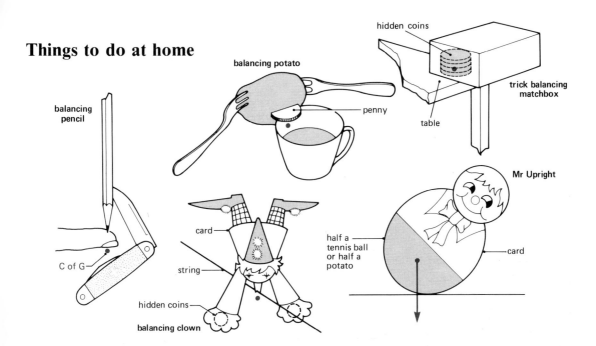

balancing pencil

C of G

balancing potato

penny

hidden coins

table

trick balancing matchbox

card

string

hidden coins

balancing clown

half a tennis ball or half a potato

card

Mr Upright

Summary

Moment of a force (newton-metre)	=	force (newton)

Moment of a force = force × perpendicular distance from the force to the pivot

(newton-metre)   (newton)   (metre)

Principle of moments:

In equilibrium,

total anti-clockwise moments = total clockwise moments

The *centre of gravity* is the point through which the whole weight of the object seems to act.
An object can be in stable, unstable or neutral equilibrium depending on the position of its centre of gravity. To be more stable an object should have a low centre of gravity and a wide base.

Discuss the physics of these cartoons.

I'LL PUT ALL THESE HEAVY THINGS IN THE TOP DRAWER

HOW CAN SHE DO THAT UP THERE BLINDFOLDED?

I THINK THE DOOR LOOKS NICER WITH THE HANDLE AT THIS SIDE!

Questions

1. Copy and complete these sentences.
   a) The turning effect (or . . . . .) of a force is equal to the force multiplied by the . . . . . distance from the . . . . . to the . . . . . Its unit is . . . . .
   b) The principle of moments states that, in . . . . ., the total . . . . are equal to the total . . . .
   c) The centre of gravity is the point through which the whole . . . . . of the object seems to act.
   d) A stable object should have a . . . . . centre of gravity and a . . . . . base.
   e) If the centre of gravity of an object
      (i) is raised by tilting, it is in . . . . . equilibrium.
      (ii) is lowered by tilting, it is in . . . . . equilibrium.
      (iii) remains at the same height, it is in . . . . . equilibrium.

2. Explain the following:
   a) A mechanic would choose a long spanner to undo a tight nut.
   b) A door handle is placed well away from the hinge.
   c) It is difficult to steer a bicycle by gripping the centre of the handlebars.

3. A mechanic applies a force of 200N at the end of a spanner of length 20cm. What moment is applied to the nut?

4. The diagrams show rulers balanced at their centres of gravity.
   a) What is the weight of X?
   b) What is the weight of Y?

5. A boy weighing 400 N sits 1·5m from the pivot of a seesaw. Where should a girl weighing 300 N sit to balance the seesaw?

6. A, B, C, D are pieces of hardboard. Copy the diagrams and mark with a cross the position of the C of G of each.

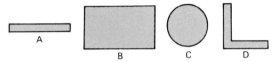

7. Sketch diagrams and mark the approximate positions of the centres of gravity of
   a) a hammer
   b) a boomerang
   c) a tea cup
   d) a human being.

8. Explain the following:
   a) A bunsen burner has a wide heavy base.
   b) Racing cars are low, with wheels wide apart.
   c) A boxer stands with his legs well apart.
   d) A wine glass is less stable when it contains wine.
   e) Lorries are less stable when carrying a load.

9. Explain what is meant by stable, unstable and neutral equilibrium, giving an example of each.

There was an old teacher called Grace,
Who often fell flat on her face,
The reason you see,
Was her C of G,
Its place was too high for her base

Further questions on page 170.

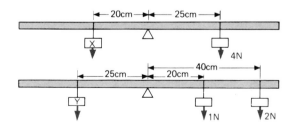

126

chapter 17
WORK, ENERGY & POWER

If the man in the picture is moving the van, we say he is doing *work*. He is doing work only if there is *movement* against an *opposing* force. The opposing force is often friction (as in the picture), or gravity. The man can only do this work if he has some *energy*.

If the man pushes with a stronger force or moves it a longer distance he will do more work. In fact,

Work done = force x distance moved
(newton)   (metre)

The distance used in this formula must be the distance moved *in the direction of the force.*

If the force is in newton and the distance is in metre then the work is measured in *joule*, often written J (see page 43).
1 joule is the work done when a force of 1 newton moves through 1 metre (in the direction of the force).

Example
A man lifts a brick of mass 5kg from the floor to a shelf 2 metre high. How much work is done?
The opposing force in this case is the *weight* of the brick. Here on Earth a mass of 5 kg weighs 50 newton (see page 87).

formula first: work done = force x distance moved
                    (J)   (N)   (m)
then put in   = 50N x 2m
the numbers: = 100 joule

Where does he get the energy to do this work?

If a man pushes a van against a friction force of 300 newton for a distance of 10 metre, how much work does he do?

Large amounts of work may be measured in kilojoules or even megajoules (see page 43).

Experiment 17.1
Use a spring balance to lift objects against gravity or to pull objects along the bench against friction. In each case measure the distance moved (using a ruler) and calculate the work done.
Where do you get the energy to do this work?

Forms of energy

We have seen already (on page 8) that there are several forms of energy.
We have studied *thermal* energy and found that it is really the movement energy of the molecules.

All moving objects have movement energy, called *kinetic* energy. A heavy lorry travelling fast has a lot of kinetic energy (if you get in the way it will use some of this energy to damage you).

Another kind of mechanical energy is stored energy, called *potential* energy.

When an object is lifted to a higher place, it is given more *gravitational potential energy*. This is stored energy which it can give out if it falls down.

A second kind of potential energy is *elastic potential energy* which is stored in the elastic of a catapult or in a stretched bow.

Chemical energy is energy stored in food and other fuels. Your body gets its energy from the food you eat.

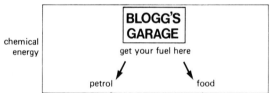

Sound energy and *light* energy are common forms of energy.

Electrical energy is common and useful. Like other forms of energy, it can be very dangerous. *Magnetic* energy is always connected with electric energy.

Nuclear energy is stored in the centre (or *nucleus*) of an atom. The Sun, like a hydrogen bomb, runs on nuclear energy.

Changing energy from one form to another

We have seen already (on page 9) that energy can be changed from one kind to another. When this happens, the *amount* of energy stays the same, because energy cannot be created or destroyed.

Energy and work

Energy is the capacity for doing work.
The amount of work done tells us how much energy has been converted from one form to another.

In the example on page 127 we calculated that the work done by the man in lifting the brick to the shelf was 100 joule. This means that 100 joule of his chemical (food) energy was converted to 100 joule of gravitational potential energy.

The diver in the diagram uses 6000 joule of his chemical energy to climb to the top where he has 6000J of gravitational potential energy (P.E.) but no kinetic energy (K.E.). As he falls down, his potential energy is converted into an *equal* amount of kinetic energy. The total amount of P.E. + K.E. is always 6000J. (See also page 158).

When a pendulum swings to and fro, its energy is constantly changing from potential energy to kinetic energy and back again.

What happens to this energy as the pendulum is gradually slowed down by friction (see page 112)?

What are the energy changes in these examples?

This fact is called:

> **the Principle of Conservation of Energy:**
> energy cannot be created or destroyed, although it can be changed from one form to another.

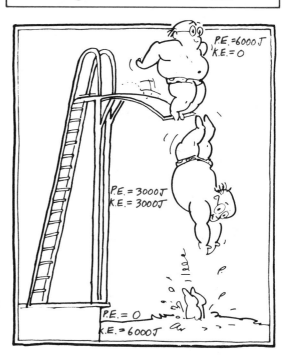

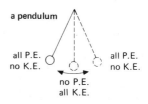

a pendulum

all P.E.
no K.E. — no P.E. all K.E. — all P.E. no K.E.

ENERGY CONVERTED TO ENERGY

ENERGY CONVERTED TO ENERGY

ENERGY CONVERTED TO ENERGY

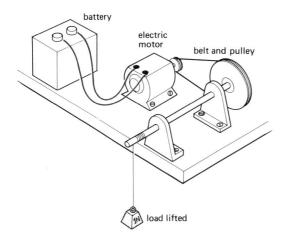

battery

electric motor

belt and pulley

load lifted

Energy experiments

Experiment 17.2

Arrange an electric motor, as in the diagram, so that it can turn an axle from which a load is hanging.

What happens when you connect the motor to a battery?

What are the energy changes here?

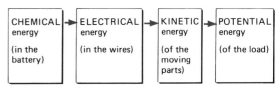

| CHEMICAL energy (in the battery) | ELECTRICAL energy (in the wires) | KINETIC energy (of the moving parts) | POTENTIAL energy (of the load) |

Experiment 17.3

Disconnect the battery and connect a lamp in its place. Start with a heavy load raised high and let it fall to drive the 'motor' which will now act as a *dynamo* and produce electricity to light the lamp.

What are the energy changes here?

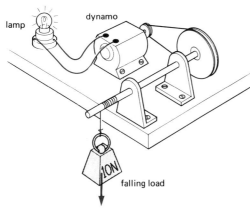

lamp

dynamo

falling load

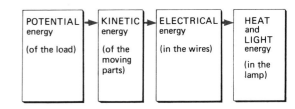

| POTENTIAL energy (of the load) | KINETIC energy (of the moving parts) | ELECTRICAL energy (in the wires) | HEAT and LIGHT energy (in the lamp) |

Experiment 17.4

Use a model steam engine to lift a load.

What are the energy changes here?

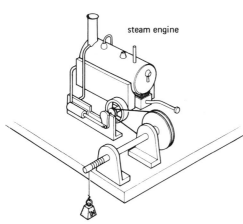

steam engine

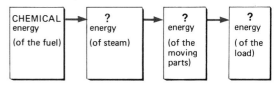

| CHEMICAL energy (of the fuel) | ? energy (of steam) | ? energy (of the moving parts) | ? energy (of the load) |

Measure the weight of the load and the height it is lifted.
Then calculate the work done on the load.

Experiment 17.5

Build a model power station using a steam engine and a dynamo. Connect them to a lamp to represent all the lights in your house.

What are the energy changes here?

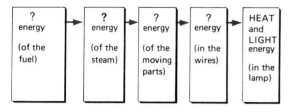

| ?
energy

(of the
fuel) | → | ?
energy

(of the
steam) | → | ?
energy

(of the
moving
parts) | → | ?
energy

(in the
wires) | → | HEAT
and
LIGHT
energy

(in the
lamp) |

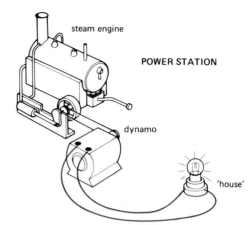

POWER STATION

steam engine

dynamo

'house'

Experiment 17.6

Build a model hydroelectric power station using the energy of falling water to turn a turbine and drive a dynamo.

What are the energy changes here?

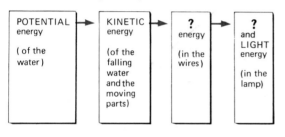

| POTENTIAL
energy

(of the
water) | → | KINETIC
energy

(of the
falling
water
and the
moving
parts) | → | ?
energy

(in the
wires) | → | ?
and
LIGHT
energy

(in the
lamp) |

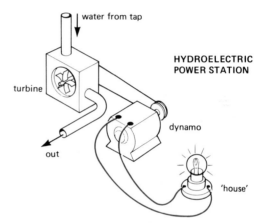

water from tap

HYDROELECTRIC
POWER STATION

turbine

out

dynamo

'house'

Sources of energy

Where does our energy come from? The answer is that it almost all comes from the *sun!*
Study the different parts of this diagram and you will see how.

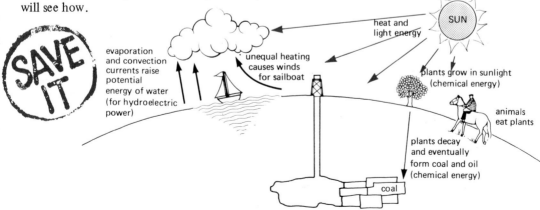

SAVE IT

evaporation
and convection
currents raise
potential
energy of water
(for hydroelectric
power)

unequal heating
causes winds
for sailboat

heat and
light energy

SUN

plants grow in sunlight
(chemical energy)

animals
eat plants

plants decay
and eventually
form coal and oil
(chemical energy)

coal

131

Power

If two cars of the *same* weight climb up the *same* hill, then they do the *same* amount of work. (See page 127.) But if car A climbs the hill in a shorter time than the other car, we say it has a greater *power*. Power is the *rate of working* (the rate of converting energy).

> Power $=$ $\dfrac{\text{work done (in joule)}}{\text{time taken (in seconds)}}$

The work done is measured in joule, the time taken is measured in seconds and so the power is measured in *joule per second* or *watt* (W)
1 watt (1 W) = 1 joule per second.

Very powerful engines may be measured in kilowatt (kW) or even megawatt (MW).

1kW = 1 kilowatt = 1 000 watt
1 MW = 1 megawatt = 1 000 000 watt.

Engines may also be measured in horsepower (h.p.) (1 h.p. is about 750W)

Example
A crane lifts a load weighing 3000 N through a height of 5m in 10 seconds. What is the power of the crane?

Calculate it in two parts:

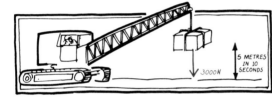

part 1.
formula first:

Work done	=	force x distance moved
		(see page 127)

= 3000 N x 5m

then put in the numbers: = 15 000 J

part 2.
formula first:

Power $=$ $\dfrac{\text{work done}}{\text{time taken}}$

then put in the numbers: $=$ $\dfrac{15\ 000\ \text{J}}{10\ \text{seconds}}$

= 1500 W = 1·5 kW

$\left(= \dfrac{1500}{750}\ \text{h.p.} = 2\ \text{h.p.}\right)$

A family car might develop a power of 40 kW, a racing car perhaps 400kW and a moonrocket 100 000 MW.

How much power can you develop with your legs?

Experiment 17.7
You need a flight of stairs to run up, so that you are doing work against the force of gravity (your weight).

First measure your weight in newton (see page 87).

Then, since you are going to lift your body against the *vertical* force of gravity, you must measure the *vertical* height of the stairs (in metre).

Then ask someone to use a stopclock to time how long it takes you to run from the bottom to the top of the stairs.

Now use the method shown on the opposite page to calculate
1.   the amount of work you did
2.   the power developed by your legs.

Who develops the most power in your class?
What are the energy changes in this experiment?

Wattsisname?
Q.   The unit of power is named after which person?
A.   Watt.
Q.   I'm asking who, not what.
A.   I mean Sir James Watt!

Electric power

Electric power is also measured in watt. Electric lamps are marked with the electrical power that they convert. Look at the writing on a lamp. '60W' means that the lamp converts 60 joule per second (from **electrical** energy to heat and light energy).

60W

Example
An electric kettle is rated at 2kW. How many joules of energy are converted in 10 seconds?

Power = 2 kW = 2000 W = 2000 joule per second
∴ In 10 seconds, energy converted to heat   = 2000 × 10 joule
                                             = 20 000 joule (= 20 kJ)

2kW

Summary

Work done = Force x distance moved
(joule)   (newton)   (metre)
1 joule is the work done when a force of 1N moves through 1 metre.
Principle of conservation of energy: energy cannot be created or destroyed but can be changed from one form to another.

Forms of energy: thermal, light, sound, kinetic and potential energy, electric and magnetic energy, chemical energy and nuclear energy.

Power = $\dfrac{\text{work done (joule)}}{\text{time taken (seconds)}}$

1 watt is a rate of working of 1 joule per second.

Questions

1. Copy and complete these sentences.
    a) The work done (measured in . . . . .) is equal to the . . . . . (in newton) multiplied by the . . . . . moved (in metre).
    b) 1 joule is the work done when a force of one . . . . . moves through a distance of one . . . . (in the direction of the . . . . .).
    c) The principle of . . . . . of energy says that energy cannot be . . . . . or . . . . . but it can be . . . . . from one form to another.
    d) The nine forms of energy are:
    e) Power is the rate of doing . . . . .

       Power (in . . . . .) = $\dfrac{\text{work done (in . . . . .)}}{\text{. . . . . (in seconds)}}$

    f) 1 watt is a rate of working of one . . . . . per . . . . .

2. A man lifts a parcel weighing 5 newton from the ground on to a shelf 2 metre high. How much work does he do on the parcel?

3. A boy weighing 500 N climbs 40 m vertically when walking up the stairs in an office block. How much work does he do against gravity? What are the energy changes here?

4. Which of the following are examples of working?
    a) a moving car
    b) a stationary car
    c) pushing a pram
    d) a brick wall
    e) a paperweight holding down paper
    f) winding up a watch.

5. Which of the following are examples of potential energy?
    a) a boy at the top of a slide
    b) a stretched catapult
    c) a hot iron
    d) a stretched bow
    e) the arrow in a stretched bow
    f) a wound-up clock
    g) a guillotine.

6. Name the forms of energy stored in the following
    a) a slice of bread
    b) a travelling bullet
    c) dynamite
    d) a stretched elastic band
    e) a torch battery
    f) a turning flywheel
    g) petrol
    h) water in a high dam.

7. Name the energy changes in each example in the list below. For example, in an electric lamp.

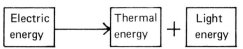

    a) a torch battery   g) a pendulum
    b) a falling stone   h) bicycle brakes
    c) a bunsen burner   i) a human body
    d) a dynamo   j) an electric bell
    e) a microphone   k) an electromagnet
    f) a loudspeaker   l) a car crash

8. What are the energy changes in these examples?

9. Name the forms of energy at each stage of the following:
   a) a car engine (see page 72)
   b) a coalfired power station
   c) a nuclear power station
   d) a hydroelectric power station

10. Draw an 'energy chain diagram' to show how the energy you use for living comes from the sun.

11. What things can be used to change
    a) electrical energy into kinetic energy
    b) thermal energy to kinetic energy
    c) sound energy to electrical energy

12. Professor Messer dreamed one night that the world had suddenly changed so that energy could not be converted from one form to another. Why did his dream turn into a nightmare?

13. A girl does 1000 joule of work in 5 seconds. What power does she develop?

14. What energy changes take place in sports events like pole-vaulting, weight-lifting, javelin-throwing, the marathon?

15. A man lifts a weight of 300 N through a vertical height of 2m in 6 seconds. What power does he develop?

16. A man weighing 1000 N runs up some stairs, rising a vertical height of 5m in 10 seconds. What power does he develop? What are the energy changes here?

17. A crane lifts a load of 3000 N through a vertical height of 10m in 4 seconds. What is its rate of working in a) watt   b) kilowatt   c) horsepower?

18. An electric lamp is marked 100W. How many joule of electrical energy are transformed into heat and light a) during each second b) during a period of 100 seconds.

19. Find out all you can about the work of
    a) James Joule   b) Sir James Watt.

20. What are the energy changes in these examples?

Further questions on page 171.

Physics at work: Building Bridges

Experiment 1   Bending a beam

Put a ruler as a bridge between two books. Then press down gently on the middle so that the 'beam' bends.
What must be happening to the molecules of the ruler?

In the lower part of the ruler the molecules are pulled farther apart (as the ruler bends). This part of the ruler is in *tension*. In the upper part of the ruler the molecules are pushed closer together. This part of the ruler is in *compression*.

Experiment 2   Building paper bridges

Using only a sheet of A4 paper, build a bridge between two supports. Try different designs and test each bridge with pennies or laboratory weights. Here are some possible designs:

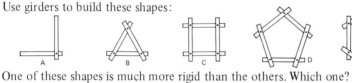

How can you devise a *fair test* to compare them?
Which is the strongest design? Why?

Experiment 3   Testing girders

For girders you can use drinking straws, with pins to fasten them (or you can use balsa wood, or wooden 'spills', LEGO, or Meccano).
Use girders to build these shapes:

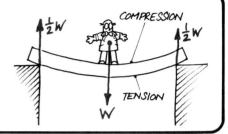

One of these shapes is much more rigid than the others. Which one?
How can you add extra girders to make the others more rigid?

Experiment 4   Building a girder bridge: a competition

Use your girders to build a bridge to cross a 10cm gap.
You can use up to 10 straws but imagine each one costs £1000.
You can test each bridge (until it collapses) by hanging a paper cup from it and gently adding weights or sand to it. How can you ensure your tests are *fair*?
Who can build the cheapest bridge to support a 500 g mass?
Who can build the strongest bridge for £10 000?

Experiment 5   A competition

Your materials are a paper cup of dry sand, some rubber bands and some straws (which 'cost £1000' each). Push a straw through the cup as shown and then make a structure that will support the cup clear of the table. The winner is the person who can make the cheapest structure.

Physics at work: Sport

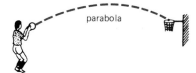

In many games, a ball is thrown or hit.
The ball does not travel in a straight line, but
in a *parabola* shape. (This is because gravity is
pulling it downwards.)
To be more accurate, it is the centre of gravity
of the ball that follows the parabola.

The centre of gravity (centre of mass) of a person
is usually behind the navel, but it depends on the
positions of the arms and legs.
When an acrobat or a diver jumps through the air,
the centre of gravity moves in a smooth parabola,
no matter how the person twists or spins.

The long jump

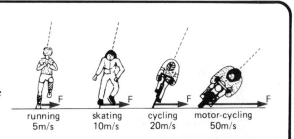

Sprinting: The athlete accelerates
because of the force F (the
friction force, see also page 112).
His weight W is balanced by the
reaction force R.
You can measure the acceleration
of a runner by using ticker-tape
(see page 149).

The take-off:
A final push to
make R and F as
large as possible.

The landing:
Now the forces F
and R bring him to
a stop.

His centre of gravity follows a parabola
between take-off and landing.
The world record is 9m.

Using friction to turn corners

In many sports you have to turn corners.
This can only happen if there is an inwards force
F provided by friction.
The faster you go, and the sharper the bend, the
bigger the force F needs to be (and the bigger the
angle you lean over).
What happens if the friction F is not big enough
(for example if the cyclist meets ice or gravel)?

running
5m/s

skating
10m/s

cycling
20m/s

motor-cycling
50m/s

Analysing movement

One way to analyse an athlete's movement is to use videotape or a
cine film (see page 213).
Another way is to use a camera in the dark, with a flashing light
called a *stroboscope*. Each flash of light shows one position of
this golfer:
Where is the club moving the fastest and the slowest?

There are many examples of Physics in other sports. For example:
football (p 162), snooker (p 163), car-racing (p 72, 124),
cycling (p 142, 147), shooting (p 115, 163), fishing (p 140) etc.

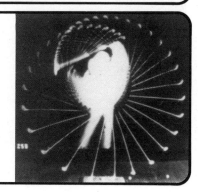

chapter 18

MACHINES

Machines help us to apply forces more easily. In this chapter we will look at some simple machines like levers, gears and pulleys.

A *Lever* is a very common simple machine. The diagram shows a crowbar being used to lever up a large stone. It is like the seesaw we looked at on page 121.

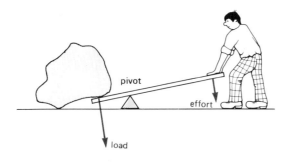

The man is exerting an *effort* force, against a *load* force (the weight of the stone).

As you can see in the diagram, the effort force of the man is *smaller* than the load force of the stone because he is pushing a *longer* distance from the pivot (see the principle of moments on page 121).

This means that the lever is magnifying the force. To see how much it magnifies the force, we calculate the *mechanical advantage (M.A.)*

$$\text{Mechanical advantage (M.A.)} = \frac{\text{load}}{\text{effort}} \qquad \cdots\cdots\cdots (1)$$

This number tells us by how much the force is magnified. For example, if the load is 50N and the effort is only 5N then

$$\text{M.A.} = \frac{\text{load}}{\text{effort}} = \frac{50}{5} = 10$$

This tells us that the force has been magnified 10 times. In other words, the man has to push only $\frac{1}{10}$th as hard as the load and so he finds it is much easier to lift the stone.

In practice, *the M.A. depends on how much friction there is* in the machine. A lot of friction means a smaller M.A.

Although the man does not have to push as hard as the load, he has to push *further*. As the diagram shows, the distance moved by the effort is greater than the distance moved by the load. The ratio of these two distances is called the *velocity ratio (V.R.)*.

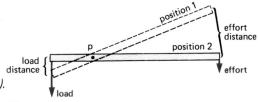

$$\text{Velocity Ratio (V.R.)} = \frac{\text{distance moved by the effort}}{\text{distance moved by the load}} \qquad \dots (2)$$

In the diagram, the V.R. = 3. This means that the effort moves three times farther than the load.
The smaller force always moves the longer distance.
The V.R. does *not* depend upon friction.

Efficiency

The *efficiency* of a machine is very important. If we could build a machine that had *no* friction in it, then it would be 100% efficient and *all* the energy put into the machine (by the effort) would be got out of the machine (and given to the load).
In practice, all machines have some friction in them and the work got out is *always less* than the work put in.

$$\% \text{ Efficiency} = \frac{\text{work got out (work done on the load)}}{\text{work put in (work done by the effort)}} \times 100\% \qquad \dots (3)$$

The efficiency is always less than 100% because of friction.
If the efficiency is 50% it means that only half the energy is given out as useful work on the load and the rest is converted to heat by friction.

From the formula for % efficiency it can be proved that

$$\% \text{ Efficiency} = \frac{\text{M.A.}}{\text{V.R.}} \times 100\% \qquad \dots (4)$$

Efficiency will soon decrease
If you forget the oil and grease

Example
In a certain machine an effort force of 10N is used to lift a load of 30N.
To lift the load by 1 metre the effort has to be moved 4 metre.
What is the efficiency of the machine?

Step 1
(use equation 1) $\text{M.A.} = \dfrac{\text{load}}{\text{effort}} = \dfrac{30N}{10N} = 3$

Step 2
(use equation 2) $\text{V.R.} = \dfrac{\text{distance moved by effort}}{\text{distance moved by load}} = \dfrac{4m}{1m} = 4$

Step 3
(use equation 4) $\% \text{ Efficiency} = \dfrac{\text{M.A.}}{\text{V.R.}} \times 100\% = \tfrac{3}{4} \times 100\% = \underline{75\%}$

Levers

The V.R. of a lever can be found by measuring the two arms of the lever (marked e and ι in the diagram). By comparing distances in the diagram,

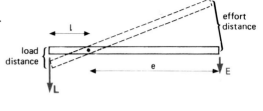

V.R. $= \dfrac{\text{length of effort arm, (e)}}{\text{length of load arm, } (l)}$

If the effort arm (e) is three times longer than the load arm (ι), then the V.R. = 3

If there is not much friction, then the M.A. will be almost 3 (and the lever will exert a force almost 3 times stronger than the effort).

Look at these diagrams of levers and work out an approximate value for the V.R. in each case:

Type 1
These levers have the pivot or *fulcrum* between the effort force and the load.

Type 2
These levers have the *load* between the effort and the fulcrum. A small effort moves a large load (the force is magnified).

Type 3
These levers have the *effort* between the load and the fulcrum. The load is smaller than the effort but moves farther (the distance is magnified).

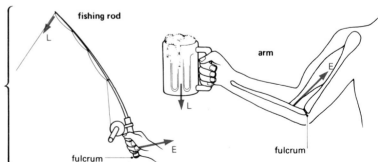

Here are some more examples of levers. For each one, work out whether it is a 'force magnifier' or a 'distance magnifier' and then decide which group on the opposite page it belongs to.

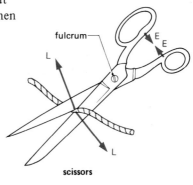

scissors

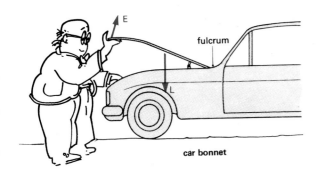

car bonnet

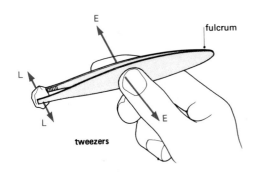

tweezers

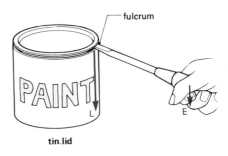

tin lid

Wheel and axle

A wheel and axle machine is really a kind of continuous lever.

A small effort force on the wheel gives a large force on the axle.

Since the wheel's radius R is the effort arm and the axle's radius r is the load arm, then from the formula on the opposite page,

$$\frac{\text{Velocity}}{\text{Ratio}} = \frac{\text{radius of wheel, R}}{\text{radius of axle, r}}$$

As always, the M.A. and the efficiency depend on how much friction there is.

There are many examples of a wheel and axle in everyday use: the winch in the diagram, the steering wheel on a car, the handlebars and foot crank on a bicycle, a door knob, a spanner turning a nut, a screwdriver, and many others.

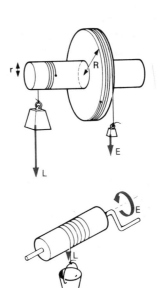

141

Gears

Like the wheel and axle, gears are a kind of continuous lever. They can be used as a force magnifier or as a distance magnifier.

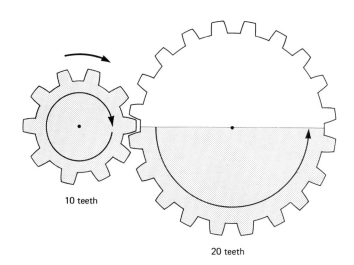

In the diagram, the small wheel has 10 teeth and the large wheel has 20 teeth. When the small wheel moves clockwise through one complete turn, the large wheel moves *anti*clockwise through only half a turn. The smaller (effort) wheel has to turn *twice* to turn the large (load) wheel *once*. This means that the V.R. = 2. In fact,

10 teeth

20 teeth

$$\frac{\text{Velocity}}{\text{Ratio}} = \frac{\text{number of teeth on } \textit{driven} \text{ wheel}}{\text{number of teeth on } \textit{driving} \text{ wheel}}$$

In this case, the M.A. will be almost 2 (depending on friction) and so the large wheel will exert almost twice the force (at half the speed).

Look at the diagram at the top of page 138. If the 'm' gear turns clockwise, which way does the 's' gear turn?

Bevel gears can also be used to change the direction.

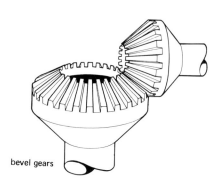

bevel gears

Remember: the large wheel always moves slower but with more force, while the smaller wheel moves faster but with less force. Exactly the same applies to the sprocket wheels on a bicycle.

Experiment 18.1
Count the number of teeth on the rear sprocket and the chain wheel of a bicycle.
What is the V.R. of that part of the bicycle?
Is it a force magnifier or a distance magnifier?

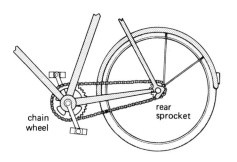

chain wheel

rear sprocket

The inclined plane

A ramp, a slope and a hill are examples of inclined planes.

In the diagram the man is lifting a heavy load into the van using a small effort force.
As in all these machines, the smaller force has to go a longer distance — in this case the effort has to move the full length of the slope while the load moves a shorter distance vertically.
This means that:

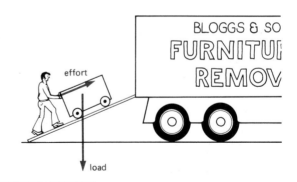

$$\text{Velocity Ratio} = \frac{\text{distance moved by the effort along the slope}}{\text{distance moved by the load vertically}}$$

Experiment 18.2
Weigh a small trolley (with a spring balance) and then pull it up an inclined plane.

Is the effort smaller than the load?
What is the M.A.?
Measure the distances.
What is the V.R.?

How did the Egyptians build the pyramids?

The screw

In days of old, so we are told,
Men built the pyramids high.
They took the strain on an inclined plane
And their motto was 'Do it or die'.

The thread of a screw is really an inclined plane wrapped round the screw.

Experiment 18.3

Cut out two pieces of paper so that they look like inclined planes, one shallow and one steep.
Wrap each piece of paper round a pencil and you will see that each one forms a screw.

Which inclined plane (A or B) would need less effort?
Which screw (a or b) would need less effort?

A screw is used to exert a large force in a nut and bolt, a vice and a car-jack.

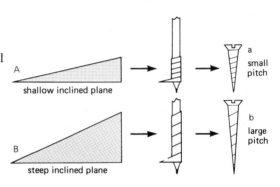

A shallow inclined plane

a small pitch

B steep inclined plane

b large pitch

Pulleys

Pulleys are very useful for lifting loads vertically. Even a single pulley is useful because it changes the direction of the force. It is easier to pull downwards (with your weight helping you) than it is to pull upwards.

The velocity ratio of a single pulley is 1 (because the effort and the load move equal distances).

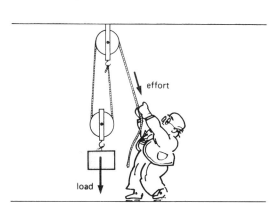

Two pulleys

It is easier to lift a load when the rope goes round two pulleys. You can see that one end of the rope is fixed and the man is pulling on the other end to lift up the bottom pulley and the load.

How far would you have to move the effort to lift the load by one metre?
Because the bottom pulley is held up by two ropes and because they *both* have to be shortened by 1 metre, the effort has to move 2 metres. This means that the velocity ratio = 2.

In fact, for any pulley system,

$$\text{V.R.} = \frac{\text{number of pulleys used}}{} = \frac{\text{number of ropes supporting the load}}{}$$

The M.A. of the pulley system would be less than 2 because the effort has to overcome friction and it has to lift the bottom pulley as well as the load.

A complex pulley system used to lift very heavy loads.

Three pulleys

The diagram shows how *three* pulleys can be connected by a rope to make the lifting even easier.

Now there are two pulleys in the top block and the rope is fixed to the bottom pulley.

(In practice the two top pulleys are of equal size, but here one is drawn smaller so that you can see where the rope goes).

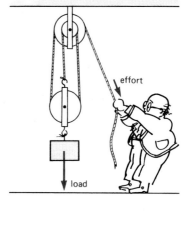

Example

If an effort of 100N is needed to lift a load of 200N with this pulley system, what is its efficiency?

Step 1 M.A. $= \dfrac{\text{load}}{\text{effort}} = \dfrac{200\text{N}}{100\text{N}} = 2$

Step 2 V.R. = number of pulleys = 3

Step 3 % Efficiency $= \dfrac{\text{M.A.}}{\text{V.R.}} \times 100\% = \dfrac{2}{3} \times 100\% = \underline{66\frac{2}{3}\%}$

Experiment 18.4

Set up a pulley system with four pulleys and hang a known weight on the bottom pulley block. Use a spring balance to measure the effort needed to lift the load.

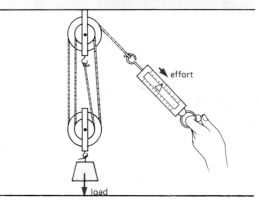

Calculate the efficiency in the same way as in the example above (an electronic calculator is useful here).

Repeat with different loads, each time measuring the effort and calculating the efficiency.

Put your results in a table and draw a graph of *efficiency* against *load*.

Some possible results:

Load (N)	effort (N)	M.A. $= \dfrac{\text{Load}}{\text{effort}}$	% efficiency
1·0	0·5	2·0	50%
2·0	0·8	2·5	63%
3·0	1·0	3·0	
4·0	1·2		
5·0			

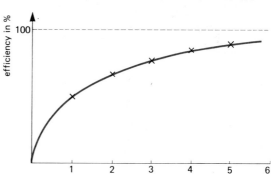

Can you explain why the efficiency increases with heavier loads?

(Hint: think of the weight of the bottom pulley block compared with the load.)

Something to do

A forceful trick to beat two strong men.

The diagram shows how you can pull two men together against their full strength.

What kind of machine is this?
What is its velocity ratio?

Summary

Mechanical Advantage (M.A.) = $\dfrac{\text{Load}}{\text{Effort}}$ (it depends on friction)

Velocity Ratio (V.R.) = $\dfrac{\text{distance moved by the effort}}{\text{distance moved by the load}}$ (it does not depend on friction)

% Efficiency = $\dfrac{\text{work got out}}{\text{work put in}} \times 100\%$ (it depends on friction and is always less than 100%)

$= \dfrac{\text{M.A.}}{\text{V.R.}} \times 100\%$

The weaker force always moves the longer distance.

V.R. of a lever = $\dfrac{\text{length of effort arm}}{\text{length of load arm}}$

V.R. of wheel and axle = $\dfrac{\text{radius of wheel}}{\text{radius of axle}}$

V.R. of gears = $\dfrac{\text{number of teeth on driven wheel}}{\text{number of teeth on driving wheel}}$

V.R. of an inclined plane = $\dfrac{\text{slope length}}{\text{vertical height}}$

V.R. of block and tackle = number of pulleys used
= number of strings lifting the bottom block

Barrows for Sale – but would you buy? Please work it out and tell me why.

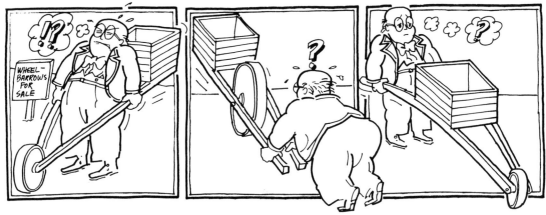

WHEEL-BARROWS FOR SALE

146

Questions

1. Copy out and complete these sentences.
   a) Mechanical advantage is . . . . divided by . . . . . Its size depends on how much . . . . . there is.
   b) Velocity ratio is the distance moved by the . . . . . divided by the distance moved by the . . . . .
   c) Efficiency is defined as the . . . . . got out of the machine divided by the . . . . . put in. Its size depends upon . . . . . and is always less than . . . % The efficiency is also equal to the . . . . . divided by the . . . . .

2. A man uses a crowbar 1·5 metre long to lift a rock weighing 600 newton. If the fulcrum is 0·50 metre from the end of the bar touching the rock, how much effort must the man apply? (Draw a diagram first).

3. Make a list of all the examples of the lever and the wheel-and-axle that you can find in your home.

4. A gear wheel A with 20 teeth is used to drive a gear wheel B with 60 teeth. What is the velocity ratio? If wheel A rotates three times every second, how many times does wheel B rotate in each second?

5. The diagram shows a trolley of weight 300N being pulled at a steady speed up a ramp by a force of 200N.

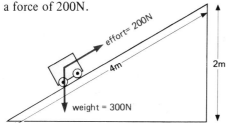

Calculate

a) the mechanical advantage
b) the velocity ratio
c) the work done on the load (see page 127).
d) the work done by the effort
e) the efficiency of the machine.

Further questions on page 171.

6. How many examples of simple machines can you find in the photograph? In each case name the type of machine and say whether it is a force magnifier or a distance magnifier. Why does a cyclist often zig-zag up a steep hill?

7. Think carefully about the following machines saying where the effort is applied and what the load is   a) a vice   b) a bolt and nut
   c) a corkscrew   d) a brace and bit
   e) a screwjack   f) an adjustable spanner
   g) a ship's propeller.

8. Look at these diagrams of pulley systems. For each one, calculate   a) the M.A.
   b) the V.R.   c) the work done in lifting the load 1 metre   d) the work done by the effort in this case   e) the % efficiency.

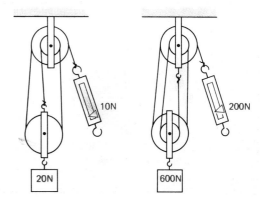

chapter 19

VELOCITY and ACCELERATION

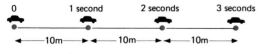

a constant speed of 10m/s

If a car travels down the road at a *constant speed* of 10 metre per second (10 m/s), it means that it travels 10 metre in every second.

The speed can be found by

Average speed	=	distance travelled (metre)
		time taken (second)

Velocity is almost the same thing as speed but it is a *vector* quantity (see page 116). Velocity has a *size* (called speed) *and* a *direction.*

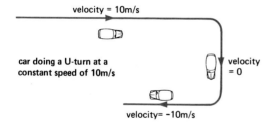

car doing a U-turn at a constant speed of 10m/s

If a car is travelling at a constant speed *in a straight line,* then it has a constant velocity. If it turned a corner then its velocity would change, even though its speed remained constant. A car travelling in the opposite direction has a negative velocity.

Acceleration

If a car-driver presses the accelerator, his car goes faster — it accelerates.

The acceleration can be found by

Acceleration	=	change in velocity (m/s)
		time taken for the change (s)

If the change in velocity is measured in m/s and the time is measured in seconds, then the acceleration is measured in m/s^2 (metre per second squared).

An acceleration of $2m/s^2$ means that the velocity increases by 2m/s every second.

If a car is slowing down then it has a *deceleration* or a *retardation* or a *negative acceleration.*

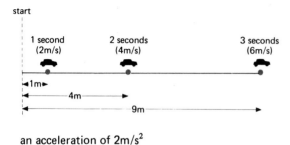

an acceleration of $2m/s^2$

Ticker-timer experiments

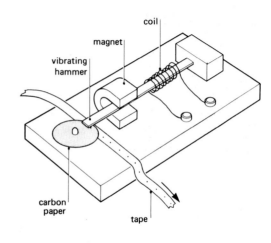

A ticker-timer is simply a vibrator that puts little black dots on to a paper tape at the rate of 50 dots in each second.

Experiment 19.1

Pull a length of tape through the ticker-timer at a constant velocity.

What do you notice about the spacing of the dots?

a) Count along the tape, marking off every fifth dot. Because there are 50 dots in each second, each 5-dot length is produced in $\frac{1}{10}$th second.
b) Use scissors, at each mark, to cut the tape into several 5-dot lengths (carefully keeping them in the right order).
c) Stick the 5-dot lengths side by side in the right order to get a strip chart.
Since each strip is the distance travelled in $\frac{1}{10}$th second, you have made a strip chart of *velocity* against *time*.

The constant height of the strips means that the tape was moving at a constant velocity.

Experiment 19.2

Repeat experiment 19.1 but make the tape accelerate as you pull it.

What do you notice about the spacing of the dots?

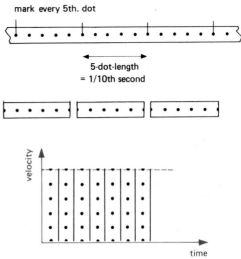

Repeat steps (a), (b) and (c) to build a velocity-time strip chart for this accelerating tape.

You can see that the velocity is increasing (because the tape was accelerating).

What would be the velocity-time graph for an object which was slowing down (decelerating)?

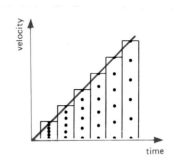

Experiment 19.3

Make a strip chart for a more complicated move-ment — for example, fix the tape to a drawer or door which is then closed gently.

Velocity-time graphs

The diagram shows a velocity-time graph for a car travelling at constant velocity.
It is not accelerating and not decelerating.

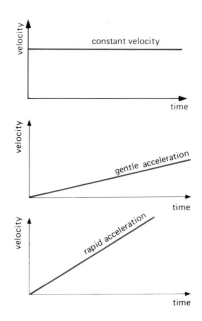

What would the graph look like if the car accelerated gently?
Because its velocity would increase, the graph would slope upwards:
In this graph, the car starts off from rest (velocity is zero) and accelerates uniformly (steadily).
A straight graph means uniform, constant acceleration.

What would the graph look like if the car accelerated more rapidly?
Since its velocity would increase more rapidly, the graph would be *steeper*. In fact,

> the *acceleration* is shown by the *slope* (or gradient) of the velocity-time graph.

Page 388 explains how to find the slope of a graph.

Here is a velocity-time graph of a car starting off at one set of traffic lights and stopping at the next set of lights.
Can you see that the car accelerates from A to C, travels at constant velocity between C and D and then decelerates (brakes) rapidly to stop at E.

At which point (A, B or C) is it accelerating most?

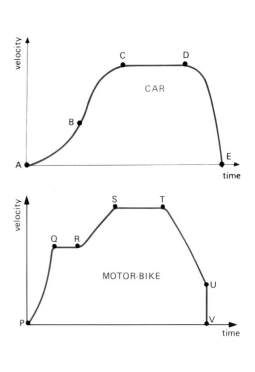

Here is a velocity-time graph of a motor-bike.

Between which points is the motor-bike accelerating most rapidly?
Which part of the graph might show the motor-bike changing gear? When is the bike travelling fastest?

When does the rider start to brake? Which part of the graph shows the bike hitting a solid brick wall?

If this was a ticker-tape strip chart you could find the total *distance travelled* by measuring the total amount of tape under the graph. In fact,

> the *distance travelled* is shown by the *area* under the velocity-time graph.

Here is a velocity-time graph of a 'stockcar' starting a race, crashing into another car, and then reversing.

How long do you think the driver has to wait before the starter waves his flag to start the race?

Is his acceleration greatest at B, C, D or E?
At which point is he travelling fastest?
What is his velocity then? How much time has passed?

When does he start to brake?
At what time does he come to a stop after hitting the other car? How long has he been travelling then?

For how long does he stay at rest before reversing?
At which point (G, H, I or J) does he start to reverse his car?
When the car is reversing, it has a *negative* velocity (remember velocity is a *vector* quantity, see page 116).

What is his maximum velocity in reverse?
For how many seconds does he reverse the car? When does he finally stop the car?
How can you tell whether the distance travelled is farther in forward gear or reverse?

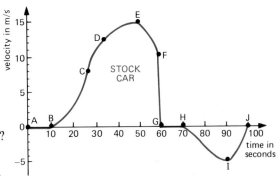

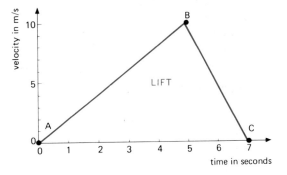

Here is a velocity-time graph of a lift climbing from the ground floor to the top of a building.

At which point (A, B or C) is it on the ground floor?
At which point (A, B or C) is it travelling fastest?
What is its maximum velocity in m/s?
How long does it take to reach this velocity?

At which point does it reach the top of the building? How long does the journey take?

We can calculate the acceleration from the formula on page 148:

$$\text{Acceleration} = \frac{\text{change in velocity (m/s)}}{\text{time taken for the change (s)}}$$

$$= \frac{10 \text{ m/s}}{5 \text{ s}} = 2 \text{ m/s}^2$$

Can you see why the deceleration is 5 m/s²?
This is more difficult: how can you calculate the distance travelled, from the numbers on the graph?

151

Displacement-time graphs

We have seen that velocity is almost the same thing as speed, but it is a vector quantity (see p 116). In a similar way, *displacement* is almost the same thing as distance, but it is a *vector* quantity. Displacement has a *size* (called distance) *and* a *direction*.

If you move to a different part of the room you might have a displacement of 2 metres *in an easterly direction*. The direction is important.

"The direction is important"

Ticker-timer measurements are sometimes used to draw **displacement-time graphs** (although they are not as useful as velocity-time graphs).

The diagram shows a displacement-time graph for a car which is *at rest* (stationary, not moving):

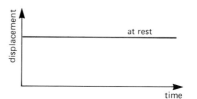

If the car moves away at constant velocity then the displacement will increase:

If the car moved away at a higher velocity then the line would be steeper. In fact,

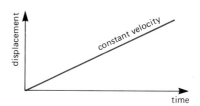

> the *velocity* is shown by the *slope* (or gradient) of the displacement-time graph.

If the car accelerates, then the velocity increases, and so the slope of the graph increases:

If the car decelerated, then the velocity would decrease, and so the slope of the graph would decrease. The graph would curve the other way.

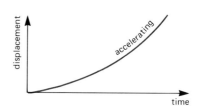

Here is a displacement-time graph for the same lift that is shown on the previous page. It has the same points (A, B, C) marked.

The lift accelerates from A to B, and is travelling fastest at B.

Where is it decelerating?

What is the total distance travelled?
How long does it take the lift to travel 25m?

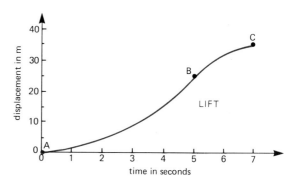

Equations of motion

These are 4 equations which can be used whenever an object is travelling with **constant, uniform acceleration** in a straight line.

We write these equations using 5 symbols: s, u, v, a, t.
Suppose an object is travelling at a velocity u and then moves with a uniform acceleration a for a time t.
Its velocity is then v and it has travelled a distance s.

s = distance travelled (metres)
u = initial velocity (m/s)
v = final velocity (m/s)
a = acceleration (m/s^2)
t = time taken (seconds)

Then from page 148: acceleration $= \dfrac{\text{change in velocity}}{\text{time taken}}$

or, in symbols: $a = \dfrac{v - u}{t}$

rearranging the equation (see page 402)

$$\boxed{v = u + at} \quad \dots (1)$$

Also from page 148: average speed $= \dfrac{\text{distance travelled}}{\text{time taken}}$

or, in symbols: $\dfrac{u + v}{2} = \dfrac{s}{t}$

rearranging the equation (see page 402)

$$\boxed{s = \dfrac{(u + v)}{2} t} \quad \dots (2)$$

Using equation (1) to replace v in equation (2): $s = \left[\dfrac{u + (u + at)}{2} \right] t$

rearranging:

$$\boxed{s = ut + \tfrac{1}{2}at^2} \quad \dots (3)$$

Using equation (1) to replace t in equation (2): $s = \dfrac{(u + v)}{2} \dfrac{(v - u)}{a}$

rearranging:

$$\boxed{v^2 = u^2 + 2as} \quad \dots (4)$$

Make sure you learn these 4 equations. You can remember the 5 symbols by '*suvat*'. If you know any *three* of 'suvat' the other 2 can be found.

In questions, 'initially at rest' means $u = 0$.
A negative number for the acceleration means that the object is slowing down (decelerating).

Example
A car starts from rest and accelerates at 2 m/s^2 for 10 seconds. Calculate a) the final velocity and b) the distance travelled.

Write 'suvat' $s = ?$
and show what $u = 0$
you know: $v = ?$
 $a = 2$ m/s^2
 $t = 10$ s

a) Use equation (1): $v = u + at$
put in numbers: $= 0 + 2 \times 10$
 $= \underline{20 \text{ m/s}}$

b) Use equation (3): $s = ut + \tfrac{1}{2}at^2$
put in numbers: $= 0 + \tfrac{1}{2} \times 2 \times 10^2$
 $= \underline{100 \text{ m}}$

Acceleration due to gravity

The force of gravity pulls down on all objects here on Earth (see page 85). If objects are allowed to fall, they *accelerate* downwards.

If there is no air resistance or friction then all objects accelerate downwards at the *same* rate. You may have heard of a famous experiment Galileo is supposed to have done from the leaning tower of Pisa: if a heavy stone and a light stone are dropped together, they accelerate at the same rate and land at the same time.

Experiment 19.4
Fasten a weight to a length of ticker-tape and let it fall from the ticker-timer.
Cut up the tape as before (page 149) to get a velocity-time strip chart.

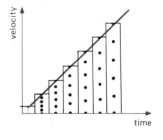

Repeat the experiment with a heavier weight and compare the graphs. What do you find?
Is the acceleration always the same?

How could you find the acceleration from your graphs?
Accurate measurements show that:

> the acceleration of free fall, g = 9·8 m/s^2
> (*the acceleration due to gravity*)

For simple calculations we often take this number as 10 m/s^2. This means that for an object falling with no air resistance, the velocity after 1 second is 10 m/s and after 2 seconds the velocity is 20 m/s and so on. If a man fell for 5 seconds with no air resistance what would his speed be?

In practice, there is usually air resistance. A parachute is designed to make the air resistance as large as possible. If a man falls a long way without a parachute then because of air friction he reaches a final or *terminal* velocity of about 50 m/s – the speed of a fast racing car.

There was a young man who had heard,
That a person could fly like a bird.
To prove it a lie
He jumped from the sky,
—His grave gives the date it occurred!

Experiment 19.5 Measuring the acceleration of free fall, g

An electric stop-clock (accurate to $\frac{1}{100}$ second) is used to measure the time taken for an object to fall through a known distance s.

When the switch is in position A, the electromagnet holds up a small steel ball.

When the switch is moved quickly to B, the electromagnet releases the ball and the clock starts timing.
When the ball hits the trapdoor, the circuit is broken at contact C, and the clock stops.

From page 151, $s = ut + \frac{1}{2}at^2$
But $u = 0$ and $a = g$, acceleration due to gravity

$\therefore s = \frac{1}{2}gt^2$ or $g = \dfrac{2s}{t^2}$

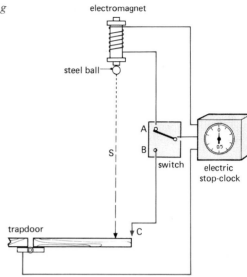

electromagnet

steel ball

A

S

B

switch

electric stop-clock

trapdoor

C

How does your value compare with the accepted value of $g = 9\cdot8$ m/s^2?

Example
A ball is thrown vertically upwards at 20 m/s.
Ignoring air resistance and taking $g = 10$ m/s^2, calculate
a) how high it goes b) the time taken to reach this height
c) the time taken to return to its starting point.

When travelling upwards it is *decelerating*, so $a = -10$ m/s^2.
At the moment when it reaches its highest point, $v = 0$.

Write $s = ?$
'suvat' $u = 20$ m/s
first: $v = 0$
          $a = -10$ m/s^2
          $t = ?$

a) From page 153: $v^2 = u^2 + 2as$
   Then numbers: $0 = 20^2 + 2(-10)s$
                  $20s = 20^2$
                  $\therefore s = \underline{20\ m}$

b) From page 153: $v = u + at$
   Then numbers: $0 = 20 + (-10)t$
                  $\therefore t = \underline{2\ seconds}$

c) It will also take 2 seconds to fall down again.
   $\therefore$ Total time $= \underline{4\ seconds}$

Vertical *and* horizontal motion

Experiment 19.6
Set up a ruler and two pennies as shown.
Press on the middle of the ruler and tap the end so that A falls vertically while B is projected sideways.

Listen to the pennies hitting the ground.
Do they hit at the same time?

The vertical accelerations of the pennies are exactly the same and unconnected to any horizontal movement.

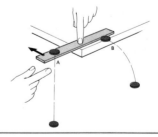

The horizontal and vertical motions of a body are independent and can be treated separately.

155

Newton's second law of motion

Experiment 19.7   Effect of force on acceleration
In this experiment a force is applied to a trolley using a length of elastic. The acceleration is measured using a ticker-timer.

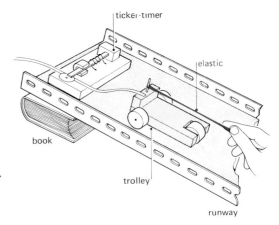

First eliminate the effects of friction by raising one end of the runway until, when tapped, the trolley runs down by itself at constant speed. The runway is then 'friction compensated'.

Now practise pulling the trolley with the elastic as shown, *always keeping the same stretch on the elastic* (judge this stretch against the front of the trolley). Then do it with the ticker-timer running.

Now apply *twice* the force to the trolley by using 2 elastics side by side, each with the *same stretch as before*.
Then apply three times the force, using 3 elastics.

From each tape make a velocity-time strip chart (see page 149). Remember that the slope of each graph shows the acceleration.

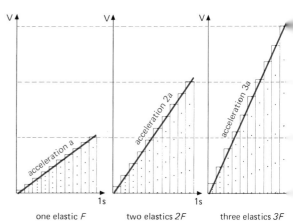

one elastic F   two elastics 2F   three elastics 3F

Can you see that doubling the force doubles the acceleration? Does three times the force give three times the acceleration?
  ∴ acceleration is *directly proportional* to the force

or,   **acceleration ∝ force**

Experiment 19.8   Effect of mass on acceleration
Now double the mass of the trolley by stacking a second trolley on top of the first one.
Repeat experiment 19.7 using 2 elastics with the *same stretch as before*.

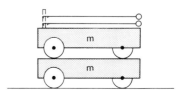

Then repeat it with a mass of 3 trolleys.
Make velocity-time strip charts as before.

Can you see that doubling the mass (with the same force) gives half the acceleration.
Does 3 times the mass have $\frac{1}{3}$ of the acceleration?
  ∴ acceleration is *inversely proportional* to the mass

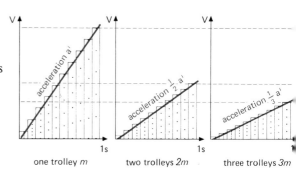

one trolley m   two trolleys 2m   three trolleys 3m

or,   acceleration $\propto \dfrac{1}{\text{mass}}$

Combining both results: acceleration $\propto \dfrac{\text{force}}{\text{mass}}$

or,   **Force ∝ mass × acceleration**

Provided we measure the force in newtons, the formula from the opposite page becomes:

Force = mass x acceleration   or $F = ma$
(in N)   (kg)   (m/s²)

This is Newton's Second Law of Motion.
Remember: (1) the force F is the resultant (or unbalanced) force on
the mass m,
(2) the force F must be measured in newtons.

One newton (1N) is defined as the force which gives to a mass of 1 kg, an acceleration of 1 m/s².

Example
When a force of 6 N is applied to a block of mass 2 kg, it moves along a table at constant velocity.
a) What is the force of friction?
   When the force is increased to 10 N, what is
b) the resultant force?
c) the acceleration?

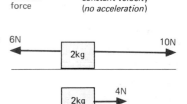

a) When the block moves at constant velocity (no acceleration), there is no resultant force on it (see Newton's First Law, page 89). This means the forces are balanced and equal.
   ∴ The frictional force = 6 N

b) When the applied force = 10 N, the frictional force is still 6 N.
   ∴ The resultant (unbalanced) force = (10 − 6)N = 4 N
c) formula first: Force = mass x acceleration
   then numbers: 4 N = 2 kg x acceleration
   ∴ acceleration = $\frac{4}{2}$ = 2 m/s²

If you wanted to find the distance travelled from rest in 10 seconds, which equation of motion would you use next? (see page 151).

Gravitational Field Strength g
The Earth is surrounded by an invisible *gravitational field*. This field exerts a force (called weight) on any mass which is in the field.

On the surface of the Earth, experiments show that this force is **9·8 N on every 1 kg** (see page 87).
This is, the **gravitational field strength, g = 9·8 N/kg.**

We now have 2 ways of thinking of g:
(1) For an object in free fall, the acceleration g = 9·8 m/s².
(2) For any object, moving or at rest, the gravitational field strength g = 9·8 N/kg.
   For m kg the force is m times as much, | ∴ **Weight = mg newton** |

We usually take g = 10 m/s² = 10 N/kg.

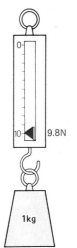

157

Gravitational potential energy

This is the *energy that a body has because of its position* (see also page 128).

To lift an object to a height h above the ground, work has to be done against the force of gravity (see also page 127).

Gravitational potential energy   = work done against gravity
                                  = force x distance moved (see page 127)
                                  = weight x vertical height
                                  = mass x g x height (see page 157)

$\therefore$ | **Gravitational Potential Energy** = **mass** x ***g*** x **height** | or | **P.E. = *mgh*** |
       (in joules)   (kg) (10 N/kg) (m)

weight = mg

Kinetic energy

This is the *energy that a body has because of its movement* (see also page 128).

If mass m is accelerated, from rest, by a force F for a distance s, then its kinetic energy will increase. Referring to the diagram:

From page 153: $v^2 = u^2 + 2as$
But $u = 0$: $\therefore v^2 = 2as$
$$\therefore s = \frac{v^2}{2a} \quad \ldots\ldots \text{(equation A)}$$

Now: Kinetic energy gained by object   = work done by the force
From page 127:   = Force F x distance s
But $F = ma$ (page 157):   = ma x s

And from equation A above: = $ma \times \dfrac{v^2}{2a} = \frac{1}{2}mv^2$

$\therefore$ | **Kinetic Energy** = $\frac{1}{2}$ x **mass** x **velocity squared** | or | **K.E** = $\frac{1}{2}mv^2$ joules |
      (in joules)   (kg)   (m/s)^2

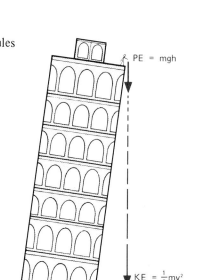

PE = mgh

KE = $\frac{1}{2}mv^2$

When the mass hits the ground 1800 J of kinetic energy is converted to 1800 J of heat and sound energy.

Example
The leaning tower of Pisa is 45 m high. A mass of 4 kg is dropped from the top.
Calculate a) the P.E. at the top,
              b) the K.E. just before hitting the ground,
              c) the velocity just before hitting the ground.

a) Formula first: Potential energy = mgh joule
    Then numbers: = 4 x 10 x 45 J =   <u>1800 J</u>

b) From the Principle of Conservation of Energy (see page 129),
                     Loss in P.E. = Gain in K.E.
        $\therefore$ K.E. just before impact = <u>1800 J</u>

$\int$) $\therefore \frac{1}{2}mv^2$ = 1800
                   $\frac{1}{2}$ x 4 x v^2 = 1800
                         v^2 = 900
  $\therefore$ Velocity just before impact, v = <u>30 m/s</u>

This last answer could also have been found from $v^2 = u^2 + 2as$.

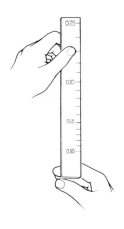

Something to make—a reaction timer

You can test a friend's reaction time by holding the top of a 30cm (1 foot) ruler so that it hangs vertically with your friend's fingers close to (but not touching) the zero mark at the bottom of the ruler. When you let go (with no warning) your friend has to grip the ruler as quickly as possible.

The slower his reaction time, the farther the ruler will fall as it accelerates downwards (at 9.8 m/s^2).

The table below shows how to mark the ruler (or a strip of card). It also shows how far you would travel in a car at 50 km/hr (30 m.p.h.) (that is, how far you would travel *before* your foot could reach the brake pedal to *start* to slow down).

At this distance (in cm) from the zero mark:	4.9	5.9	7.1	8.3	9.6	11.0	12.5	14.2	15.9	17.7	19.6	21.6	23.7	25.9	28.2	30.6
mark this time (in seconds) on your ruler:	0.10	0.11	0.12	0.13	0.14	0.15	0.16	0.17	0.18	0.19	0.20	0.21	0.22	0.23	0.24	0.25
Distance travelled (in metre) at 50 km/hr (30 m.p.h.)	1.4	1.5	1.7	1.8	1.9	2.1	2.2	2.4	2.5	2.6	2.8	2.9	3.1	3.2	3.3	3.5

Summary

$$\text{Average Speed (m/s)} = \frac{\text{distance travelled (metres)}}{\text{time taken (seconds)}} \qquad \text{Acceleration} = \frac{\text{change in velocity (m/s)}}{\text{time taken for change (s)}}$$

Velocity (a vector quantity) is speed in a particular direction.

Acceleration is shown by the *slope* of a velocity-time graph.
Distance is shown by the *area* under a velocity-time graph.

For constant acceleration: $v = u + at \qquad s = \frac{(u + v)}{2} t$

$$s = ut + \tfrac{1}{2}at^2 \qquad v^2 = u^2 + 2as$$

The acceleration of free fall (due to gravity) = 9.8 m/s^2 (10 m/s^2)
The horizontal and vertical motions of a body are independent.

Newton's Second Law: **Force = mass × acceleration**
  (N)   (kg) (m/s^2)

1 N = force which gives to a mass of 1 kg an acceleration of 1 m/s^2.
Weight = mg newton

Gravitational Potential Energy = mgh joule. Kinetic Energy = $\tfrac{1}{2}mv^2$ joule.

Questions

(Take $g = 10 \text{ m/s}^2 = 10 \text{ N/kg}$ where necessary)

1. Write down formulae (showing all units) for the following: a) average speed b) acceleration c) four constant acceleration equations d) Newton's Second Law e) weight f) gravitational potential energy g) kinetic energy.

2. a) A car travels 100 m in 5 seconds. What is its average speed?
   b) A car accelerates from 5 m/s to 25 m/s in 10 seconds. What is its acceleration?

3. Describe in as much detail as possible, the motion of a car which has this graph.

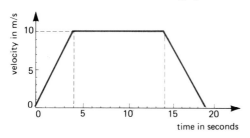

4. Sketch a displacement-time graph for the car in question 3.

5. A sports car accelerates from rest at 4 m/s² for 10 seconds. Calculate a) the final velocity (use equation 1) and b) the distance travelled.

6. A girl on a bicycle accelerates uniformly from rest to 10 m/s in a distance of 50 m. Find a) her acceleration (use equation 4) and b) the time taken.

7. A stone is dropped down a well. The splash is heard after 2 seconds. How deep is the well? (Use equation 3.)

8. The Eiffel Tower is 300 m high. A boy at the top slips on a banana skin and falls over the side. How long has he got to live?

9. A ball is thrown upwards with a velocity of 10 m/s. a) How high does it go? b) How long is it in the air?

10. A film stunt man dives off a cliff 20 m high, so that initially his *vertical* velocity is zero but his horizontal velocity is 5 m/s. Find a) his vertical velocity as he hits the water b) the time taken and c) the horizontal distance travelled during this time.

11. A car travelling within the town speed limit, at 13 m/s, hits a brick wall. For a passenger without a seat belt, this is like falling from the top of a house of height h metre. Find the value of h.
    (Most houses are about 7 m high.)

12 Explain why a parachutist, a feather and a leaf do *not* fall with a constantly accelerating motion.

13. A mass of 4 kg is accelerated by a force of 20 N. What is the acceleration if there is no friction? If a frictional force of 8 N is acting, what is the acceleration?

14 A Saturn V moon rocket has a mass at lift-off of 3.0×10^6 kg. The thrust at lift-off is 3.3×10^7 N. Find a) the weight of the rocket on Earth b) the resultant (unbalanced) force at lift-off c) the acceleration at lift-off d) the apparent weight of the rocket in orbit.

15. A fighter plane lands on an aircraft carrier at 60 m/s and is brought to rest in 2 seconds by an arrester wire. a) What is the deceleration? b) How far does it travel in this time? c) If the mass of the plane is 3000 kg, what is the braking force?

16. An astronaut has a mass of 100 kg. What is his weight a) on Earth, b) on the moon where the gravitational field strength = 1·7 N/kg?

17. In question 8, if the boy has a mass of 50 kg, find a) his P.E. at the top b) his P.E. at the bottom c) his K.E. just before impact and d) his velocity just before impact. What happens to his K.E. on impact?

160

Further questions on page 172.

Mechanics crossword

(Do not write on this book — copy the grid with tracing paper first)

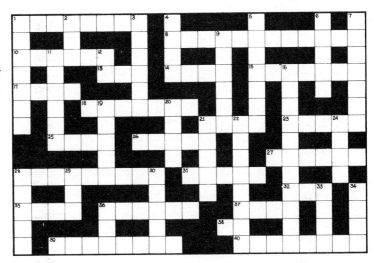

Clues across:

1. The pull of the Earth on one ——— = 9·8 newton.
8. Change in velocity divided by time taken.
10. A fresh weight for Sir Isaac?
13. One joule is the energy needed to move 1N through ——— metre.
14. Velocity equals distance travelled divided by ———.
15. Source of wisdom!
17. Like action or reaction?
18. Velocity and force are ———.
21. Your most valuable equipment in experiments.
23. Physics is ——— special!
25. Power is work done divided by ———.
26. A ——— converts electric energy to kinetic energy.
27. Not the first unit of time?
28. A wound-up spring has ——— energy.
31. It might make you move.
32. Shortened unit of electric current.
35. What is the unit!
36. You have been, all term.
37. Newton's first law: a stationary object, with no resultant force acting on it, remains at ———.
38. 3·14159265358979323846264338327950288 41971693993751058 . . . .
39. Speed? Not quite.
40. If a velocity-time graph is a straight line, the acceleration is ———.

Clues down:

1. and 9. A heavy lorry travelling at high speed has a lot of this.
2. Newton's third law: apply a force in, the reaction is———.
3. In a little while you have the turning effect of a force!
4. Newton's third law: apply a force to the west; the reaction is to the ———.
5. Force is measured in———.
6. A man walking at 2m/s for 4 seconds covers a distance of———metres.
7. It keeps you going when the car crashes.
9. See 1 down.
11. The pull of gravity.
12. A joule is a unit of power – yes or no?
16. When there is a resultant force on a body, it —
19. Push or pull to ——— a force.
20. When you divide by 'time', you are calculating a ———.
21. An experimental inaccuracy.
22. A dynamo converts kinetic energy to ——— energy.
24. Cold place with little friction.
28. Rate of working.
29. Take in fuel.
30. An electric fire converts electric energy to heat and ——— energy.
33. A city where you might be inclined to experiment!
34. An engine converts ——— energy to kinetic energy.
36. If two ropes support the bottom block of a pulley system, the velocity ratio equals ———.

chapter 20

MOMENTUM

The *momentum* of an object depends on its *mass* and its *velocity*.
In fact:

Momentum = mass × velocity	or	Momentum = *mv*

(kg)   (m/s)

Momentum is a *vector* quantity. It is measured in units of **kg m/s**.

A bicycle of mass 10 kg moving at 5 m/s has a momentum of 50 kg m/s.
What is the momentum of a 4 kg mass moving at 10 m/s?

Consider a force F acting on a mass m for a time t so that it accelerates from velocity u to velocity v.
The quantity (force × time) is called *impulse*. **Impulse = $F \times t$.**

From page 148, acceleration a $= \dfrac{v - u}{t}$

∴ Newton's Second Law (page 157) is: $F = ma = m\left(\dfrac{v-u}{t}\right) = \dfrac{mv - mu}{t}$

or in words: **Force** = $\dfrac{\textbf{change in momentum}}{\textbf{time taken}}$

Multiplying both sides by time, we get:

Impulse = **Force × time = change in momentum**	or	$Ft = mv - mu$

(N)   (s)   (kg m/s)

Example 1

Consider first a boy kicking a **stone** of mass 1 kg and accelerating it
from rest to 10 m/s. Because the stone is rigid, the force of his foot
acts for only $\frac{1}{100}$ second. Calculate this force.

formula first: Force × time = $mv - mu$
then numbers: Force × $\frac{1}{100}$ $= 1 \times 10 - 1 \times 0$

             Force × $\frac{1}{100}$ $= 10$

                ∴ Force = <u>1000 N</u> (painful!)

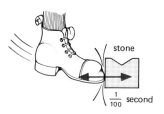

Example 2

Now consider the same boy kicking a **football** of the same mass
(1 kg) to give it the same speed (10 m/s) and therefore the same
momentum (10 kg m/s).
Because the ball is soft and he follows through with his foot, this
force is applied for $\frac{1}{10}$ second (10 times longer than before).
Is this force larger or smaller than before?

Using the same formula with time = $\frac{1}{10}$, gives force = <u>100 N</u>
Which kick hurts less? Why?

The *longer* the time of a collision, the *smaller* the force.

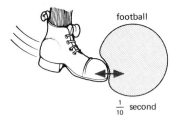

Collisions

Here are two balls rolling towards each other so that they collide:

When they collide they exert a force F on each other for a short time t and so the momentum of each ball changes.
From Newton's Third Law (page 114) the forces are equal and opposite. Therefore (from the equation on the opposite page) the changes in momentum are equal and opposite.

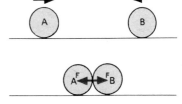

That is, the momentum *gained* by one ball is equal to the momentum *lost* by the other ball.
This is the **Principle of Conservation of Momentum**:
When two or more bodies act on each other, their total momentum remains constant, providing there is no external force acting.
That is:

Total momentum before collision	=	Total momentum after collision

The same is true for an explosion (as an explosion is the opposite of a collision)

Example 3
A bullet, mass 10 g (0·01 kg), is fired into a block of wood, mass 390g (0·39 kg), lying on a smooth surface. The wood then moves at a velocity of 10 m/s. a) What was the velocity of the bullet?
b) What is the kinetic energy before and after the collision?

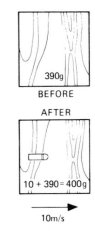

a) formula first: Total momentum = Total momentum
                          before collision   after collision

$$(\text{mass} \times \text{velocity})_{\text{before}} = (\text{mass} \times \text{velocity})_{\text{after}}$$

then numbers: $0{\cdot}01 \times V = (0{\cdot}01 + 0{\cdot}39) \times 10$
                  $0{\cdot}01 \times V = 0{\cdot}4 \times 10$
                  $V = \underline{400 \text{ m/s}}$

b) Total K.E. before collision $= \frac{1}{2}mV^2 = \frac{1}{2} \times 0{\cdot}01 \times 400^2 = \underline{800 \text{ J}}$
Total K.E. after collision $= \frac{1}{2}Mv^2 = \frac{1}{2} \times 0{\cdot}4 \times 10^2 = \underline{20 \text{ J}}$
The difference, 780 J, is converted to heat energy and sound energy during the collision.

Note: **momentum is conserved in a collision, but kinetic energy is not.**

Other examples
On page 74 and page 115, jets, rockets and the recoil of a gun were explained using Newton's Third Law. To calculate the velocity of the moving parts, the momentum equation must be used (see example 4 on the next page).

163

Momentum experiments

The conservation of momentum can be verified by experiments:

Experiment 20.1   Inelastic collisions

Set up a runway which is 'friction-compensated'
(see page 156).
Use two trolleys as shown here:
Trolley **A** is connected by ticker-tape to a ticker-timer. **A** has a pin at the front which sticks into a
cork on trolley **B**, to give an *inelastic* collision.

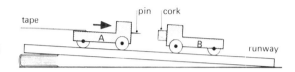

Start **A** moving, with **B** at rest.
Afterwards, mark off 5-dot lengths of tape (see p. 149)
to find the velocity before and after the collision.

Find the masses of the trolleys and so calculate
the momentum before and after the collision.

What do you find?
Repeat with weights added to trolley **A** or trolley **B**.

Experiment 20.2   Elastic collisions

This time use a trolley with a spring-loaded buffer
to give an *elastic* collision.
You will need two tapes to find the velocities of
both trolleys after the collision.

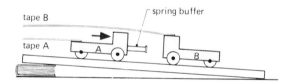

Is the total momentum constant?

Experiment 20.3   Explosion

On a level runway, use a trolley with a spring
buffer which "explodes" when a peg is tapped.
Again use ticker-tapes to find the velocities.
Momentum is a *vector*, so remember to take one
direction as positive and the other direction as negative.
Is the total momentum constant (= zero)?

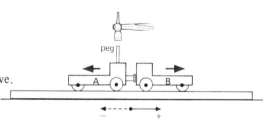

Example 4

A bullet of mass 10 g (0·01 kg) is fired at 400 m/s
from a rifle of mass 4 kg. What is v, the recoil
velocity of the rifle? (see also page 115).

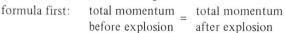
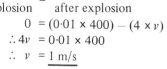

Remembering that momentum is a vector, let the
positive direction be to the right.

formula first: $\dfrac{\text{total momentum}}{\text{before explosion}} = \dfrac{\text{total momentum}}{\text{after explosion}}$

then numbers:
$$0 = (0{\cdot}01 \times 400) - (4 \times v)$$
$$\therefore 4v = 0{\cdot}01 \times 400$$
$$\therefore \ v = \underline{1 \text{ m/s}}$$

164

Summary

Momentum = mass x velocity.   It is a vector quantity.
(kg m/s)   (kg)   (m/s)

Newton's Second Law: Force = $\dfrac{\text{change in momentum}}{\text{time taken}}$ = rate of change of momentum

Impulse = Force x time = change in momentum = $mv - mu$
(Compare: Force x *distance* = change in K.E. = $\frac{1}{2}mv^2 - \frac{1}{2}mu^2$ from page 156)

Principle of Conservation of Momentum: Total momentum before = Total momentum after
If no external force is acting,   collision (or explosion)   collision (or explosion)

Questions

1. Copy and complete:
   a) Momentum = ... x . . . . .
   b) It is a . . . . . quantity
   c) Its unit is . . . . .
   d) Newton's Second Law can be written as
      Force = mass x . . . . . or
      Force = ... of change of . . . . .
   e) Impulse = ... x . . . . .
   f) Impulse = change in . . . . .
   g) The principle of conservation of
      momentum states:
   h) . . . . . is conserved in any collision but
      . . . . . is not.

2. a) What is *impulse*?
   b) Explain the physics of this limerick:
      A dashing young footballer, Paul,
      Scored a goal with a one-kilo ball,
      But a similar kick
      To a one-kilo brick
      Made Paul bawl, and then fall, and then
      crawl.
   c) When you jump down from a table, why
      do you bend your legs rather than keep
      them rigid? Why do paratroopers roll on
      landing? Why would you prefer to fall on
      to a bed than on to concrete?
   d) Why are seat belts designed to stretch in a
      collision? Why is the front of a car
      designed to collapse in a serious collision?

   e) Why does a cricket batsman, a tennis
      player or a golf player 'follow through'?
   f) Professor Messer invents a foam-rubber
      hammer ("to make less noise"). Will it
      work? Why?

3. A truck of mass 2 kg travels at 8 m/s
   towards a stationary truck of mass 6 kg.
   After colliding, the trucks link and move off
   together. a) What is their common velocity?
   b) What is the K.E. before and after the collision?
   c) Explain the apparent loss in energy.

4. In a sea battle, a cannonball of mass 30 kg
   was fired at 200 m/s from a cannon of mass
   3000 kg. What was the recoil velocity of the gun?

5. A heavy car A, mass 2000 kg, travelling at
   10 m/s has a head-on collision with a sports
   car B, mass 500 kg. If both cars stop dead on
   colliding, what was the velocity of B?

6. A man wearing a bullet-proof vest stands
   still on roller skates. The total mass is 80 kg.
   A bullet of mass 20 gram is fired at 400 m/s.
   It is stopped by the vest and falls to the
   ground. What is then the velocity of the man?
   How does this compare with what you see in
   TV films?

7. A Saturn V moon rocket burns fuel at the
   rate of 13 000 kg in each second. The
   exhaust gases rush out at 2500 m/s.
   a) What is the change in momentum of the
      fuel in each second?
   b) What is the thrust? (see also question 14,
      page 160)

Further questions on page 173.

Multiple choice questions on mechanics

Most examination papers include questions for which you are given answers (labelled **A, B, C, D, E**) and you have to choose the *one* correct answer.

Although the answers are given, the questions are not always as easy as they look. Take care to read *all* the answers before you choose the one that you think is correct.

Example
The kilogram is a unit of

A   speed
B   time
C   force
D   length
E   mass

The correct answer is **E**.

Now write down the correct letter for each of the following questions (but do not mark these pages in any way).

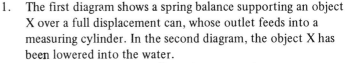

1. The first diagram shows a spring balance supporting an object X over a full displacement can, whose outlet feeds into a measuring cylinder. In the second diagram, the object X has been lowered into the water.
    What is the volume, in cm³, of the object X?
    A   2
    B   8
    C   20
    D   100
    E   120

2. The density of a certain gas is 3kg/m^3.
    What is the mass of 2m^3 of the gas?

    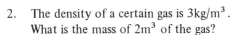

    A 0·67 kg   **C** 1·5 kg   **E** 6·0 kg
    B 1·0 kg   **D** 5·0 kg

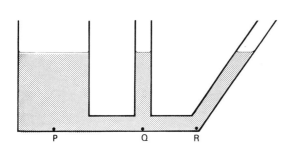

3. The diagram shows three containers, connected together and containing a liquid.
    Which of the following is correct?
    A The greatest pressure is at P.
    B The greatest pressure is at Q.
    C The greatest pressure is at R.
    D The least pressure is at R.
    E The pressures at P, Q and R are equal.

4. The diagram shows five objects placed on a slope. The dots show the position of each centre of gravity. Which object is unstable?

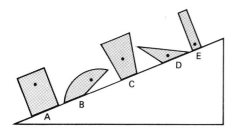

Questions 5, 6 and 7

A boy weighing 600 newton takes 4 seconds to run upstairs to a floor which is 3 metre higher.

5. The work done against gravity is
   A   150J
   B   200J
   C   1800J
   D 2400J
   E 7200J

6. His power (rate of working against gravity) is
   A   50W
   B 150W
   C 450W
   D 600W
   E 800W

7. The most useful energy change in this case is
   A chemical to potential
   B heat to potential
   C heat to kinetic
   D kinetic to heat
   E chemical to sound

8. A girl weighing 400N sits on a see-saw at a distance of 4m from the pivot as shown in

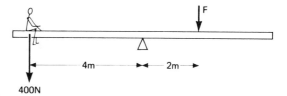

the diagram. What force is needed at F to just balance the see-saw?
   A   100N
   B   200N
   C   600N
   D   800N
   E 1600N

9. A pulley system has five pulleys and five strings supporting the bottom block. When the effort moves 10m the load moves 2m. When the load is 100N and the effort is 25N, the efficiency is 80%. What is the mechanical advantage?
   A 0·20
   B 0·25
   C 0·80
   D 4
   E 5

10. The diagram shows a velocity-time graph for a car.

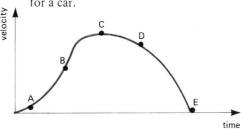

At which point is the car decelerating most rapidly?

When you have finished,
a) check all your answers carefully
b) discuss your answers (and why you chose them) with your teacher.

More questions can be found in *Multiple Choice Physics for You.*

Hooke's Law

1. A spiral spring of natural length 1·5 m is extended 0·005 m by a force of 0·8 N. What will be its length when the applied force is 3·2 N?

    A   1·48 m   D   1·52 m
    B   1·50 m   E   1·70 m
    C   1·55 m

                                   (WAEC)

2. The table shows the extension of a steel wire spring when increasing loads are hung on it.

Load N	0	5	10	15	20	25	30	35
Extension cm	0	2	4	6	8	11	17	26

    a) Up to what load, according to the table, does Hooke's Law appear to apply?
    b) What load should produce an extension of 5 cm?   (S16+)

3. a) State Hooke's law and describe an experiment to verify it.
    b) (i) Sketch a graph showing the extension of a spring versus the applied load and comment on the shape of the graph.
         (ii) Show how the weight of a piece of stone can be determined from the graph.   (WAEC)

4. A light spring which obeys Hooke's law hangs vertically from a fixed support, and its unstretched length is 20 cm. When a load of 0·4 kg is attached to its lower end, the length increases to 28 cm. [Take g, the acceleration due to gravity to be 10 m/s^2.]
    a) What is the tension, in newtons, in the spring when the load is at rest?
    b) What is the force, in newtons, required to extend the spring by 1 cm?
    c) A student takes hold of the mass, and pulls it down until the length of the spring is 32 cm. What is now the tension in the spring?
    d) What force does he have to exert in order to keep the mass in this new position?
                                       (O)

5. Theory suggests that, when a beam is loaded at the centre, the deflection x is directly proportional to the cube of L, the distance between the supports, provided that the same force W is applied for each value of L.

    In other words $x = kL^3$ where k is constant

    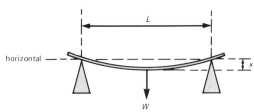

    You are required to carry out an experiment to confirm that $x = kL^3$, using a metre rule as a beam. From initial checks you know that when a metre rule rests on edge supports 0.90 m apart, its centre is deflected by about 1 cm when a load of 8 N is hung from the centre. Describe:
    a) how you would set up the apparatus   [6 marks]
    b) how you would make your measurements,   [4]
    c) the number and range of measurements you would make,   [5]
    d) how you would use your measurements to test the theory.   [5]
                                       (SEG)

Density

6. Draw labelled diagrams to show how you would find the volume of a large metal nut. If the density of the material of the nut is to be found what further measurement would be required?
    How would the measurements be used to find the required density?   (C)

7. a) What is meant by the expression 'density of a substance'?
    b) Describe, with a diagram if helpful, an experiment by which you could find the density of some gravel (small stones).
    c) A cube has a mass of 8 kg and each side is 0·2 m. What is its density?
                                       (Middx)

Pressure

8. (i)   Draw a labelled diagram of a simple mercury barometer.
    (ii) What is the purpose of a barometer?
    (iii) Indicate clearly on your diagram the measurement that is taken and label it h.
    (iv) The standard barometric pressure is 760 mmHg. If, having set up a simple barometer, you measured h and found it to have the value 680 mm, what would you suppose had happened?
          Explain how you would test to see if your conclusion was correct.
          (S16+)

9. A concrete paving slab measures 0·6 m × 0·5 m × 0·05 m.
    a) What is its volume?
    b) Taking the density of concrete to be 2 500 kg/m³, what is
       (i)   the mass of the slab;
       (ii) its weight in newtons?
    c) What is the average pressure the slab exerts on the ground beneath it when it is laid horizontally?
    d) What is the average pressure the slab exerts on the ground when it stands upright with the 0·6 m edges vertical and the 0·5 m edges horizontal?
       (O)

10. The diagram below illustrates the mode of operation of a hovercraft.

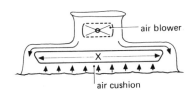

If the section marked X is circular and of radius 12 m and the pressure of the air-cushion is 2×10^5 Pa (N/m²) above atmospheric pressure, estimate the total weight of the hovercraft.
       (AEB)

Floating and sinking

11. A body of mass 2 kg is suspended from a spring balance, calibrated in newtons, which reads 17 N when the body is completely submerged in water. The intensity of the Earth's gravitational field is 10 N/kg (i.e. the acceleration due to gravity is 10 m/s²).
    a) What is the upthrust (in newtons) of the water on the body? b) What is the mass (in kg) of water displaced by the body?
    c) If the density of water is 1000 kg/m³, what is the volume (in m³) of water displaced?
    d) What is the density (in kg/m³) of the material of which the body is made?
       (N16+)

12. A block of wood measures 2·0 m by 0·5 m by 0·8 m: its density is 800 kg/m². Where necessary assume that the acceleration of free fall = 10 m/s² (10 N/kg).
    i)   What is the volume of the block?
    ii) What is the mass of the block?
    iii) If the block floats in water (density 1000 kg/m³) what will be the upthrust on it?
    iv) What volume of the floating block is submerged (i.e. underneath the water)?
    v)   A piece of iron is placed on the upper surface of the floating block and as a result the whole of the wood is submerged but none of the iron is under the water. What does the iron weigh?
    vi) The iron is removed and the block of wood is placed in a liquid of density 750 kg/m³ What does it appear to weigh?
       (N16+)

13. When a solid object is placed in 30·0 cm³ of water in a measuring cylinder, it floats and the level of the water goes up to 38·1 cm³. When the same solid is placed in 30·0 cm³ of paraffin oil in the same measuring cylinder, it sinks to the bottom of the cylinder and the level of the oil goes up to 40·0 cm³. Calculate i) the relative density of the solid, ii) the level of the liquid which would result if the solid were placed in 30·0 cm³ of a liquid of relative density 0·90.
       (AEB)

Vectors

14. Two ropes are attached to a block at the same point. The ropes are horizontal, at an angle of 90° to each other, and forces of 400 N and 600 N act along them.

a) Draw to scale a labelled diagram to show how the resultant of the two forces on the block can be found.

b) Find the size of the resultant force.

c) Measure, and record, the angle between the resultant force on the block and the direction of the 400 N force.

(C)

15. a) Distinguish between a vector quantity and a scalar quantity, giving an example of each.

b) A girl can row a boat in still water at 6 km/h. She is at a point A on one bank of a river, which is 3 km wide and which flows due South at a constant speed of 2 km/h.

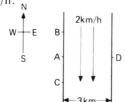

Find

(i) the speed relative to the bank which the girl would achieve when rowing directly upstream towards B,

(ii) the speed relative to the bank which the girl would achieve when rowing directly downstream towards C.

The girl wishes to travel due East across the river from A to D. **Either** by scale drawing **or** by calculation, find

(iii) how far from D the girl would meet the other bank of the river, if she mistakenly steered due East,

(iv) the direction in which she must steer the boat in order to reach D,

(v) the speed relative to the point A which she will achieve,

(vi) the time which it will take her to reach the point D.   (AEB)

Moments

16. A uniform metre rule of mass 100 g balances at the 40-cm mark when a mass X is placed at the 10-cm mark. What is the value of X?

A   33·3 g   D   400 g
B   133·3 g   E   1000 g
C   300 g

(WAEC)

17. By giving **three** examples, one for each, of bodies in different states of equilibrium, explain what is meant by (i) stable equilibrium, (ii) unstable equilibrium, (iii) neutral equilibrium. In each case, state what happens to the centre of gravity of the body when the body's equilibrium is disturbed.

(JMB)

18. A uniform metre rule balances horizontally on a knife-edge placed at the 58 cm mark when a mass of 20 g is suspended from one end.

a) Draw a diagram of the arrangement.

b) What is the mass of the rule?

c) What weight is supported by the knife-edge? [Take the force of gravity on 1 kg to be 10 N.]

(C)

19. a) Define the *moment of a force about a point*.

b) State the principle of moments.

A uniform plank PQ weighing 300 N and of length 3·0 m rests on two narrow supports C and D. C is 0·5 m from P and D is 1·0 m from Q.

c) Draw a neat diagram of the arrangement; on this diagram draw arrows at the points where the forces act **on the plank**, clearly marking the distances. The arrows should show the directions of the forces. Label the known force with its value.

d) A weight W causes the plank to tip over when placed at one end, but does not do so when placed at the other end. At which end is it placed when it causes tipping? Give the reason for your answer.

e) What is the least value of W which will cause tipping when placed

(i) at end P,

(ii) at end Q?   (N16+)

Energy and power

20. a) When a simple pendulum is oscillating, energy is changing from the form of . . . . . energy to the form of . . . . . energy and then back into the form of . . . . . energy.
   b) In a gas fire, ... energy is converted into energy in the form of . . . . . and . . . .
   c) In a dynamo, ... energy is converted into . . . . . energy.
   d) When a spring-driven toy motor-car is running, ... energy is converted into . . . . . energy.
   e) In an explosion, ... energy is converted into mechanical energy and also into . . . . . and . . . . .
   f) In order to propel an electric train, ... energy is converted into . . . . . energy.
   g) In a battery, energy is stored in the form of . . . . . energy ready to be released in the form of . . . . . energy.
   h) When a catapult is used, ... energy is converted into . . . . . energy.   (AEB)

21. A motor is used on a building site to raise a block of stone. The weight of the block is 720 N and it is raised 20 m in 24 s. Calculate
   a) the work done on the block,
   b) the useful power supplied by the motor. (C)

22. Define a) energy, b) power.
   An electric pump whose efficiency is 70% raises water to a height of 15 m. If water is delivered at the rate of 350 dm^3 per minute, what is the power rating of the pump? [Mass of 1 dm^3 of water = 1 kg] What is the energy lost by the pump? (WAEC)

23. An electric drill bores a hole in a sheet of steel in 3·0 s. The electrical input to the motor driving the drill has a power of 1·20 kW, and the efficiency of the drill is 70%.
   a) How much energy is expended in driving the motor?
   b) How much energy is expended by the drill in making the hole?
   c) If the cutting edge of the drill moves 4·0 m against friction whilst cutting the hole, what is the average force of friction between the cutting edge and the steel? (C)

Machines

24. Draw a labelled diagram of a single-string pulley system (block-and-tackle) with a velocity ratio of *five*. The following results were obtained in an experimental investigation of the performance of such a system.

Load (N)	50	100	200	300	400
Effort (N)	30	40	60	80	100

Calculate the mechanical advantage and the efficiency of the system for each value of the load and show your results clearly in a table.
Plot a graph of percentage efficiency against load, using 1 cm to represent 5% and 50 N respectively. Use your graph to determine
   a) the efficiency obtained when raising a load of 250 N,
   b) the load for which the efficiency would be 60%.
Assuming that the loss of efficiency was due only to the raising of the lower pulley block calculate the weight of the lower block. (AEB)

25. a) Draw a diagram of a block and tackle pulley system which has a velocity ratio of 2. Show where the load and effort are applied to the machine.
   b) The pulley system described in (b) was used to find the relation between load and minimum effort required to raise the load. The results obtained are shown below.

Load	N	1·0	2·0	3·0	4·0	5·0	6·0
Effort	N	1·0	1·5	2·0	2·5	3·0	3·5
Mechanical advantage				1·33		1·67	1·71
Efficiency	%			66·5		83·5	85·5

   i) Complete the table by calculating the missing values of mechanical advantage and efficiency.
   ii) Plot a graph of load against efficiency.
   iii) Use the graph to estimate the maximum useful efficiency of this machine for large loads.   (JMB)

Velocity and acceleration

26. a) (i) For a body to be given an acceleration, a . . . . . must be applied to it.
 (ii) The distance moved by a body travelling at constant velocity is calculated
by multiplying its . . . . . by the . . . .
 (iii) The acceleration of a body is given
by the expression

$$acceleration = \frac{change\ in . . . . .}{. . . . .}$$

b) The velocity, v, of a car varies with time,
t, according to the following table

t in s	0	5	10	15	20	25	30	35	40	45	50
v in m/s	0	5	10	15	15	15	15	11	7·5	3·5	0

 (i) Plot a graph of v against t.
 (ii) Describe the motion of the car during
each of the following periods of time:
0–15 s   15 s–30 s   30 s–50 s
 (iii) Calculate the acceleration of the car
during the period 0–15 s.
 (iv) How far did the car travel in the
following periods of time?
0–15 s   15 s–30 s   0–50 s
 (v) State what forces would be acting on
the car when it is travelling at
constant velocity. What is the resultant
of these forces equal to?
 (vi) Why does the engine have to work
at a higher rate when the car is
accelerating than when it is travelling
at constant speed?
 (vii) Explain why the car's fuel consumption
in kilometres per litre would be more
when travelling at 30 m/s than at 15 m/s.
(S16+)

27. A spacecraft travelling on a straight course at
75 km/s fired its rocket motors for 6·0 s. At
the end of this time its speed was 120 km/s
in the same direction. What was the spacecraft's average acceleration while the motors
were firing?
How far did the spacecraft travel in the first
10 s after the rocket motors were started, the
motors having been in action for only 6·0 s? (C)

28.

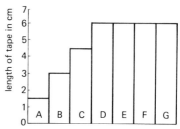

A tape attached to a trolley moving along a
runway, passes beneath a vibrator making 50
dots on the tape per second. The tape is then
cut into successive lengths containing 10
spaces between the dots on each tape length.
These lengths are then used to produce the
tape chart shown in the diagram above.
a) What time does each tape length represent?
b) Calculate the average speed for length A.
c) Calculate the average speed for length B.
d) What is happening to the trolley between
A and D?
e) Describe briefly, with the aid of a diagram, how you would set up the trolley
and runway apparatus to produce the
type of motion represented by tapes A to D.
f) What is happening to the trolley between
D and G?
g) What is the total force acting on the
trolley between D and G?
h) What is the total time from the start of
the motion until the trolley reaches its
maximum speed?   (EM)

29. The following readings were taken in an
experiment to determine the acceleration of
free fall g.

Height of fall from rest h (cm)	200	180	160	140	120	
Time taken t (s)	0·64	0·61	0·54	0·53	0·50	0
t^2 (s^2)		0·37	0·29	0·28	0·25	

Complete the table by inserting the two
missing values of t^2 corrected to two signifi-
cant figures. Plot a graph of h against t^2.
Which pair of readings appears to have been
measured or recorded incorrectly?
Ignoring that pair of readings, draw the best
straight line through the other points on
your graph.
What is the gradient of the straight line obtained?
Hence determine an approximate value for g
stating clearly the units of your answer. (AEB)

30. A man runs a race against a dog. Here is a graph showing how they moved during the race.

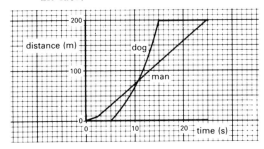

a) What was the distance for the race?
b) After how many seconds did the dog overtake the man?
c) How far from the start did the dog overtake that man?
d) What was the dog's time for the race?
e) Use the equation $v = \dfrac{s}{t}$ to calculate the average speed of the man.
f) After 8 seconds is the speed of the man increasing, decreasing or staying the same?
g) What is the speed of the dog after 18 s? (NEA)

Force, acceleration and momentum

31. A body of mass 2 kg initially at rest is acted upon by a force of 6 N for a time of 7 s. The body is not subject to any force for the next 5 s but is then brought to rest in a further 10 s by a constant retarding force. Calculate
a) the acceleration produced by the first force.
b) the distance travelled by the body during the first 12 s,
c) the deceleration of the body during the last 10 s,
d) the force producing that deceleration.
(AEB)

32. a) Define the newton.
b) A mass of 5 kg changes its velocity from 4 m/s to 24 m/s in 10 s. Calculate
(i) the acceleration,
(ii) the force needed to cause this acceleration,
(iii) the momentum at the start,
(iv) the kinetic energy at the start,
(v) the kinetic energy after 2 s.

33.

Slow motion photography (see above) shows that a jumping flea pushes against the ground for about 0.001 s, during which time its body accelerates upwards to a maximum speed of 0.8 m/s.
a) Calculate the average upward acceleration of the flea's body during this period. [3 marks]
b) If the flea then moves upwards with constant downward acceleration of 12 m/s^2
i) how long it will take, after leaving contact with the ground at a speed of 0.8 m/s, to reach the top of its jump, [3]
ii) how high it will jump after leaving contact with the ground. [2]
c) Why is the acceleration of the flea after leaving the ground not equal to g? [2]
(LEAG)

34. A trolley having a mass of 2 kg and velocity 4 m/s collides with another trolley having a mass of 1 kg which is at rest. If the two trolleys stick together after the impact calculate
a) the total momentum before the collision,
b) the common velocity after the collision,
c) the kinetic energy before the collision,
d) the kinetic energy after the collision.
Why are the answers to (c) and (d) different?
(JMB)

Further Reading

Physics Plus: *Karate, Physics And Sport, The Bicycle, Artificial Hip Joints* (CRAC)
The Science of Movement – Hancock (Macmillan)
The Physics of Human Movement – Page (Wheaton)

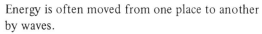

chapter 21

Waves Waves Waves

Q. What does the ocean say when it sees the beach?

A. Nothing, it just waves!

Energy is often moved from one place to another by waves.

You can see this if you throw a stone into a pond — waves spread out from the splash and soon the water at the edge of the pond moves as it gets some of the energy of the splash.

There are *two* kinds of waves:

Transverse waves

Experiment 21.1
To see a transverse wave, stretch a long 'slinky' spring along a smooth bench or floor and then give one end a quick wiggle *at right angles* to the spring.

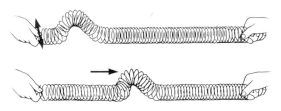

What happens?

You can see that energy is moving down the spring although each bit of the spring is only vibrating. It is vibrating at right angles (*transversely*) to the spring.

In this chapter we will be investigating transverse water waves. Light waves are also transverse waves.

Longitudinal waves

Experiment 21.2
Stretch the same 'slinky' spring along the floor and this time give it a quick wiggle to and fro *along the length* of the spring.

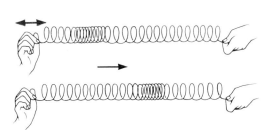

What happens?

If you look closely you can see that again, energy is passing down the spring although each bit of the spring is only vibrating. It is vibrating lengthways or *longitudinally*.

We will investigate longitudinal waves in chapter 30 (page 238). Sound waves are longitudinal waves.

The ripple tank

We can study transverse waves by using water waves in a ripple tank.

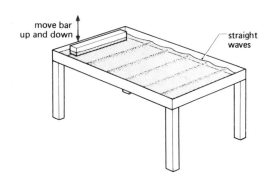

Experiment 21.3

Put some small pieces of cork on the water in a ripple tank and move the wooden bar up and down to make straight waves.
As the waves move down the ripple tank, which way do the pieces of cork move?
Are these transverse or longitudinal waves?

Here is a side view of a transverse water wave:

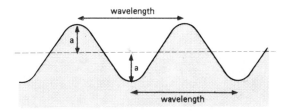

The high parts are called *crests* and the low parts are called *troughs*.
The distance marked *a* is called the *amplitude*. It is the height of a crest above the average water level (or the depth of a trough below it).

The distance between two successive crests (or troughs) is called the *wavelength*.

If the wooden bar (or the pieces of cork) in the experiment moves up and down through two complete vibrations in each second, then the *frequency* of the vibrations is two cycles per second (this is also written as 2 hertz or 2Hz).

Experiment 21.4

Move the wooden bar up and down at a low frequency first and then at a higher frequency. What do you notice about the velocity of the waves and their wavelength?

You should be able to see that
a) the velocity stays the same and
b) when the frequency increases, the wavelength decreases.

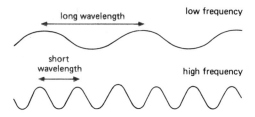

Example

A wave has a frequency of 2 hertz (2 cycles per second) and a wavelength of 10 cm.
What is the velocity of the wave?

In 1 second:

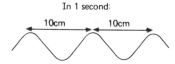

In each second, two waves are produced, each of length 10 cm.
Therefore the wave travels forward 20 cm in each second. That is, the velocity of the wave is 20cm/s.

In fact, for any wave:

velocity	=	frequency	x	wavelength
(m/s)		(Hz)		(m)

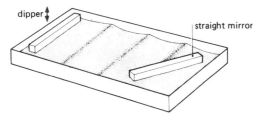

More experiments in a ripple tank (or your bath)

Experiment 21.5   Reflection at a straight mirror
Place a straight metal or plastic barrier in the tank so that it reflects the waves (like a mirror).
Send single waves across the ripple tank and look carefully for the waves reflected off the 'mirror'.

What do you notice about the angle at which the wave hits the mirror and the angle at which the reflected wave leaves the mirror?

What do you know about the angles when light is reflected from a mirror? (See page 184.)

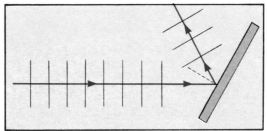

Experiment 21.6   Reflection at a curved mirror
Place a curved barrier in the ripple tank. Again send single waves across the tank and look carefully for the reflected waves. What do you see?

The waves are *converged* to a point called the *focus*. (See also page 189.)

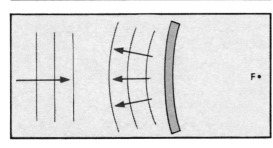

Experiment 21.7
Repeat experiment 21·6 with the curved mirror turned round the other way.

Can you see that the waves are spread out or *diverged* by the mirror? (See also page 189.)

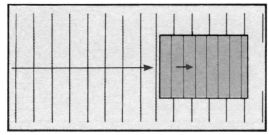

Experiment 21.8   Refraction
Place a thick sheet of glass in the tank and adjust the depth so that the water is very shallow over the glass sheet.
Again send waves across the tank and look at the waves when they travel through the shallow water.

What do you see?
What happens to a) the velocity and b) the wavelength of the waves when they move into the shallow water?

Experiment 21.9   Refraction

Now place the sheet of glass at an angle and repeat experiment 21.8.

What happens to the *direction* of the waves as they slow down in the shallow water?

When waves change their speed and direction like this, we say they have been *refracted*.
Light waves can also be refracted (see page 194).

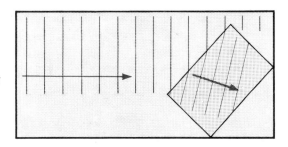

Summary

Waves move energy from one place to another (although each part of the wave only *vibrates* to and fro). In *transverse* waves (like water waves and light waves) this vibration is at right angles to the direction of the wave.

In *longitudinal* waves (like sound waves) the vibration is along the direction of the waves.

For *all* waves: velocity = frequency x wavelength

Waves can be reflected and refracted.

Questions

1. Copy out and complete.
    a) If the vibrations of a wave are at right angles to the direction of the wave, it is called a . . . . . wave. If the vibrations of a wave are along the direction of the wave, it is called a . . . . . wave. A water wave is a . . . . . wave and a sound wave is a . . . . . wave.
    b) The wavelength of a wave is the . . . . . between two successive . . . . . (or . . . . .) The frequency of a vibration is measured in . . . . . or . . . . .
    c) The velocity of a wave (measured in . . . . .) is equal to the frequency (measured in . . . . .) multiplied by the . . . . . (measured in . . . . .).
2. A certain sound wave has a frequency of 170 hertz (cycles per second) and a wavelength of 2 metres. What is the velocity of sound?
3. A certain radio wave has a frequency of 200 000 hertz (cycles per second) and a wavelength of 1500 metres. What is the velocity of radio waves?

Further questions on page 252.

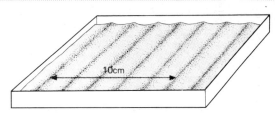

4. The diagram shows some waves in a ripple tank travelling at a velocity of 6cm/s.
    a) What is the wavelength?
    b) What is the frequency?

5. The diagram shows straight waves approaching a straight barrier at an angle of 45°.
    Draw a diagram to show how the waves are reflected from the 'mirror'.

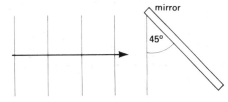

6. Some sea waves approach a gently sloping beach at an angle. Draw a diagram to show what happens to the direction of travel of the waves.

chapter 22

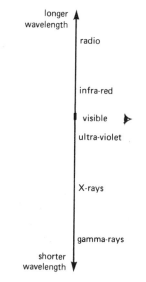

Visible light is a form of energy that we can detect with our eyes.

Objects that produce their own visible light are called *luminous sources* (for example: the Sun, other stars, electric lamps, television screens, glow-worms).

Other objects are *illuminated* by this light and *reflect* it into our eyes (for example: this page and most objects in this room, the Moon).

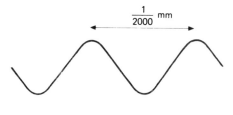

$\frac{1}{2000}$ mm

Light waves

Light is a wave motion, rather like the water waves that you may have seen on a pond or in a ripple tank (see page 175).

The light that we can see has a *wavelength* of only about $\frac{1}{2000}$ th of a millimetre.

Some waves have a wavelength even shorter than this, but then we cannot see them – for example: ultra-violet waves, X-rays and gamma-rays.

longer wavelength

radio

infra-red

visible

ultra-violet

X-rays

gamma-rays

shorter wavelength

Other waves have a wavelength longer than visible light. Again we cannot see them – for example: infra-red (heat rays) and radio waves.

All these waves are called *electromagnetic waves* (see page 224).
In a vacuum, they all travel at the same speed.

The speed of light

Light travels at a very high speed — about a million times faster than the speed of sound.
This is fast enough to travel 8 times round the Earth in one second or fast enough to cross this room in about one-hundred-millionth of a second.
The speed of light in air is 300 million metres *per second.*

A racing car travelling non-stop would take about 100 years to travel from the Sun to the Earth.
Light does the journey in about 8 minutes!
This means that you see the Sun as it was 8 minutes ago.

Other stars are much farther away.
From some stars, the light has taken 1000 million years to reach us, so we are seeing them as they were 1000 million years ago.
How long would you have to wait to see them as they are now?

A *light year* is a distance — how far is it?

Rays of light

For experiments with visible light, we often use a very narrow beam of light called a *ray*.

Experiment 22.1
Use a *raybox* to shine a narrow beam of light across a piece of paper.
What do you notice about the edges of the beam of light?

Why can't you see round corners?

Light travels in straight lines.

Sometimes the word *rectilinear* is used because it means 'in a straight line'.

Mary had a little lamp
It shone so straight for her.
The rays of light so straight and true
Were rectilinear.

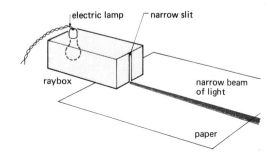

Shadows

Shadows are formed when some rays of light continue to travel in straight lines, while other rays are stopped by an object.

The kind of shadow that is formed depends on the size of the source of light:

1. Shadow due to a small source of light

Experiment 22.2
Hold a table-tennis ball between a *small* electric lamp and a white screen.

small lamp   ball   screen   dark shadow (umbra)   bright

Can you see a dark shadow with sharp edges? This shadow is called the *umbra*.

2. Shadow due to a large source of light

Experiment 22.3
Change the small lamp of experiment 22·2 for a *large* pearl lamp.

large lamp   ball   screen   umbra   penumbra

Does this shadow still have sharp edges?
Is all the shadow equally dark?

The dark centre of the shadow is still called the umbra. No light from the lamp reaches the umbra.
It is surrounded by a partial shadow or *penumbra*. The penumbra varies in brightness — from very dark near the umbra (where not much light can reach it) to very bright at the outer part where more light can reach it.

What kind of shadow is formed on a clear sunny day?
What kind of shadow is formed on a cloudy day?

Eclipses

Eclipses occur because the Moon and the Earth cast large shadows.

1. Eclipse of the Sun (solar eclipse)

This happens when the Moon comes between the Sun and the Earth, so that the Earth (in the shadow) darkens during the day.
At the time of total eclipse, only the flames of the outer edge of the Sun can be seen.

Solar eclipse

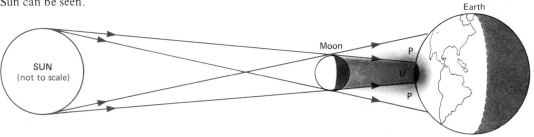

Experiment 22.4

Set up a model of an eclipse of the Sun (in a darkened room) using a large pearl lamp for the Sun, a table-tennis ball for the Moon and a white football for the Earth.

Total eclipses of the Sun do not occur often – the next one to be seen in Britain will be on 11th August 1999.

2. Eclipse of the Moon (lunar eclipse)

This happens when the Moon moves into the Earth's shadow. When the Moon is in the Earth's umbra, no sunlight can reach the Moon and it is seen to go dark.

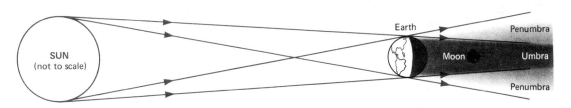

Experiment 22.5

Use the same apparatus as in experiment 22.4 to show an eclipse of the Moon.
Eclipses of the Moon can be seen much more often than eclipses of the Sun. Try to find out the date of the next eclipse of the Moon to be seen from where you live.

Experiment 22.6

Use the same apparatus as in experiments 22·4 and 22·5 to show the different *phases* of the Moon as it moves round the Earth.
How long does it take for the Moon to move once round the Earth?

The pinhole camera

A simple pinhole camera consists of a closed box with a screen at one end and a small pinhole at the other end. It uses the fact that the light going in through the pinhole travels in straight lines to the screen.

The diagram shows one way to make a pinhole camera.

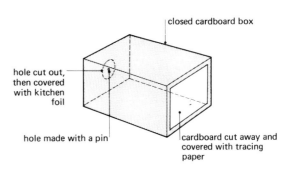

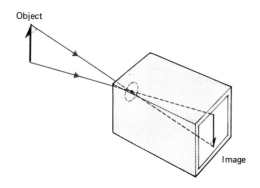

Experiment 22.7
Use a pinhole camera in a darkened room and point the pinhole at a lamp or at the outside world through a gap in the curtains.

What do you see on the screen?

The picture that you see is called an *image*. In this case because the picture is formed on a screen, it is called a *real* image.
In the pinhole camera the image is *inverted* – it is turned upside-down and turned left-to-right.

Experiment 22.8
Look carefully at the image on your screen and notice how sharp and dim it is.
Then use a pin or a pencil to make the pinhole larger.

Is the image sharper or more blurred?
Is the image dimmer or brighter?
Why does this happen?

A pinhole camera is not often used to take a photograph because to get a sharp picture, a small pinhole must be used and this means a long exposure time for the film.

Something to do   Improve your eye-sight without spectacles

If you normally wear glasses you can also make things look sharper by looking through a small pinhole in a piece of card.
In this way you are converting your eye (which is like a lens camera, see page 210) to a pinhole camera.

Summary

Shadows and eclipses occur because light travels in straight lines.
A small source of light causes a sharp shadow, called the umbra.
A large source of light causes a blurred shadow with two parts, the umbra and the penumbra.

In a pinhole camera, the image is real and inverted.
A small hole gives a sharp, dim image.
A large hole gives a blurred, bright image.
Visible light travels at the same high speed as gamma-rays, X-rays, ultra-violet, infra-red and radio waves.

Questions

1. Copy out and complete:
   a) Light travels in . . . . . lines.
   b) A shadow formed by a small source of light has . . . . . edges and is called an . . . . .
      A large source of light causes a shadow with two parts called the . . . . . and the . . . . .
   c) In an eclipse of the Sun, the . . . . . comes between the Sun and the . . . . .
      In an eclipse of the Moon, the . . . . . comes between the . . . . . and the . . . . .
   d) In a pinhole camera, the image is . . . . . and . . . . .
      If the pinhole is made larger, the image becomes . . . . . and . . . . .

2. How could a detective tell from a photograph, without seeing the sky, whether it was a bright clear day or cloudy?

3. Professor Messer thinks he's clever,
   Put him right or he's gone for ever.

4. Which lighting would you prefer in your class-room — a single small lamp or a long fluorescent lamp? Explain why. Draw diagrams to show, for each case, the shadow cast on the floor by a ball held in mid-air.

5. Draw a diagram to explain an eclipse of the Sun. Label all parts of the diagram, including the umbra and penumbra. Why is your diagram not drawn to scale?

6. Draw a diagram of a pinhole camera with two rays passing from the object through the pin-hole. How can you make the image a) sharper b) brighter c) larger?

7. Professor Messer fell asleep and dreamed that light no longer travelled in straight lines. Why was his dream a nightmare?

Further questions on page 252.

chapter 23

REFLECTION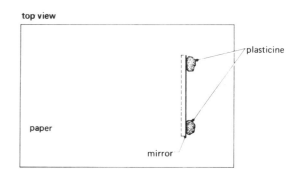

Have you ever reflected sunlight from a mirror to dazzle someone or to flash messages to a friend?

Experiment 23.1   Investigating reflection
Get hold of a *plane* (flat) glass mirror and look at it carefully.
Where is the light reflected from – the glass at the front of the mirror or the 'silvering' at the back?

Place the plane mirror on a sheet of white paper, holding it vertical with plasticine or a block of wood.

Draw a line along the *back* of the mirror (where the light is reflected).

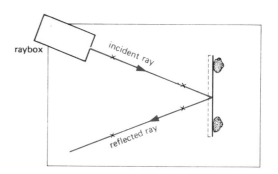

Use a ray-box to shine a ray of light on to the mirror. Can you see the reflected ray of light? Mark the position of each ray with crosses as shown in the diagram.

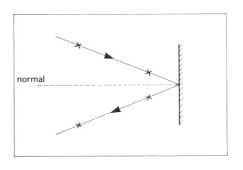

Now take off all the apparatus and use a ruler to join up the crosses, along the lines of the rays of light.

Use a protractor or a set-square to draw a line at right angles (90°) to the mirror, as shown. This line is called the *normal.*

Now use a protractor to measure the angles on each side of the normal.
What do you find?

Now try the experiment again with different angles.

What do you find each time?

Here are the scientific words and symbols used to describe experiment 23.1.
You should learn all of these.

The **law of reflection**:

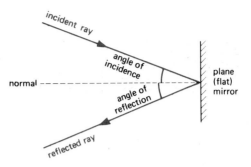

The angle of incidence	=	the angle of reflection

Water waves are reflected in the same way as light waves (see the ripple tank experiments on page 176).

If the angle of incidence is 45°, what is the angle of reflection and what is the angle turned by the ray of light?

The periscope

Periscopes are used in submarines so that the men below the surface can see what is happening above the surface.

The diagram shows how light is reflected by the two mirrors.

Remember that the angle of incidence = the angle of reflection and the rays should be turned through 90° each time. At what angle should each mirror be placed?

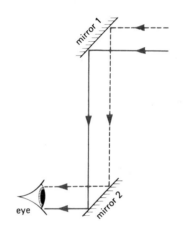

Building a periscope
You can use two handbag mirrors and a plastic bottle (like a washing-up-liquid bottle) to build a periscope which you can use to see over people's heads or to spy round corners.

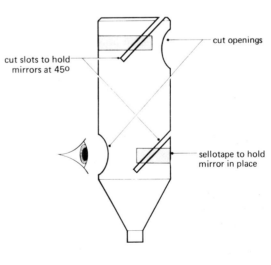

cut openings

cut slots to hold mirrors at 45°

sellotape to hold mirror in place

The image in a plane mirror

When you look into a plane (flat) mirror, you can see a picture or *image* of yourself. Whereabouts do you think this image is — on the mirror or behind the mirror?

Experiment 23.2   To find out where the image is
Use a vertical sheet of clear glass as a mirror and place a lighted candle (or bunsen burner) on the bench in front of it. (Take care).

Then move an <u>unlighted</u> candle or bunsen burner (of the same height) into a position behind the mirror so that it *appears* to be burning.

When the image of the flame appears to be on the unlighted candle, we know where the image is — it must be in the same position as the unlighted candle.

Measure the distance of the image *from the mirror* and the distance of the lighted candle (called the object) *from the mirror*. What do you notice?

The image is as far behind the mirror as the object is in front.

This means that if your face is 30cm in front of a mirror, then your image is 30cm behind the mirror.

Imagine a line joining the object to the image in experiment 23.2.
What angle would it make with the mirror?

The line joining the object to the image is at right angles (90°) to the mirror.

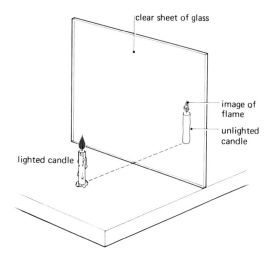

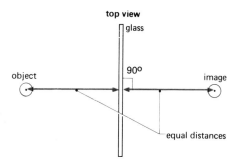

'Imagic' tricks
Magicians sometimes use images in their tricks.
If you put your finger on the <u>unlighted</u> candle in experiment 23.2, it looks as if your finger is burning!
Another trick is to fix the unlighted candle in a tall beaker which you then fill with water so that it appears to burn underwater. Can you think of some other tricks like these?

Virtual images

If you go behind the mirror to look for the image, you cannot see it.

This *virtual* image only *appears* to be behind the mirror (because the rays of light are reflected to your eye as though they came from behind the mirror). This virtual image cannot be formed on a screen (as a real image can, see page 182).

Look at the *size* of the real flame compared with its virtual image.
What do you notice?

The image in a plane mirror is virtual and is the same size as the object.

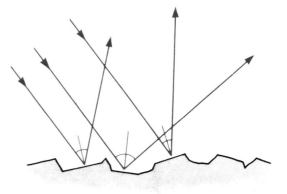

Regular and diffuse reflection

We have been looking at *regular* reflection from mirrors and other shiny flat surfaces.
Other surfaces, like this sheet of paper, also reflect light but because the surface is rough, the light is reflected off in all directions.
This is *diffuse* reflection.
If a surface does not reflect any light at all, it looks dull black, like this black ink.

Experiment 23.3

Look into a mirror and hold your *right* ear.
Which ear is your image holding?

Write your name on a piece of paper and pin it to your chest as a label. What do you notice when you look in the mirror?
This is an example of *lateral inversion*.

The image in a plane mirror is laterally inverted.

A tattooist once wrote on my throat
A short but mysterious note
And so to inspect it
I tried to reflect it
And found I'd got ТАОЯНТ on my throat!

diffuse reflection from a rough surface

Summary

The law of reflection:

the angle of incidence = the angle of reflection

The image in a plane mirror is:

a) as far behind the mirror as the object is in front, with the line joining the object and the image at right angles $(90°)$ to the mirror.

b) the same size as the object

c) virtual

d) laterally inverted

Questions

1. The angle between an incident ray and the mirror is $30°$.
   a) What is the angle of incidence?
   b) What is the angle of reflection?
   c) What is the total angle turned by the ray of light?

2. Complete this verse:

   A periscope's a useful thing,
   As everyone agrees.
   It has . . . . . mirrors fixed inside,
   At . . . . . degrees.

3. A boy with a mouth 5cm wide stands 2m from a plane mirror. Where is his image and how wide is the image of his mouth?

4. **AMBULANCE** Explain where you would see this sign and explain why it is written like this.

5. Explain how to read the following message which was found on some blotting paper:

   Meet me at midnight

6. A friend wishes to draw a vase of flowers life-size on a piece of paper. Explain with a labelled diagram how he could use a sheet of clear glass to copy the object.

7. Draw a labelled ray diagram to show the following arrangement:

   A periscope builder called Fred,
   Fixed two mirrors to the head of his bed.
   The light was reflected,
   And so he inspected,
   The hair on the top of his head.

8. Professor Messer fell asleep and dreamed that all objects stopped reflecting light. Why was his dream a nightmare?

9. A witness says that he saw a murder committed when the hands on the clock showed 5-12 p.m., but the chief suspect has an alibi for this time. Later, you, as Detective-in-charge, realise that the witness saw the clock reflected in a mirror. What would you investigate next?

10. Explain, with a diagram, how 2 plane mirrors could be fixed at a dangerous T-junction so that motorists approaching the junction could see all other traffic.

11. Explain why you cannot see an image of yourself in this page.

12. A room with only one window is dark even at mid-day. Explain how you could use wallpaper, furnishings and mirrors to brighten the room.

188

Further questions on page 253.

chapter 24

Have you ever looked into a polished spoon to see an image of your face?
Have you noticed that the front and back of a spoon give different images?

Two kinds of curved mirror:
A mirror that curves in (like a cave) is called a *concave* mirror.
A mirror that curves outwards is called a *convex* mirror.

Experiment 24.1   A concave mirror
Use a ray-box (on a sheet of white paper) to produce several rays of light to shine on to a *concave* mirror. It is better if you can adjust your ray-box to make parallel rays of light. Look for the rays reflected off the mirror.

Can you see that the rays coming off the concave mirror are getting closer together – we say they are *converging*. (See also page 176.)

A concave mirror is a converging mirror.

If the rays of light from the ray-box are *parallel* rays, the rays reflected off the mirror converge to one point called the *principal focus* (marked F on the diagram).

Parallel rays of light are reflected through the principal focus of a concave mirror.

The distance from F to the mirror is called the *focal length* of the mirror.

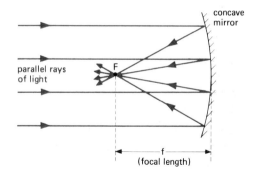

Experiment 24.2   A convex mirror
Repeat experiment 24.1 but with a *convex* mirror. Can you see that the light rays are spread out or *diverged* by this mirror?

A convex mirror is a diverging mirror.
Parallel rays of light are reflected so that they appear to come from the principal focus (F) of a convex mirror.

Where have you seen curved mirrors used?

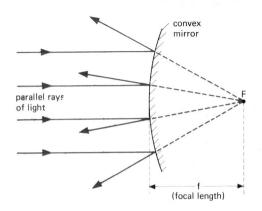

Concave (converging) mirrors: ray diagrams

The ray diagrams on the opposite page show the different images that are produced when an object is placed at different distances in front of a concave mirror.
To draw these diagrams you need to know 3 constructions:

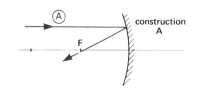

Construction A: parallel rays of light are reflected through the principal focus F. This was shown in experiment 24.1.

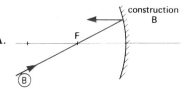

Construction B: rays of light passing through the principal focus F are reflected parallel. Rays of light can always be reversed to travel back along the same path, so this is just the reverse of construction A.

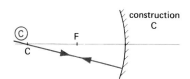

The *centre of curvature* (C) is the point at the centre of the large sphere from which the small mirror is formed.

Construction C: a ray of light passing through C will meet the mirror at right angles (90°) and so be reflected back along its own path. Measurement shows that the centre of curvature (C) is *twice* as far from the mirror as F.

Drawing a ray diagram

Step 1. On graph paper draw a central line (called the principal axis) with the mirror drawn (near one end) as a straight line (with curved ends) to represent a small central part of the mirror.

Step 2. Where distances are given, choose scale(s) for the object's size and position (often $\frac{1}{2}$ scale horizontally and full size vertically).

Step 3. Mark in the object and the positions of F and C.

Step 4. Start from the top of the object and draw at least two of the constructions.

Step 5. The position of the top of the image is where the reflected rays cross.

Step 6. Measure the position and size of the image using the scales. Say whether the image is:
  inverted or erect;
  magnified or diminished (smaller);
  real or virtual.

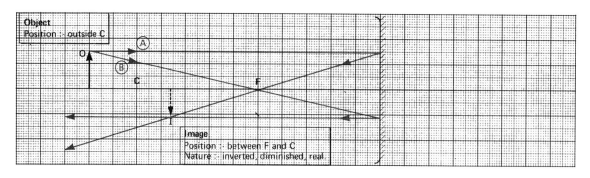

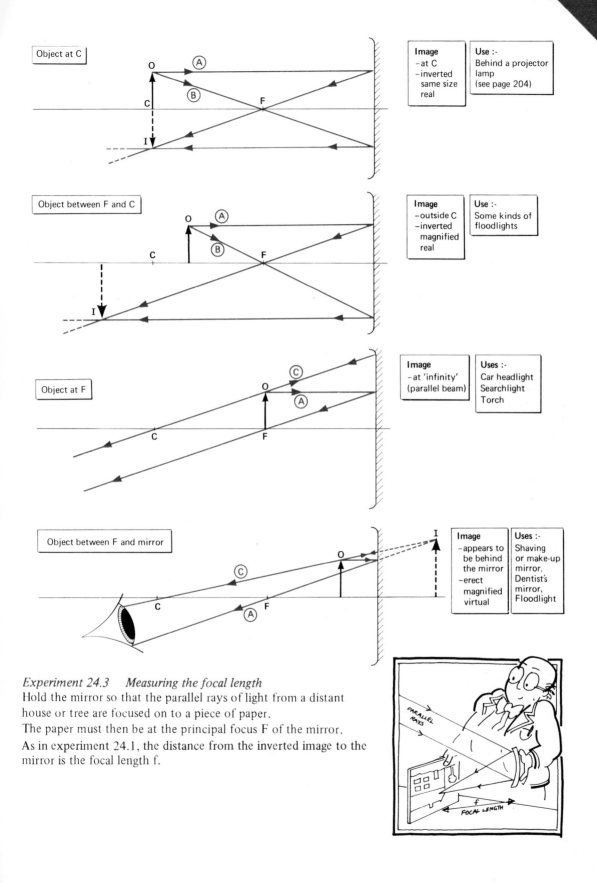

Object at C

@A
@B

O
C
F
I

Image
- at C
- inverted same size real

Use :-
Behind a projector lamp
(see page 204)

Object between F and C

@A
@B

O
C
F
I

Image
- outside C
- inverted magnified real

Use :-
Some kinds of floodlights

Object at F

@C
@A

O
C
F

Image
- at 'infinity' (parallel beam)

Uses :-
Car headlight
Searchlight
Torch

Object between F and mirror

@C
@A

O
C
F
I

Image
- appears to be behind the mirror
- erect magnified virtual

Uses :-
Shaving or make-up mirror,
Dentist's mirror,
Floodlight

Experiment 24.3   Measuring the focal length
Hold the mirror so that the parallel rays of light from a distant
house or tree are focused on to a piece of paper.
The paper must then be at the principal focus F of the mirror.
As in experiment 24.1, the distance from the inverted image to the
mirror is the focal length f.

PARALLEL RAYS

f
FOCAL LENGTH

Convex (diverging) mirror : ray diagram

The constructions used here are not quite the same as for a concave (converging) mirror. Here, the light is diverged *away from* the principal focus which is *behind* the mirror (see experiment 24.2).

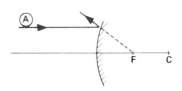

Construction A: parallel rays of light are reflected so that they appear to come from the principal focus F.

Construction B: rays of light travelling towards the centre of curvature C hit the mirror at 90° and are reflected back along their own path. The centre of curvature C is behind the mirror and twice as far as F.

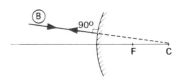

A convex mirror gives only one kind of image, so only one diagram need be drawn. The 6 steps used in drawing this ray diagram are the same as before (page 190).

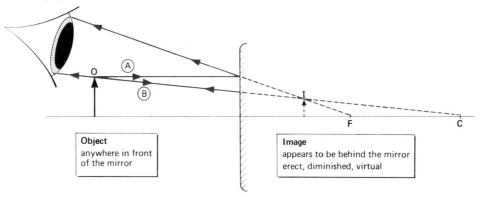

Object
anywhere in front of the mirror

Image
appears to be behind the mirror erect, diminished, virtual

Uses

A convex mirror gives an erect, virtual image like a plane mirror. But the image is smaller and so the field of view is wider. Because of this, convex mirrors are often used as driving mirrors on cars and as staircase mirrors on double-decker buses.

A hairy young student called Raven
Used a <u>d</u>iverging mirror for shaving,
His image was diminished,
And when he had finished,
His shaving looked more like engraving!

Summary

Concave mirrors converge parallel rays of light towards a principal focus in front of the mirror. Concave mirrors produce different images when the object is placed in different positions. (See ray diagrams on page 191).

Convex mirrors diverge parallel rays of light away from the principal focus which is behind the mirror.
The image in a convex mirror is always erect, virtual and diminished.
Convex mirrors have a wide field of view.

Questions

1. a) A concave mirror . . . . . rays of light, whereas a convex mirror . . . . . rays of light.
   b) Parallel rays of light are reflected by a concave mirror to a point called the . . . .
   The focal length is the distance from the
   . . . . to the mirror.
   c) For a convex mirror, parallel rays of light are
   reflected . . . . a point called the
   . . . .

2. Draw up two lists headed 'Concave mirrors' and
   'Convex mirrors' from the following: shaving
   mirror, car headlight mirror, searchlight mirror,
   driving mirror, dentist's inspection mirror,
   torch mirror, projector lamp mirror, staircase
   mirror on a double-decker bus, make-up mirror,
   reflecting telescope mirror, solar furnace mirror.

Position of the object	Position of the image	Real or virtual	Inverted or erect	Magnified or diminished	Uses
At infinity					
Outside C.					
At C.					
Between F&C					
At F.					
Between F & the mirror					

3. Draw up a table with the headings shown
   above and fill in the sections by studying the
   ray diagrams for a *concave* mirror.

4. The image in a convex mirror is always . . . . . ,
   . . . . . , and . . . . . Why is a convex mirror
   often used as a driving mirror?

5. Describe an experiment to measure the focal
   length of a concave mirror. Should you measure
   to the front or to the back of a glass mirror?
   Why?

6. An object 4cm high is placed 24cm from a
   concave mirror of focal length 8cm. Draw a
   ray diagram on graph paper (full size or $\frac{1}{2}$ scale)
   and find the position, size and nature of the
   image.

7. In the same way as in question 6, find the
   position, size and nature of the image formed
   when an object 2 cm high is placed a) 12cm from
   a concave mirror of focal length 8cm b) 5cm from
   a concave mirror of focal length 10cm c) 10cm
   from a *convex* mirror of focal length 10cm.

8. A car headlight bulb contains two filaments
   labelled A and B in the diagram. Filament B is
   at the principal focus of the concave mirror.
   Copy the diagram and draw in the rays of light
   from each filament. Why is this arrangement
   used in car headlights?

9. Professor Messer's in a dither,
   He's seeing things in his driving mirror:

Further questions on page 253.

chapter 25

REFRACTION

How do telescopes work? Why do swimming pools appear shallower than they really are?
Before we can answer questions like these, we must find out what happens when light travels from one substance into another.

Experiment 25.1
Place a glass block on a sheet of white paper and draw a line all round it. Use a raybox to shine a ray of light from the air into the glass.

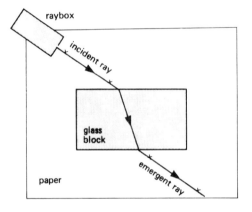

Look carefully for the ray inside the glass and where it comes out again.
Has the ray of light travelled straight into the glass or has it been bent?

Mark the incident ray and the emergent ray with crosses as shown in the diagram.
Then take off all the apparatus and use a ruler to join the crosses and re-construct the path of the ray of light.

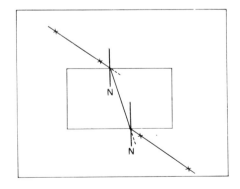

You can see that the ray was bent or *refracted* where it entered the glass and where it left the glass.

Use a protractor or a set-square to draw lines at right angles (90°) to the glass surface, as shown. Each of these lines is called a *normal* (N).

Where is the light bent towards a normal and where is it bent away from a normal?

Rays of light travelling from air into glass are bent or refracted towards the normal.
Rays of light travelling from glass into air are refracted away from the normal.

Here are some of the scientific words used to describe experiment 25.1.

Angle i is called the angle of incidence.

Angle r is called the angle of refraction.

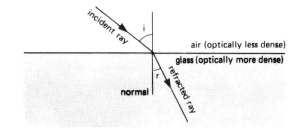

Refractive index

We saw on page 179 that the speed of light in air is about 300 million metres/second. But light travels more slowly in glass – in fact, only about $\frac{2}{3}$rds as fast. This is the reason why light is refracted – in just the same way as a squad of soldiers marching quickly on concrete are 'refracted' as they move on to muddy ground where they have to walk more slowly.

Experiments 21.8 and 21.9 with a ripple tank showed how water waves are refracted when they are slowed down (see page 177).

A number called the *refractive index* of a substance is the ratio of the two speeds:

> **Refractive index of a substance** $=$ **speed of light in air** / **speed of light in the substance**

The refractive index of most kinds of glass is about 1.5.

The refractive index of water is less (about 1.33) which means that the light is not bent as much when it enters water because it is not slowed as much.

Experiment 25.2 (more difficult)

In experiment 25.1, measure i, the angle of incidence and r, the angle of refraction.

Then use your calculator or a book of mathematical tables to find 'sine i' and 'sine r'.

The refractive index of the glass can be found from another formula, called *Snell's law*:

> **Refractive index** $= \dfrac{\text{sine i}}{\text{sine r}}$

A motorist drove down a straight country road,
When the curb gave the steering a twitch,
One wheel was slowed by the edge of the road,
And refracted him into a ditch.

Substance	Refractive index
Water	1.33
Perspex	1.5
Glass	about 1.5
Diamond	2.4

Real and apparent depth

Have you noticed that swimming pools and buckets of water always appear to be shallower than they really are?

Experiment 25.3
Place a thick glass block or a beaker of water on top of some writing.
What do you notice?

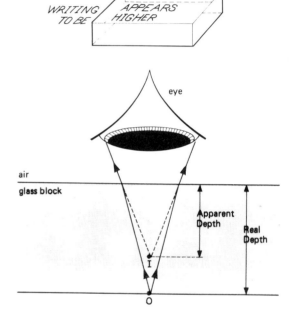

The apparent depth is *less* than the real depth because rays of light are refracted away from the normal as they leave the glass or the water. When they are received by your eye, they *appear* to come from the point marked I (a virtual image).

The apparent depth can be calculated from the formula:

$$\frac{\text{Real depth}}{\text{Apparent depth}} = \frac{\text{refractive index}}{\text{of the substance}}$$

This means that water is really 1.33 times as deep as it appears to be; and glass is about 1.5 times as deep as it appears to be.

Experiment 25.4   An 'imagic' trick
Put a penny in the bottom of an empty cup and position your head so that the penny is just out of sight.
How can you bring it into view without moving your head or the cup or the penny (and without using a mirror)?

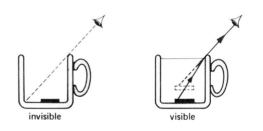

invisible   visible

Simply pour in water. Rays of light from the penny are refracted away from the normal and into your eye so that you can see it. Its apparent position is higher than its real position.

Experiment 25.5   The 'broken' pencil
Hold a pencil at an angle in a beaker of water.
What do you see?

Use the diagram to explain why the pencil appears to be broken.

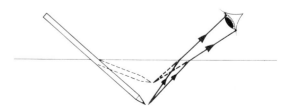

Total internal reflection

Experiment 25.6
Place a semi-circular glass block on a sheet of
white paper. Draw a line round it and mark the
mid-point of the straight side.
Use a ray-box to shine a ray of light at right-angles
to the curved surface so that it goes straight through
to the mid-point of the straight side.

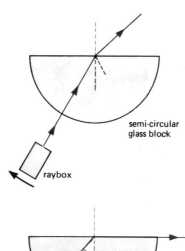

semi-circular
glass block

raybox

Can you see the ray of light refracted away from
the normal as it comes out of the glass block?
Can you also see that only a small amount of light
is reflected back?

Now move the ray-box round so that the ray of
light still goes straight in to the mid-point, but
with the angle of incidence increased until the
refracted ray comes out *along the edge of the block*.
When this happens, the angle of incidence in the
glass is called the *critical angle* (C).

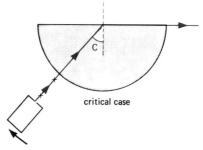

critical case

Mark the ray with crosses so that you can take the
apparatus off and re-construct the diagram.
Use a protractor to measure the critical angle (C).
Is it the same size as the angle in the diagram?

Now move the ray-box to increase the angle of
incidence so that it is *greater* than the critical
angle.
What happens?

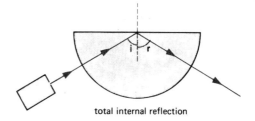

total internal reflection

We call this *total internal reflection:* 'total' because
all the light is reflected and 'internal' because it
only takes place *inside* the more dense substance.
As with ordinary reflection, angle i = angle r.

Total internal reflection takes place only
when (i) the rays are travelling in a dense medium
towards a less-dense medium.
and (ii) the angle of incidence is greater than the
critical angle.

The critical angle is about 42° for glass, and 49° for
water.

Using the formula on page 195 and referring to
the middle diagram above:

$$\frac{\text{refractive index}}{\text{of glass}} = \frac{\text{sine } 90°}{\text{sine C}}$$

$$\therefore \boxed{\text{refractive index} = \frac{1}{\text{sine C}}}$$

Uses of total internal reflection: totally reflecting prisms

Experiment 25.7

Put a prism of glass with angles measuring 45°–45°–90° on a sheet of white paper.
Use a ray-box to shine in a ray of light as shown in the diagram.

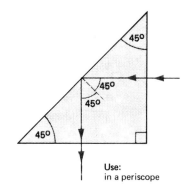

Use:
in a periscope

Notice that the ray of light goes straight through the first surface. Inside the glass it meets the second surface at an angle of incidence of 45°.
This is *greater* than the critical angle of 42° and so total internal reflection takes place.
The long side of the prism acts as a mirror and turns the ray of light through 90°.

In periscopes, prisms are better than ordinary plane mirrors (see p. 185). This is because ordinary mirrors give multiple images due to multiple reflection in the glass at the front of the mirror. This is shown in the diagram.
Two other reasons are that the silvering on a mirror is easily damaged and that a mirror reflects less light than a totally reflecting prism.

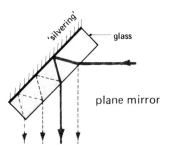

'silvering'

glass

plane mirror

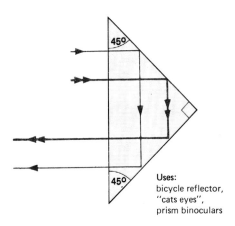

Uses:
bicycle reflector,
"cats eyes",
prism binoculars

Experiment 25.8

Use a 45°–45°–90° prism in a different way, as shown in the diagram.
Use your ray-box to show that the rays are totally reflected. This is because the angle of incidence (45°) is greater than the critical angle (42°).

The rays of light are turned through 180° so that they return to the direction they came from.
Bicycle reflectors and "cats' eyes" reflectors on roads are built like this so that they reflect back the light from car headlights.
Diamonds are cut in a similar way to make use of total internal reflection.

Notice that the 2 rays of light are inverted when they come out of this prism. This is used in prism binoculars (see page 219) where prisms are used to invert the image and shorten the binoculars.

More examples of total reflection

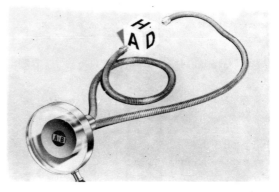

Pictures through optical fibres

The Light Guide (or optical fibre)

Experiment 25.9
Shine a lamp into one end of a curved glass or
perspex rod.
What do you see at the other end?

This is possible because of total internal reflection
of the rays of light inside the rod (as long as the
angle of incidence is greater than the critical angle).

If the rod is thin and flexible it can even be tied
in knots or pushed down your throat to take
photographs of the inside of your lungs!
Light guides are sometimes used to illuminate the
dials on radios and in cars. The streams of water
in illuminated fountains also act as light guides.
See also the examples on page 202.

Pictures through optical fibres

Mirages

Sometimes, on a hot day, you might see what look
like pools of water on a long stretch of road. This
happens because there are hot layers of air near
the hot road and cooler (denser) layers of air
higher up.

A ray of light is gradually refracted more and more
towards the horizontal. Eventually it meets a hot
layer near the ground at an angle greater than the
critical angle and total internal reflection takes
place.

A mirage: *Concorde* and its reflection

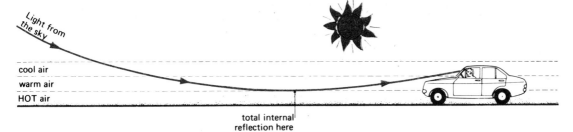

To the driver of a car, this total reflection looks
like the reflection from a pool of water: this is a
mirage.

A fish's-eye view

A fish or a diver under water can see everything above the surface, but their view is squeezed into a cone with an angle of 98° (twice the critical angle for water).

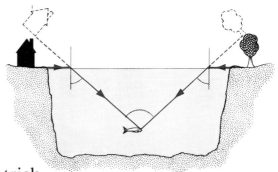

Outside this cone, the surface looks like a silvered mirror and reflects the light from objects inside the pond (by total internal reflection).

Something to do   Another 'imagic' trick

Place a beaker on top of a penny. Cover the top of the beaker so that you can only look through the side.
Can you see the penny?
Now fill the beaker full of water.
Why has the penny disappeared?
Why does the bottom of the beaker look silvery?

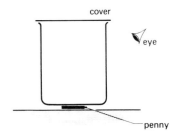

Now challenge a friend to bring the penny into view without moving the beaker, the penny or the cover.
The solution: simply run some water into the space *under* the beaker and the penny appears!

Summary

Rays of light travelling into a 'denser' medium (e.g. from air into glass) are refracted *towards* the normal; when leaving the denser medium (e.g. glass to air) the rays are refracted *away from* the normal. This happens because light travels slower in glass than in air.

The apparent depth of an object is less than its real depth. Total internal reflection takes place if (i) the rays are travelling in a dense medium towards a less-dense medium *and* (ii) the angle of incidence is greater than the critical angle C. (C = 42° for glass, C = 49° for water).

$$\frac{\text{Refractive index}}{\text{of a substance}} = \frac{\text{speed of light in air}}{\text{speed of light in substance}} = \frac{\text{sine } i}{\text{sine } r} = \frac{\text{real depth}}{\text{apparent depth}} = \frac{1}{\text{sine C}}$$

Questions

1. a) A ray of light travelling from air into glass is . . . . . or bent . . . . . the normal. A ray of light travelling from water to air is . . . . . or bent . . . . . the normal.

   b) The speed of light in water is . . . . . than the speed of light in air.

   c) A swimming pool looks . . . . . than it really is, because light from the bottom is refracted . . . . . the normal on passing into air.

   d) Total internal reflection takes place in a glass prism if the angle of incidence in the . . . . . is . . . . . than the . . . . . angle. The angles of a totally reflecting prism are . . . . . degrees, ... degrees, ... degrees.

2. Copy and complete the diagrams:

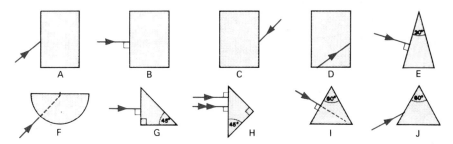

3. Explain with the aid of diagrams:
   a) A pond appears to be shallower than its real depth.
   b) An Indian who is spear-fishing does not aim his spear at where the fish appears to be.
   c) Diving for a penny is more difficult than it seems to be from the side of the swimming pool.
   d) In a photograph of a man rowing a boat, the oars seem to be broken where they enter the water.

4. Explain with the aid of diagrams:
   a) To a diver under water, most of the surface looks silvery.
   b) Bubbles of air rising from a diver look silvery.
   c) Cracks in a sheet of glass often look silvery.
   d) Light can be shone in inaccessible places by means of a 'light guide'.

5. Explain with diagrams, how $45°-45°-90°$ prisms can be used to reflect light through a) $90°$ b) $180°$.

6. Professor Messer goes out of his depth again:

7. Draw a diagram showing a periscope made with $45°-45°-90°$ prisms. What are the advantages of prisms over ordinary plane mirrors?

8. Professor Messer fell asleep and dreamed that light could not be refracted by any substance. Why was his dream a nightmare?

9. a) From the diagram, calculate the angle of incidence and the angle of refraction.

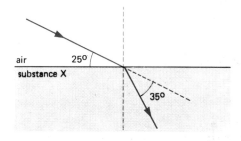

   b) What is the refractive index of substance X?

10. A swimming pool appears to be 3m deep. If the refractive index of water is $\frac{4}{3}$, what is the real depth?

Further questions on page 254.

Physics at work: Fibre Optics

An optical fibre is a very narrow *light-guide* (see page 199). It is as thin as a human hair and as flexible. It has a very narrow core made of very pure glass. This glass is so clear that if the seas were made of it, you could see the sea-bed easily.

The core is surrounded by a 'cladding', also made of a very pure glass. However, this glass has a slightly lower refractive index (see page 195). Because the cladding (like air) has a lower refractive index than the core, the light rays inside the core are *totally internally reflected.*

The edges of the core act like a perfect mirror and the light rays can travel a long way inside the glass. Optical fibres can be as long as 200 km.

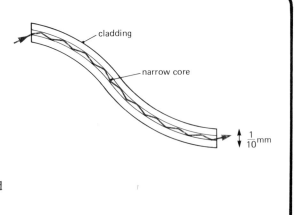

Communications

Optical fibres are being used to replace the copper wires in the telephone system.

The telephone conversations are sent down the optical fibre by switching on and off a LED (Light-Emitting Diode, page 350) or a laser (see opposite page). Infra-red rays are often used.

At the other end, the signals are converted back to electricity by a *photo-diode* which acts rather like an LDR (Light Dependent Resistor, page 351).

Over 10 000 telephone conversations can be carried at the same time by one fibre! Fibre optics can also be used to send data from one computer to another, or to carry up to ten TV channels.

Optical fibres have many advantages over copper cables — they are thinner, cheaper, carry more signals, have no cross-talk between two telephone conversations and are almost impossible to 'bug'.

The new and the old – optical fibres to replace copper wires.

Medicine

Doctors use optical fibres to look at the inside of people's lungs and stomachs:

A large number of optical fibres are held together in a bundle which goes down the patient's throat. Light is sent down some of the fibres to illuminate the stomach.

Reflected light comes back up the bundle of fibres to form an image of bright and dark spots of light.

If an ulcer is found, a laser beam can be sent down the fibres to burn and seal the ulcer neatly.

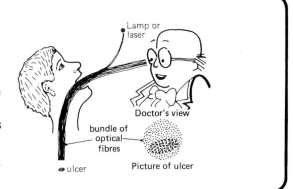

Physics at work: Lasers

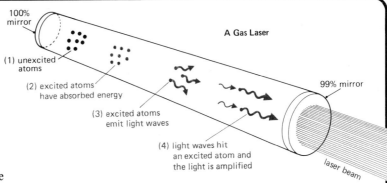

A laser is a device for producing a very intense beam of light. It is an *amplifier* of light.
The word LASER stands for Light Amplification by Stimulated Emission of Radiation.

A Gas Laser

100% mirror
(1) unexcited atoms
(2) excited atoms have absorbed energy
(3) excited atoms emit light waves
(4) light waves hit an excited atom and the light is amplified
99% mirror
laser beam

It consists of a narrow tube containing a gas:
Electricity is passed through the gas so that it starts to glow (just like the fluorescent lamp on your classroom ceiling).
This is because the electricity 'excites' the atoms of the gas by giving them energy which they absorb.
Later they give out this energy, but in the form of a light wave.
If this light wave hits an atom which is already 'excited' then it *also* gives out light (this is 'stimulated emission'). The light has been *amplified* and is brighter.
If most of the atoms are excited by the electricity, than the light waves build up like an avalanche to make a very bright light. The mirrors help by reflecting the light up and down the tube many times. One of the mirrors is not a perfect reflector so some of the light escapes. This is the laser beam.

Laser light is a special kind of light. Ordinary light is rather like the people in a football crowd: it moves in different directions with different wavelengths. Laser light is more like a column of soldiers: it all moves in the same direction, with exactly the same wavelength and with all the waves exactly in step or '*in phase*'.

Lasers can be made from solids and liquids as well as gases. Some are smaller than a pencil, others are as big as a room.

Uses of lasers

Because laser beams are perfectly straight, they are used by surveyors and engineers to ensure that roads, pipes, railway lines, oil tankers and buildings are built accurately.

Laser beams are very intense (*never* look up the beam) and they can melt or even vaporise substances. Laser beams are used to 'drill' holes and 'cut' sheets of metal. They can do this more quickly and more accurately than metal drills and saws.
The cloth for your jeans was probably cut by a laser.

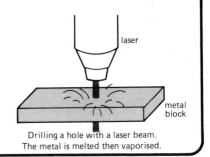

laser
metal block

Drilling a hole with a laser beam.
The metal is melted then vaporised.

Lasers are used to weld pieces of metal together. The laser melts the metal, which mixes and later solidifies.
Lasers are used by doctors to 'weld' skin. For example, if the retina in a person's eye becomes loose, a laser is used to weld it back on. Optical fibres can be used to guide the laser beam (see opposite page).

Lasers are also used in 'compact disc' players (see page 346) and laser-TV's can project big TV pictures on to a wall.

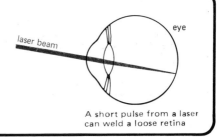

laser beam
eye

A short pulse from a laser can weld a loose retina

chapter 26

LE**NSE**S

Lenses are very useful — in fact, at this moment you are using the lens in your eye to focus on these words.

Two kinds of lenses
A lens which is thicker at the centre than at the edges is called a *convex* lens.
A lens which is thinner at its centre is called a *concave* lens.

Experiment 26.1   A convex lens
Use a ray-box (on a sheet of white paper) to produce several rays of light to shine on to a convex lens. (If necessary, rest the paper on two books and push the lens half-way through the paper).
It is better if you can adjust your ray-box to make *parallel* rays of light. Look for the rays refracted through the lens.

Can you see that the rays after the convex lens are getting closer together? We say they are *converging*.

A convex lens is a converging lens.

If the rays of light from the ray-box are *parallel* rays, the rays refracted through the lens converge to one point called the *principal focus* (marked F on the diagram).

Parallel rays of light are refracted through the principal focus of a convex lens.

The distance from F to the centre of the lens is called the *focal length* of the lens.

Experiment 26.2   A concave lens
Repeat experiment 26.1, but with a concave lens.

Can you see that the light rays are spread out or *diverged* by this lens?

A concave lens is a diverging lens.

Parallel rays of light are diverged so that they appear to come from the principal focus (F) of a concave lens.

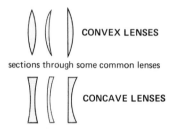

CONVEX LENSES

sections through some common lenses

CONCAVE LENSES

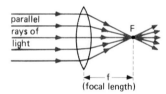

Convex (converging) Lens
(used like this as a burning glass or the objective lens of a telescope)

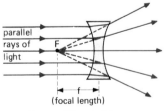

Concave (diverging) Lens

Experiment 26.3   Focal length of a convex lens (1)

Hold a convex (converging) lens so that the parallel rays of light from a distant house or tree are focused on a piece of paper.

The piece of paper must then be at the principal focus F. As in experiment 26.1, the distance from the (inverted) image to the centre of the lens is the focal length *f*.

Does a fatter lens give a shorter or a longer *f*?

Light can travel into a lens from the left or from the right. It therefore has *two* principal foci (at equal distances on opposite sides of the lens).

Experiment 26.4   Focal length of a convex lens (2)

This method uses an illuminated object **O** and a plane mirror, as shown.

Move the lens until a *sharp* image **I** (of the object) is formed on the screen next to the object. This happens when the light travels back along its own path because it is hitting the mirror normally.

Then the distance shown is the focal length. Why?

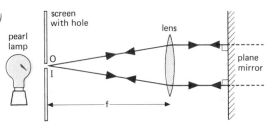

Experiment 26.5   Images formed by a convex lens

Use the apparatus shown in the diagram to investigate the different images formed when the object is placed at different distances from the lens.
For each position of the object, move the screen until you get a sharp image.

Fill in a table to show your results (you can copy the table shown in question 3 on page 209). Compare these results with the ray diagrams on the next two pages.

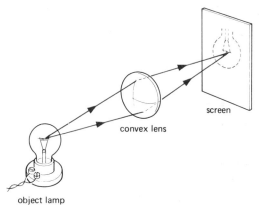

Images

The images can be *magnified* larger than the object (that is, the magnification is greater than 1) or *diminished* smaller than the object (the magnification is less than 1).
What does magnification = 1 mean?

$$\text{Magnification} = \frac{\text{height of image}}{\text{height of object}}$$

Images can also be *real* (light rays go through them, so they can be shown on a screen) or *virtual* (the light rays only appear to come from that position). To see virtual images you must remove the screen and place your eye in its place.

A short-sighted student called Menzies,
Wore his spectacles only on Wednesdays.
He said, "It's not bliss,
But the question is this:
Are these glasses for keeps or just lenses?"

205

Convex (converging) lenses: ray diagrams

The ray diagrams on the opposite page show the different images that are produced when an object is placed at different distances from a convex (converging) lens.
To draw these diagrams, you need to know 2 constructions:

Construction A: parallel rays of light are refracted through the principal focus F.
This was shown in experiment 26.1.

Construction B: rays of light passing through the centre of the lens travel straight on. This is true for a thin lens because, at the centre, its sides are parallel.

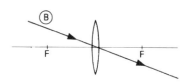

Drawing a ray diagram:
Step 1. On graph paper draw a long central line (called the principal axis). Draw a shorter line at right angles to represent the thin lens.
Step 2. Where distances are given, choose scale(s) for the object's size and position (often $\frac{1}{2}$ scale horizontally and full size vertically).
Step 3. Mark in the object and the position of F on each side of the lens. It is also useful to mark points called '2F' at twice the focal length from the lens.

Step 4. Starting from the top of the object, draw the two constructions.
Step 5. Where the refracted rays cross is the position of the top of the image.
Step 6. Measure the position and size of the image (using the scales) and say whether it is:
    inverted or erect;
    magnified or diminished;
    real or virtual.

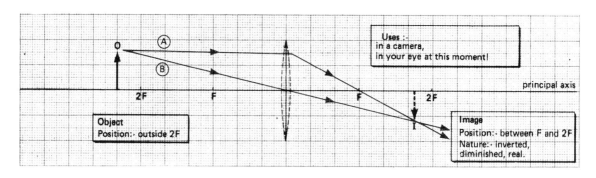

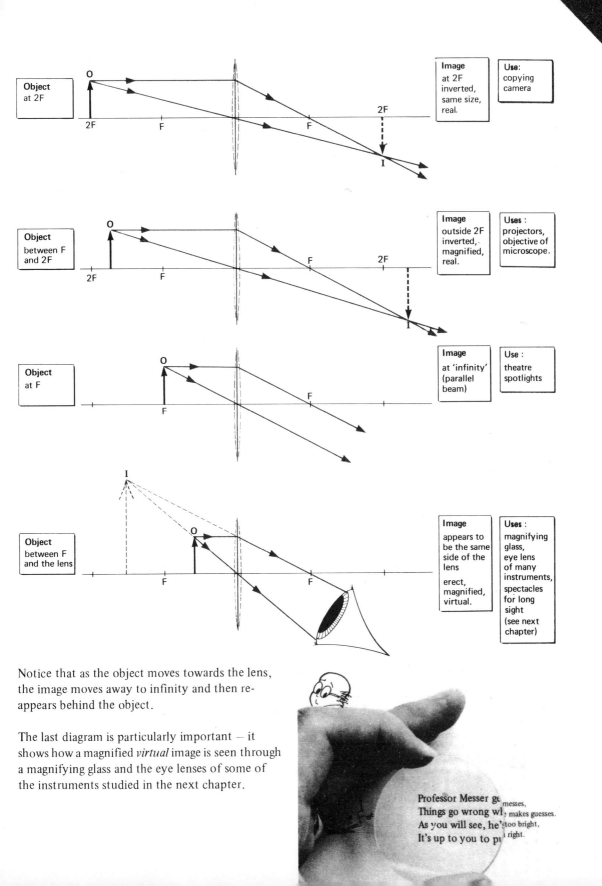

Object at 2F

O
2F · F · 2F
Image at 2F inverted, same size, real.
Use: copying camera

Object between F and 2F

O
2F · F · F · 2F · I
Image outside 2F inverted, magnified, real.
Uses: projectors, objective of microscope.

Object at F

O
F · F
Image at 'infinity' (parallel beam)
Use: theatre spotlights

Object between F and the lens

I · O
F · F
Image appears to be the same side of the lens erect, magnified, virtual.
Uses: magnifying glass, eye lens of many instruments, spectacles for long sight (see next chapter)

Notice that as the object moves towards the lens, the image moves away to infinity and then re-appears behind the object.

The last diagram is particularly important – it shows how a magnified *virtual* image is seen through a magnifying glass and the eye lenses of some of the instruments studied in the next chapter.

Professor Messer ge messes,
Things go wrong wh makes guesses.
As you will see, he' too bright,
It's up to you to p right.

Concave (diverging) lens : ray diagram

These constructions are not quite the same as for a convex (converging) lens. Here, the light is diverged *away from* the principal focus which is on the object's side of the lens (see experiment 26·2).

Construction A: parallel rays of light are refracted away from the nearer principal focus F (see experiment 26.2).

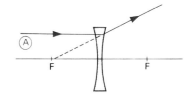

Construction B: a ray of light passing through the centre of the thin lens travels straight on.

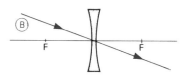

A concave lens gives only one kind of image, so only one diagram need be drawn. The 6 steps used in drawing this ray diagram are the same as before (page 206).

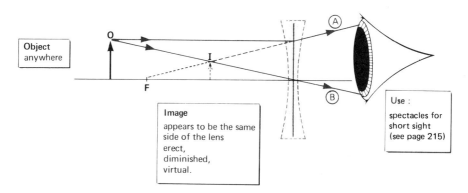

Object anywhere

Image
appears to be the same side of the lens
erect,
diminished,
virtual.

Use :
spectacles for short sight
(see page 215)

Summary

Convex lenses converge parallel rays of light to a (real) principal focus. Convex lenses produce different images when the object is placed in different positions (see the ray diagrams for details and uses).

Concave lenses diverge parallel rays of light away from a (virtual) principal focus. The image in a concave lens is always erect, virtual and diminished.

$$\text{Magnification} = \frac{\text{height of image}}{\text{height of object}}.$$

Questions

1. Copy out and complete:
   a) A convex lens . . . . . rays of light, whereas a concave lens . . . . . rays of light.
   b) Parallel rays of light are refracted by a convex lens to a point called the . . . . The focal length is the distance from the . . . . to the lens.
   c) For a concave lens, parallel rays of light are refracted . . . . a point called the . . . .
   d) The image in a concave lens is always . . . . . , ... , and . . . . .

2. Draw up two lists headed 'Convex lenses' and 'Concave lenses' from the following: camera lens, eye lens, spectacles for short sight, magnifying glass, slide projector lens, burning glass, spotlight lens.

3. Draw up a table with the following headings and fill in the sections by studying the ray diagrams for a *convex lens*.

Position of the object	Position of the image	Real or virtual	Inverted or erect	Magnified or diminished	Uses
At infinity					
Outside 2F					
At 2F					
Between 2F & F					
At F					
Between F & the lens					

4. a) Describe an experiment to measure the focal length of a convex lens.
   b) Your eye contains a convex lens – why is it unwise to look at the sun? Why is it unwise to leave glass bottles in a forest?

5. Use the diagram to explain why a convex lens converges rays of light.
   Draw a similar diagram for a diverging lens, showing clearly what happens at each surface.

6. An object 4cm high is placed 15cm from a convex lens of focal length 5cm. Draw a ray diagram on graph paper (fullsize or $\frac{1}{2}$ scale) and find the position, size and nature of the image.

7. In the same way as in question 6, find the position, size and nature of the image formed when an object 2cm high is placed
   a) 7.5cm from a convex lens of focal length 5cm
   b) 4cm from a convex lens of focal length 12cm
   c) 10cm from a *concave* lens of focal length 10cm.

8. Detective Messer looks for clues
   But hasn't a clue which lens to choose.

Further questions on page 254.

chapter 27

ptical instruments

First we will consider optical instruments that form *real* images:-
the camera, your eye, the film projector.
Then we will consider optical instruments that form *virtual* images
for your eye to look at:- microscopes and telescopes.

The lens camera

A camera consists of a light-tight box with a convex (converging)
lens at one end and the film at the other end.

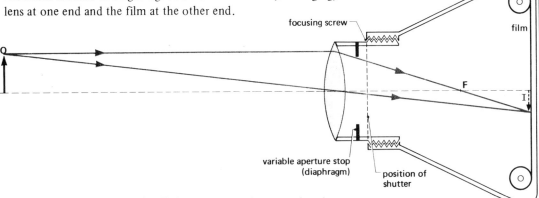

Because the object is outside 2F the image is real, inverted and
diminished, as shown in the diagram (see also the ray diagram at the
bottom of page 206).

Experiment 27.1

Convert the pinhole camera (described on page 182) into a lens
camera by enlarging the hole at the front of the box and holding a
lens over the hole (choose a lens with a focal length about as long as
the box).

Adjust the position of the lens for either near or far objects to make
a sharp image on the screen.

Is the image erect or inverted?
If the object is coloured, is the image coloured?
Is the image brighter or dimmer than in a pinhole camera?

I knew a famous photographer chap,
He snapped everyone in the town,
But his clients weren't happy,
– their tempers would snap,
When he made them all stand
upside down.

The controls on a camera

1. Focusing

A camera is focused by moving the lens. If the object moves *nearer* the lens, its image moves farther away (see page 207).

So to keep the image in focus on the film, the lens must be moved farther *out*.

1/30th s at f/2, fast film

2. The shutter

The amount of light entering the camera can be controlled by the length of time that the shutter is open. Fast moving objects will appear blurred unless the exposure time is very short (perhaps only $\frac{1}{500}$th second).

1/500th s at f/4, fast film

3. The aperture stop or diaphragm

The amount of light entering the camera can also be controlled by varying the size of the hole in the diaphragm which is just behind the lens. To take a photograph in dim light, a large hole is needed.

The size of this hole also controls the *depth of focus*. A large depth of focus means that both near and far objects will appear to be in focus at the same time. This is obtained by having a *small* hole in the diaphragm.

1/10th s at f/8, slow film

The control which varies the size of the hole is marked in *f-numbers*.

A value of f/8 means that the diameter of the hole is $\frac{1}{8}$th of the focal length. f/11 is a *smaller* opening (it lets in less light but gives a greater depth of focus).

The shutter time and the f-number which should be used, depend upon
a) the brightness of the object
b) the sensitivity or 'speed' of the film
c) the kind of effect that is wanted.

1/50th s at f/16, fast film

Look at the values given with each photograph.
Can you decide why the photographer chose the different values for each photograph?

1/200th s at f/2, slow film

The human eye

Study the different parts of the diagram of your eye.

The light enters your eye through the transparent *cornea*, passes through the *lens* and is focused on the *retina*. The retina is sensitive to light and sends messages to your brain by way of the *optic nerve*.

The *iris* changes in size to vary the amount of light that enters through the *pupil*.
When is your pupil small and when is it large?

Notice that the image on your retina is inverted (see also the ray diagram at the bottom of page 206). Although the image is inverted, your brain has learned to interpret this correctly.

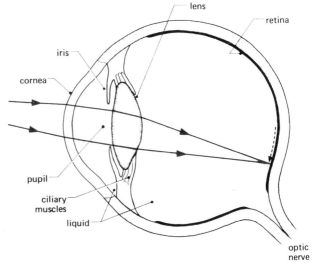

top view of your right eye

Focusing your eye
Most of the bending of the light rays is done by the curved cornea, but your lens can change its shape to alter its focal length slightly.
To focus distant objects the lens is *thin* (with the *ciliary muscles* relaxed).

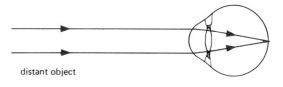

distant object

To focus on a near object (like this book) the ciliary muscles tighten (rather like the draw-strings on a duffel bag) and squeeze the lens. The lens is then fatter (with a shorter focal length) and can focus the light from near objects.

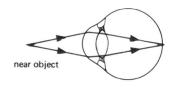

near object

Experiment 27.2 Find your eye's blind spot
Cover your *left* eye, hold the page at arm's length and stare at Professor Messer's magic wand. Then move the page slowly towards your eye until the rabbit disappears. This happens because the light rays from the rabbit are then arriving at a place where there is a gap in your retina. This gap is where the optic nerve leaves your eye.

Experiment 27.3 Binocular vision

Hold a pencil in one hand and close one eye. Then, with another pencil in your other hand, try to make the two pencil points touch.

Now try this with both eyes open. Which is easier? Why is this?

Having two eyes open allows you to see an object from two slightly different angles and so you can judge distances more accurately.

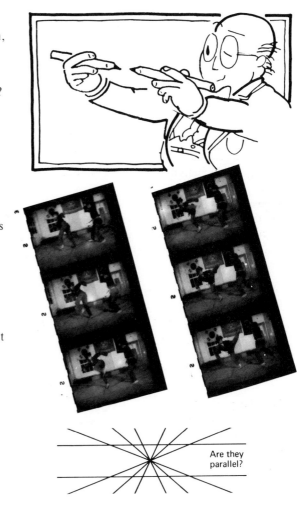

Experiment 27.4 Persistence of vision

Look at the little man at the bottom corner of this page and the next few pages.
What appears to happen if you flick over the corners of the pages quickly?

The effect of the image on your retina lasts for about $\frac{1}{10}$ th of a second — so that you can appear to see an image after it has disappeared. This effect is used in television and in the cinema where you are shown about 25 pictures every second, each picture changing slightly from the one before.

Experiment 27.5 Optical illusions

Judge each of the diagrams with your eyes and then answer each of the questions by measuring with a ruler.

Are they parallel?

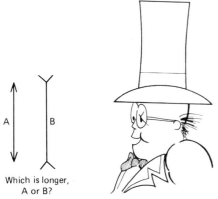

Which is longer, A or B?

Which is greater, the height of the hat or the width of the brim?

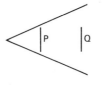

Which is longer, P or Q?

All equal in length or one shorter?

Defects of vision

People with normal vision can focus clearly on very distant objects at 'infinity'. We say their *far point* is at infinity.

People with normal vision can also focus clearly on near objects such as this book. The closest point at which they can see an object clearly is called the *near point*. The near point for adults is often about 25cm from the eye (less for teen-agers).

Experiment 27.6
Hold a book at arm's length and move it closer to find the nearest distance that you can focus it clearly without straining your eyes.
What is the distance to your near point?

Long sight
A person who can see distant objects clearly but cannot focus near objects is said to be *long-sighted*.

This is because his eye-ball is too short or his eye-lens is too thin even though his ciliary muscles are fully squeezed.

This means that rays of light from a close object O are focused towards a point *behind* the retina.

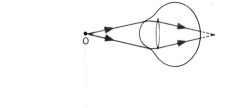

Rays of light from this person's near point N can just be focused by his eye.

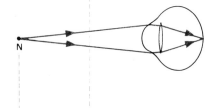

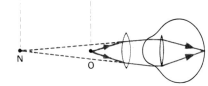

This person should wear a *convex (converging)* spectacle lens. Then rays of light from a close object are converged so that, as far as his eye is concerned, the rays appear to come from his near point N. Then they can be focused by his eye.

Did you hear about the optician who fell into his lens-grinding machine and made a spectacle of himself?

Short sight

A person who can see near objects clearly but cannot focus distant objects is *short-sighted*.

This is because the eye-ball is too long or the eye-lens is too strong. (The lens is too thick even though the ciliary muscles are relaxed.)

This means that rays of light from a distant object (at 'infinity') are focused *in front* of the retina.

Rays of light from this person's far point F can just be focused.

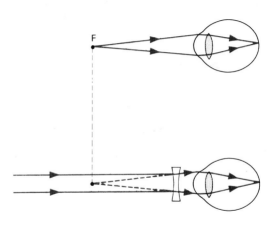

This person should wear a *concave* (*diverging*) spectacle lens.
Then rays of light from a distant object are diverged so that they appear to come from his far point F and so they can be focused by his eye.

How can you tell by inspecting a person's spectacles whether he is long-sighted or short-sighted?

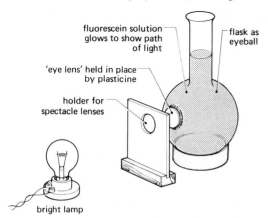

fluorescein solution glows to show path of light

flask as eyeball

'eye lens' held in place by plasticine

holder for spectacle lenses

bright lamp

A short-sighted teacher called Rose,
Confused the optician she chose,
And so in her specs,
Each lens was con<u>vex</u>,
And her far point was the end of her nose.

Experiment 27.7
The ray diagrams on these two pages may be demonstrated by your teacher using the apparatus shown in the diagram.

Experiment 27.8   Colour-blindness
If some specially coloured charts are available, your teacher may be able to test you for colour-blindness (the commonest kind is when red and green look the same to the colour-blind person).

215

The projector (for slides or for ciné film)

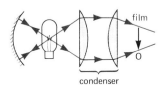

concave lamp
mirror

A projector contains a *lamp* and a *concave mirror* to make the image brighter. The lamp is placed at the centre of curvature of the mirror (see page 191) so that the rays of light are reflected back along their own path.

If the film was placed just after the lamp, the image on the screen would be poorly illuminated because the rays of light are diverging. To give a bright picture, a *condenser* is included. It is usually made of 2 plano-convex lenses, as shown.

Now the light is converging towards the screen and so the film is illuminated both brightly and evenly.

The light is then scattered by the film and focused by a convex (converging) *projection lens* on to the screen:

condenser

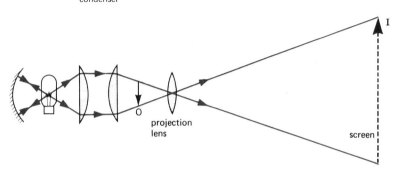

The film O is placed between F and 2F of the projection lens (see the second diagram on page 207), so that the image I is real, inverted and magnified. The film is put in the projector upside-down so that the picture is seen the right way up.

How is the projector focused?
Why does a projector usually contain a fan?

Experiment 27.9   Making a projector
Make a projector on the bench using a ray-box lamp; a single convex lens (focal length about 5cm) for the condenser; a piece of ciné film or a slide; a convex lens (focal length 5 or 10cm) as the projection lens and a sheet of white paper for the screen.

Is the image inverted?
By how much is it magnified?

216

The compound microscope

We have seen how a single converging lens (a 'magnifying glass' or 'simple microscope') can be used to magnify objects (see the ray diagram and photograph at the bottom of page 207).
However to give a higher magnifying power we need *two* lenses in a *compound microscope*.

eye lens

object placed here

objective lens

mirror to shine light on object

Fe

I₁

Fe

Fo

O

Fo

I₂

The Ray Diagram

a) The first part of the microscope is like a projector (see opposite page and page 207). The object is placed just outside Fo, the principal focus of the objective lens and so a real, inverted *magnified* image I₁ is formed.

b) Then the magnified I₁ acts as an object for the eye lens, which is used as a magnifying glass (see the bottom diagram on page 207). Fe is the principal focus of the eye lens.

The dotted lines are construction lines used to find the position of the final image I₂.

c) The final image I₂ is virtual and magnified still further. It is inverted compared with the object. I₂ may appear 1000 times larger than the object.

Experiment 27.10   Making a compound microscope
Use two lenses, of focal lengths 5cm and 10cm, together with a half-metre rule and some plasticine as shown in the diagram.

Move the object to-and-fro until it appears in sharp focus. (A plastic ruler makes a good object.)

What do you notice about the image?
Is it distorted?
Is it coloured differently in any way?

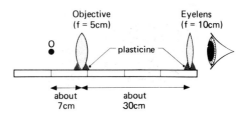

Objective (f = 5cm)   Eyelens (f = 10cm)

O

plasticine

about 7cm   about 30cm

Telescopes

A. The astronomical refracting telescope

A simple refracting telescope contains 2 lenses:
1. an *objective lens* which has a *long* focal length
2. an *eye lens* which has a *short* focal length.

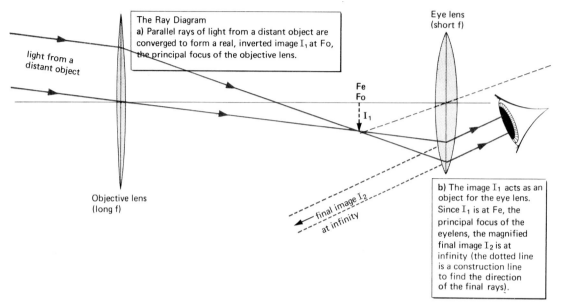

light from a distant object

The Ray Diagram
a) Parallel rays of light from a distant object are converged to form a real, inverted image I_1 at Fo, the principal focus of the objective lens.

Eye lens (short f)

Fe
Fo
I_1

Objective lens (long f)

final image I_2 at infinity

b) The image I_1 acts as an object for the eye lens. Since I_1 is at Fe, the principal focus of the eyelens, the magnified final image I_2 is at infinity (the dotted line is a construction line to find the direction of the final rays).

Astronomers use telescopes to look at very faint stars. Why is it an advantage to have a very large objective lens?

The final image is inverted — is this a disadvantage to an astronomer?

Experiment 27.11   Making a refracting telescope
Use two lenses, of focal lengths 20cm and 5cm, with a half-metre rule and some plasticine, as shown in the diagram. (Can you remember how to find the focal length of each lens? See page 205.)

Adjust the distance between the lenses until the image is in sharp focus. (Take care not to drop the lenses.)

Is the image magnified?
Is the image erect or inverted?

What happens to the magnifying power if you change the objective lens to one with a *longer* focal length?

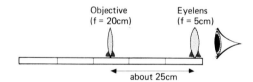

Objective (f = 20cm)   Eyelens (f = 5cm)

about 25cm

B. Prism binoculars

Binoculars are made of two refracting telescopes side by side, one for each eye.

Each telescope is built like the one on the opposite page but has 2 prisms added.

In experiment 25.8 on page 198, we saw that $45°-45°-90°$ prisms can use total internal reflection to *invert* rays of light.

By including two of these prisms in each telescope, the final image I_2 can be turned the right way up and the right way round.

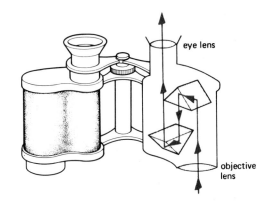

eye lens

objective lens

The prisms do two things:
1. They invert the rays of light so that the final image is seen the correct way.
2. By reflecting the light up and down the inside of the binoculars, they shorten the length of the instrument so it is more compact.

C. The reflecting telescope

The *reflecting* telescope (invented by Sir Isaac Newton) has a large concave mirror to collect the light and to converge it towards a small plane mirror.

The plane mirror reflects the light sideways to an eye lens. This acts as a magnifying glass in the same way as the eye lens on the opposite page.

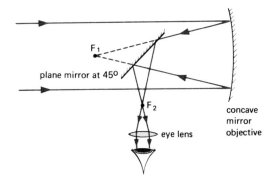

F_1

plane mirror at 45°

F_2

eye lens

concave mirror objective

Large telescopes

Large lenses tend to sag, so all the very large telescopes are the reflecting type. (The largest objective lens in a telescope is 1 metre in diameter; the largest mirror is 6 metres in diameter). This means that the larger reflecting telescopes can collect more light and let astronomers see fainter stars in the sky.

Radio telescopes are the reflecting kind; a solar furnace (see page 56) is a kind of reflecting telescope.

A radio telescope

Things to do

1. Make a microscope

Use a pin or a nail to make a hole about 2mm in diameter in a piece of kitchen foil.

Carefully let a drop of water fall on to the hole so that it stays there and acts as a tiny lens with a short focal length.

You can use it as a simple magnifying glass. Or, if you have a glass or plastic magnifying glass, you can make a compound microscope as shown.

Experiment with holes of different sizes.

Use your microscope to look at a grey part of a newspaper photograph. Can you see the individual dots clearly?

2. Make a movie 'film'.

Use a book of cloakroom tickets to make a flicker-book. Plan the movements of the figures carefully so that you get the right apparent speed.

Summary

In a camera, the real, inverted image is focused by moving the lens. The brightness of the picture depends on
a) the exposure time
b) the size of the aperture (f-number)
c) the 'speed' of the film.

In the eye, the real, inverted image is focused by changing the shape of the lens.
With long sight, a person cannot see near objects. It is corrected by a convex (converging) spectacle lens.
With short sight, a person cannot see distant objects. It is corrected by a concave (diverging) spectacle lens.

A projector produces a real, inverted, magnified image because the film is placed just outside the focal length of the projection lens. A condenser is used so that the film is illuminated evenly and brightly.

A compound microscope has two convex (converging) lenses, each with a short focal length. The image is virtual, inverted and magnified.

A refracting telescope has a long focal length objective lens and a short focal length eye lens. The image is inverted, unless prisms are used to turn it the right way up, as in prism binoculars.
A reflecting telescope has a wide concave (converging) mirror to collect a lot of light.

Things aren't always what they seem.
Is the Professor in a dream?

220

Questions

1. A camera has the following controls: a) a variable-speed shutter b) a variable aperture c) a focusing control. Explain the purpose of each of these controls.

2. Look again at the information given with the photographs on page 211. For each photograph, , explain why you think the photographer chose those values for the controls.

3. The lens of a camera has a focal length of 10cm. What must be the distance between the lens and the film in order to photograph objects at a) infinity b) 20cm from the lens.

4. Comparing the eye and a camera, make lists of a) similarities and b) differences.

5. Draw a large labelled diagram of a human eye. Explain the function of a) ciliary muscles b) the iris c) the retina. Where is the blind spot? Why are two eyes better than one?

6. Use a ruler to check the illusions on page 213. In order to look slimmer, should a girl wear a dress with vertical or horizontal stripes? What optical illusions are used by artists?

7. Among animals, the 'hunters' usually have their eyes facing forward at the front of their heads, whereas the 'hunted' usually have their eyes at the sides of their head. Why is this?

8. In a certain murder investigation, it was important to discover whether the victim was long-sighted or short-sighted. How could a detective decide by examining the spectacles?

9. With the aid of ray diagrams, explain what is wrong with the eyes of a person with a) long sight b) short sight. Use ray diagrams to explain how each fault can be corrected.

10. Explain, with a ray diagram, the purpose of the different parts of a side projector. A piece of fluff shows up at the top left corner of the screen. Where exactly would you look for it?

11. Describe, with a diagram, the action of a compound microscope. Describe the final image compared with the object.

12. Describe, with a diagram, the action of an astronomical refracting telescope. How can the final image be made erect?

13. Explain why astronomers use reflecting telescopes rather than refracting telescopes.

14. Professor Messer fell asleep and dreamed that light could not be refracted. Why was his dream a nightmare?

15. Another night, Professor Messer dreamed that none of this optical instruments in this chapter (except the eye) had been invented. In what ways would our lives be different if this was so?

16. Discuss the physics of this cartoon:

Further questions on page 255.

chapter 28

COLOUR

What causes the colours in a rainbow or in a diamond? To answer this, we must investigate the *spectrum* of white light.

Experiment 28.1   Newton's experiment
Use a raybox (or a slide projector) with a narrow slit. Shine a ray of white light through a glass prism and on to a white screen, as shown.

What do you see on the screen?
How many different colours can you count?

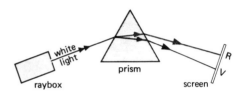

It seems that white light is really a mixture of several colours and can be split up by a prism. We say that the white light from the raybox has been *dispersed* by the prism to form a *visible spectrum*.
The colours, in order, are **R**ed, **O**range, **Y**ellow, **G**reen, **B**lue, **I**ndigo, **V**iolet. (Remember them by forming them into a boy's name: **ROY G. BIV.**)

The colour that is deviated (bent) least by the prism is red; violet is deviated through a larger angle, as shown in the diagram.

A name you should remember
As long as you may live.
The colours of the spectrum,
Are known by **ROY G. BIV.**

Experiment 28.2   An almost pure spectrum
In the last experiment, the spectrum was 'impure' because the colours overlapped each other on the screen. To form a 'pure' spectrum, a lens must be used.
Use a narrow slit and move the screen (or the lens) until the colours do not overlap and the spectrum is (almost) pure.

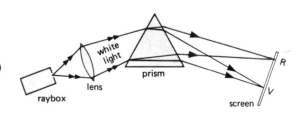

The colours of a rainbow or a diamond or a cut-class necklace are caused by the refraction and dispersion of white light into a spectrum.

Experiment 28.3   Newton's colour wheel

If white light is composed of the colours **ROY G BIV**, then we should be able to get white light by adding these colours together.

Spin a colour wheel quickly so that the colours appear to be mixed together.

What colour does it appear?
What do we really mean by white light?

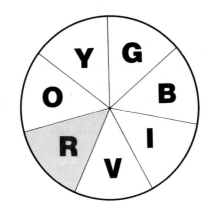

How does a prism disperse white light into a spectrum?
Different colours of light have different wavelengths (rather like waves at sea have a different wavelength from ripples on a pond — see page 175).

All colours of light travel at the same speed in a vacuum. When they enter a transparent substance like glass, they all slow down but by *different* amounts.
Because they slow down, they are refracted (see page 195) but because they slow down by *different* amounts, different colours are refracted through *different* angles.

Violet (the shortest wavelength) is slowed down the most and so is refracted through the largest angle. Red (a longer wavelength) slows down less and so is deviated through a smaller angle.

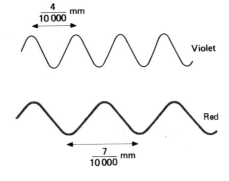

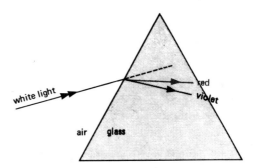

The full electro-magnetic spectrum
The visible spectrum is only a small part of a much larger spectrum containing many other wavelengths which we cannot see with our eyes (see also page 178).

The full electromagnetic spectrum is shown in more detail on the next two pages. Study each part of it carefully.

The electro-magnetic spectrum

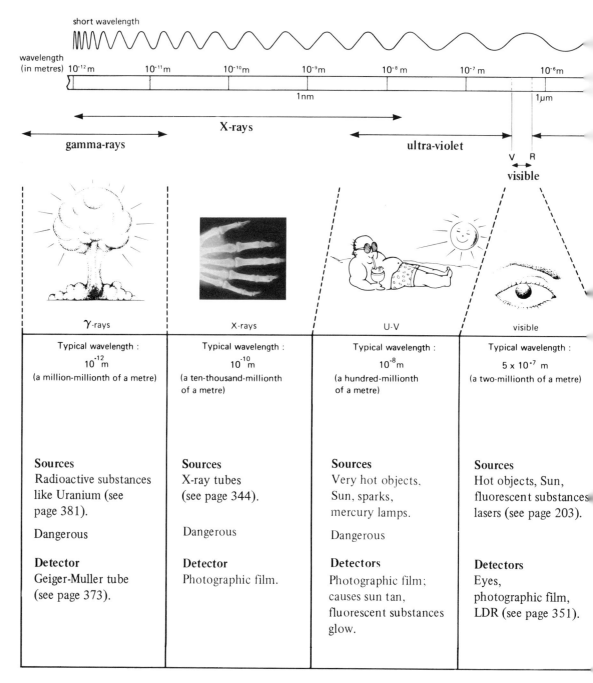

short wavelength

wavelength (in metres) 10^{-12} m 10^{-11} m 10^{-10} m 10^{-9} m 10^{-8} m 10^{-7} m 10^{-6} m

1nm   1μm

X-rays

gamma-rays   ultra-violet

V   R
visible

γ-rays	X-rays	U-V	visible
Typical wavelength : 10^{-12} m (a million-millionth of a metre)	Typical wavelength : 10^{-10} m (a ten-thousand-millionth of a metre)	Typical wavelength : 10^{-8} m (a hundred-millionth of a metre)	Typical wavelength : 5×10^{-7} m (a two-millionth of a metre)
Sources Radioactive substances like Uranium (see page 381). Dangerous	**Sources** X-ray tubes (see page 344). Dangerous	**Sources** Very hot objects, Sun, sparks, mercury lamps. Dangerous	**Sources** Hot objects, Sun, fluorescent substances lasers (see page 203).
Detector Geiger-Muller tube (see page 373).	**Detector** Photographic film.	**Detectors** Photographic film; causes sun tan, fluorescent substances glow.	**Detectors** Eyes, photographic film, LDR (see page 351).

All travel at the same **speed**, in a vacuum, of 3×10^8 m/s (300 million metres per second).

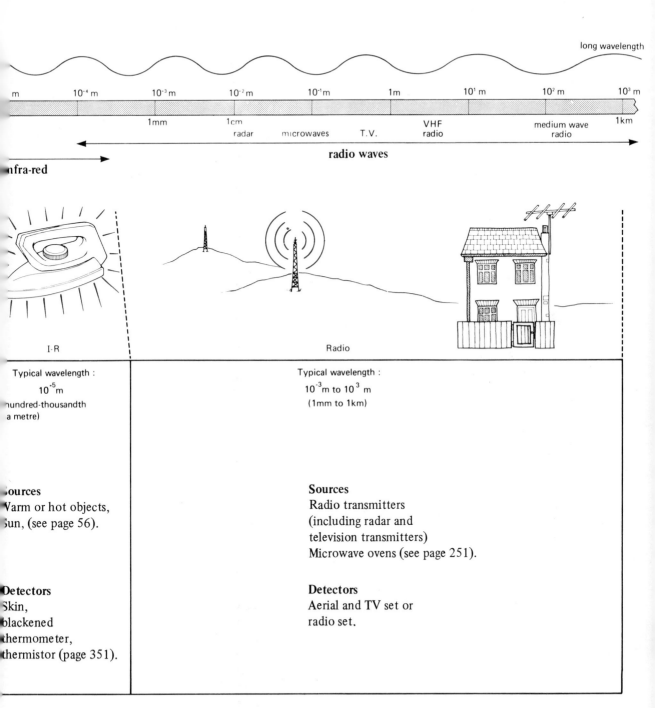

long wavelength

| m | 10^{-4} m | 10^{-3} m | 10^{-2} m | 10^{-1} m | 1m | 10^{1} m | 10^{2} m | 10^{3} m |

1mm   1cm   VHF   medium wave   1km
                   radar   microwaves   T.V.   radio   radio

radio waves

infra-red

I-R   Radio

Typical wavelength :
10^{-5} m
(a hundred-thousandth
of a metre)

Typical wavelength :
10^{-3} m to 10^{3} m
(1mm to 1km)

Sources
Warm or hot objects,
Sun, (see page 56).

Sources
Radio transmitters
(including radar and
television transmitters)
Microwave ovens (see page 251).

Detectors
Skin,
blackened
thermometer,
thermistor (page 351).

Detectors
Aerial and TV set or
radio set.

Velocity = frequency × wavelength (see page 175)

225

Subtraction of colour (absorption)

(In some of the diagrams on the next three pages, where a colour is marked *, it will help you if you use a felt-tipped pen to colour it the right colour.)

Filters

Experiment 28.4
Look through a red plastic filter. What happens to white light as it passes through the red filter?

Remember white light is really a mixture of several colours (ROY G BIV).
Which one of these colours passes through a red filter? Which colours are *absorbed* (and subtracted) by a red filter?

What happens as white light passes through a *blue* filter?

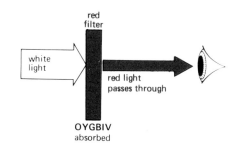

Experiment 28.5   Two filters
Place a red filter and a green filter together and look through them.
What happens to the white light? Why?

Try the same experiment with red and blue filters and then green and blue filters. (In practice most filters are not perfect, so some light may get through both.)

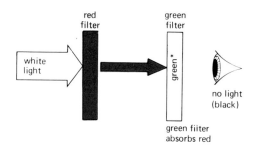

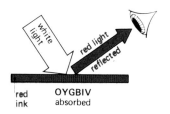

Coloured objects in white light
When white light shines on this white paper, the white light is reflected to your eye. When white light shines on this black ink, all the light is absorbed and none is reflected.

What happens when white light shines on this red ink (or red paint or red cloth)?

White light (really ROY G BIV) shines on the red ink which absorbs OYGBIV leaving only red to be reflected to your eye and so the ink appears red.
(In practice, red paint usually reflects small amounts of other colours as well.)

What happens when white light shines on to *blue* ink?

What happens if you mix red, green and blue inks together? Why?

226

Coloured objects viewed through filters

Experiment 28.6

Look at white paper through a green filter.
Look at a green book through a green filter.
What do you see?

Look at a red book through a green filter.
What do you see?
Explain why this is so.

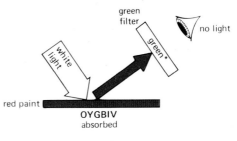

Use other filters and brightly coloured objects to fill in a copy of this table:

Colour of object	Filter	Appearance
white	green	green
green	green	green
red	green	black
blue	red	

Look at the cartoon through a red filter and then through a green filter. Explain what you see.

Coloured objects in coloured light

Experiment 28.7

Place a red filter over a raybox or a torch and shine the red light on to a bright blue object (in a dark room).

What do you see?
Explain why this is so.

Use different colours and fill in a table like the one above.

Look at the cartoon in red light and in green light and explain what you see.

Mixing paints

When an artist mixes two paints, the final colour is the one which is not absorbed by either of the paints.

Use the diagrams to explain why yellow + blue = green.

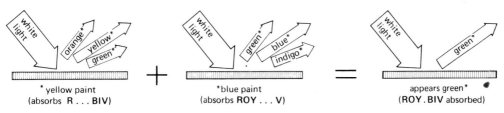

227

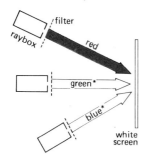

Addition of colours

Experiment 28.8   Primary colours

Use 3 rayboxes (or slide projectors) fitted with a red, a green and a blue filter and shine them on to the same white screen so that the colours overlap and add together.

If you adjust the brightness of the lamps or alter the distances correctly, you can make the three colours add up to make the screen appear *white*.
Because they add together to make white light, **Red, Green and Blue are called the primary colours of light**.

Experiment 28.9   Secondary colours

Switch on the lamps in pairs so that you see the result of adding *two* primary colours. These colours are called *secondary* colours.
Copy out and complete the table:

Primary colours added together	Secondary Colour
red + green	
red + blue	
green + blue	

The 3 secondary colours are *yellow*; a purple colour called *magenta* and a greeny-blue colour called *cyan*.

Switch on all 3 lights and put your hand in front of the screen to cause shadows.
Can you explain what you see?

Look closely at the screen of a colour TV set.
Can you see the red, green and blue dots?
Why are these three colours used?

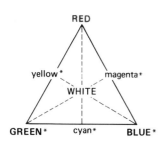

The colour triangle diagram shows the primary and secondary colours and also pairs of *complementary* colours which add together to form white. e.g. red + cyan = white; green + magenta = white.
Can you name two more colours that are complementary?

Summary

The spectrum of white light can be remembered by ROY G BIV. (Red has the longest wavelength and is deviated least by a prism.)

The full electromagnetic spectrum in order of increasing wavelength is: gamma-rays, X-rays, ultra-violet, visible light, infra-red, radio waves.

Subtraction of colours (absorption): filters, paints, pigments and inks subtract colours. e.g. a blue flower absorbs ROYG, IV and reflects blue light. In red light it looks black.

Addition of colours: the three primary colours (R,G,B) add to give white.
The secondary colours are yellow, magenta, cyan.

Questions

1. Copy out and complete:
   a) White light is composed of . . . . . colours, in order: red, ... , ... , ... , ... , ... , ... , which form the visible . . . . . The colour with the longest wavelength is . . . . . The colour deviated through the largest angle by a prism is . . . . .
   b) The full electromagnetic spectrum, in order, is: gamma-rays, ... , ... , ... , ... , ... The section with the longest wavelength is . . . . . ˙
   c) A red filter allows . . . . . light to pass through but . . . . . the other colours. Blue paint reflects . . . . . light but . . . . . the other colours.
   d) The three primary colours of lights are . . . . . , ... , ... and when added together they give . . . . . light.
   Red and green light added together give . . . . . light, called a . . . . . colour. The other . . . . . colours are . . . . . and . . . . .

2. The diagram shows white light being dispersed by a prism.
   a) What colour would you see (i) at X (ii) at Y?
   b) What may be detected (i) above X (ii) below Y?

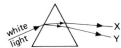

3. On the chart on pages 224–225 find the wavelength of your favourite radio station.

4. Why does lipstick look unpleasant under yellow street lighting?

5. Copy out and complete the table:

Colour of object in white light	Colour of filter used to view object	Appearance of object
white	red	
green	green	
green	red	
blue	magenta*	
red	cyan	
yellow	green	

(* a magenta filter passes red & blue)

6. Copy out and complete the table:

Colours of lights added together	Appearance
red+green+blue	
red + green	
red + blue	
blue + green	
blue + yellow	
green + magenta	
red + cyan	

7. Write essays on the following topics (or discuss them in class).
   a) The use of coloured spot lights for different stage effects.
   b) The choice of colours for road signs, advertisements, clothes and rooms.
   c) The use of colour in language e.g. feeling blue/red with rage/rose-coloured spectacles/ green beginner etc.

8. His face is red, he's feeling blue,
   Can't see his green mistake — can you?

Further questions on page 256.

229

Multiple choice questions on light

Most examination papers include questions for which you are given 5 answers (labelled **A, B, C, D, E**) and you have to choose the *ONE* correct answer.

Although the answers are given, the questions are not always as easy as they look. Take care to read *all* the answers before you choose the one that you think is correct.

Example
The metre is a unit of

    A   speed
    B   force
    C   mass
    D   length
    E   time

The correct answer is **D**.

Now write down the correct letter for each of the following questions (but do not mark these pages in any way).

1. A pin-hole camera is aimed at a tree. The image seen in the camera is found to be blurred. This is because
    A   the tree is too near the camera.
    B   the screen is too near the pin-hole.
    C   not enough light is coming through the pin-hole.
    D   the pin-hole is too large.
    E   the pin-hole is too small.

2. Which of the following statements, about the image in a plane mirror, is NOT correct?
    A   It is real.
    B   It is erect.
    C   It is laterally inverted.
    D   It is the same size as the object.
    E   It is as far behind the mirror as the object is in front.

3. Which one of the following statements about curved mirrors is NOT correct?
    A   A convex mirror is often used as a driving mirror.
    B   A convex mirror is often used as a shaving or make-up mirror.
    C   A concave mirror is often used as the objective of a reflecting telescope.
    D   A concave mirror is often used in a searchlight or a torch.
    E   A concave mirror is often used as the reflector behind the lamp in a projector.

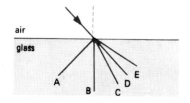

4. The diagram shows a ray of light passing from air into glass. Which line shows the correct path of the ray?

Questions 5, 6, 7

In each of the diagrams below, two rays of light are shown as they are reflected by a mirror or refracted by a lens. The mirror or lens is shown by a dotted line.

The mirrors and lenses available are

    A   plane mirror.

    B   concave mirror

    C   convex mirror.

    D   converging lens.

    E   diverging lens.

For each of the questions 5, 6 and 7, choose the correct mirror or lens.

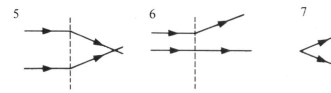

8. A boy is short-sighted, but not wearing spectacles. Which of the following statements is NOT correct?

    A   He cannot see distant objects clearly.

    B   Images of distant objects are formed in front of his retina.

    C   This is because his eye lens is too strong (or his eye-ball is too long).

    D   The image is inverted.

    E   The defect can be corrected by wearing converging lenses.

9. Which of the following shows the colours in order of increasing wavelength?

    A   Violet, green, yellow, red.

    B   Violet, yellow, green, red.

    C   Yellow, red, green, violet.

    D   Red, green, yellow, violet.

    E   Red, yellow, green, violet.

10. A blue dress with red spots is viewed in red light. It will appear

    A   completely red.

    B   completely magenta.

    C   blue with red spots.

    D   magenta with red spots.

    E   black with red spots.

When you have finished,

    a) check all your answers carefully

    b) discuss your answers (and why you chose them) with your teacher.

Have you started revising for your examinations yet?
See page 392.

More questions can be found in *Multiple Choice Physics for You.*

chapter 29

more about

Waves Waves Waves

Diffraction

Do waves change shape as they go through a gap in a wall?

Experiment 29.1   Wide gap
Use a ripple tank (see page 175) to send straight waves across water to a wide gap between two barriers.

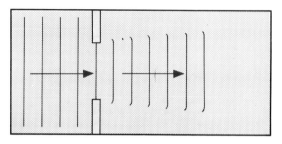

Look at the waves after they have passed through the gap. Are they still perfectly straight?

Experiment 29.2   Narrow gap
Now send straight waves through a narrow gap. What happens to the shape of the waves?

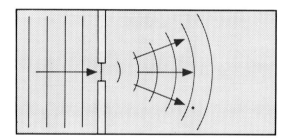

In the first diagram, with a wide gap, the waves travel almost straight on (*rectilinear propagation*). There is a clear shadow behind each barrier.
In the second diagram, the waves spread out after passing through the gap. This is called *diffraction*.

Diffraction is most obvious when the *width* of the gap is about the same size as the *wavelength* of the waves.

What happens when sea waves pass through the narrow entrance of a harbour?

Sound waves are diffracted through doorways because the wavelengths of sound are about the same size as doorways. (This is one reason why you can hear people without seeing them.)

Radio waves can also be diffracted. Your teacher may be able to show you this with 3 cm radar waves.

Light is seen to be diffracted only if it passes through a very narrow slit. This shows us that light is a wave motion with a very small wavelength.

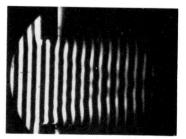

Ripple tank photographs

Interference of water waves

What happens when two waves are *superposed* – in the same place at the same time.

Experiment 29.3
In a ripple tank, use two dippers to make two circular waves that spread out and overlap.

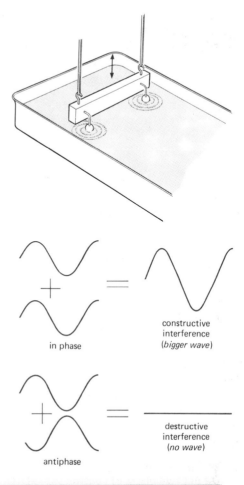

Look carefully where the waves overlap.
What happens when the *crest* of one wave overlaps the *crest* of another wave?
What happens when the *crest* of one wave overlaps the *trough* of another wave?

We call this effect *interference*.

At some places in the ripple tank, the two waves are arriving in step or *in phase* so that the two crests overlap. At these places (called *antinodes*) the waves add up to give a bigger wave.
This is called **constructive interference**.

in phase — constructive interference (*bigger wave*)

At other places in the ripple tank, the waves are exactly out of step (out of phase or in *antiphase*). At these places (called *nodes*) one wave is trying to push the water up while the other wave is trying to push it down and so the water does not move. This cancelling effect is called **destructive interference**.

antiphase — destructive interference (*no wave*)

The photograph shows two water waves interfering. The sources of the waves (the dippers) are marked S_1 and S_2. Look at the photograph carefully:

Close to the dippers you can see the circular waves spreading out and then overlapping.

Can you see that there are lines of places where there is constructive interference (marked C).
In between, there are other lines where the water is not moving (destructive interference, marked D).

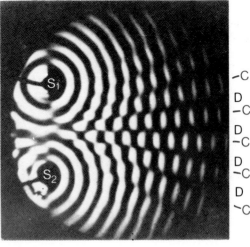

Interference can occur with water waves, sound waves or light waves.

Interference of light waves

Experiment 29.4
Use a ripple tank to send waves through a wall
with *two* gaps or slits (S_1 and S_2 in the diagram).

Do you get the same pattern as in experiment 29·3?
This time the waves are diffracted through each
slit (as in experiment 29.2) and then interfere (as
in experiment 29.3).

Experiment 29.5   Young's slits
This experiment is like the last, but uses light
waves.

Use a double slit made by scratching two fine
parallel lines on a piece of painted glass. The
distance between the slits should be $\frac{1}{2}$ mm or less.

Hold the slits vertically,
close to one eye and look
through them carefully
at a lamp with a thin
vertical filament.
What do you see?

You should see a group of vertical lines called *interference fringes*.
The bright lines are places of constructive interference (C), with
places of destructive interference (D) between them.

This experiment proves that light is a wave motion.
It was first done by Thomas Young in 1801.

Hold a red filter in front of the double slit and look at the distance
between the red fringes.
Then use a green filter instead. Are the fringes closer together?
What happens with a blue filter?

As we will see on the opposite page, the red fringes are farther apart
because red light has a longer wavelength than blue light (see p. 224).

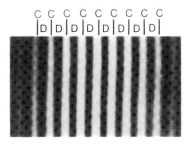

Interference fringes using red light

What does *monochromatic* mean?
When you used white light, why were the fringes blurred with
coloured edges?

Formula for Young's slits

In the diagram (not to scale) S_1 and S_2 are the slits. B is the mid-point of S_1S_2 and C is the central bright fringe. A is the first bright fringe next to the central one.

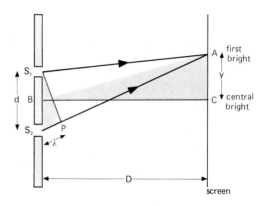

At A, the two rays S_1A and S_2A overlap to give constructive interference.
For the waves to arrive crest-on-crest, in phase, S_2A must be exactly one wavelength longer than S_1A.

The symbol for wavelength is λ (lambda).
Point P is marked so that $S_1A = PA$, and so $S_2P = \lambda$.

The two pink triangles (ABC and S_2S_1P) can be shown to be almost equi-angular (similar triangles).

Therefore:

$$\frac{S_2P}{S_1S_2} = \frac{AC}{BA}$$

from the diagram:

$$\frac{\lambda}{d} = \frac{y}{D} \qquad \therefore \boxed{\lambda = \frac{y \times d}{D}}$$

Experiment 29.6   Measuring the wavelength of light

With this formula, you can find the wavelength of light using this apparatus:

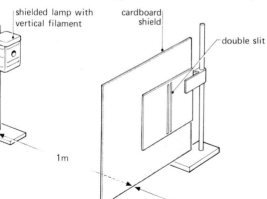

You must use a very dark room and shade the screen from any stray light.

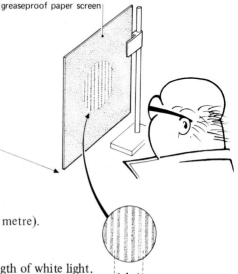

There are three measurements to take:

1. *The separation of bright fringes (y)*
   Mark two places on the screen which are 4 bright fringes apart, as shown.
   In daylight, measure this distance $(4y)$ and so find y (in metre).

2. *The slit separation (d)*
   In daylight, put the double slit and a $\frac{1}{2}$ mm rule side by side and view them through a magnifying glass. Estimate the value of d (in metre).

3. *The distance of the screen from the slits (D).* Use a metre rule.

Put your values in the formula and get a value for the average wavelength of white light. Does it agree with the value on page 224?

235

Diffraction grating

We have seen what happens with one slit (page 232) and with two slits (page 234). Now see what happens when we have a large number of slits placed very close together.

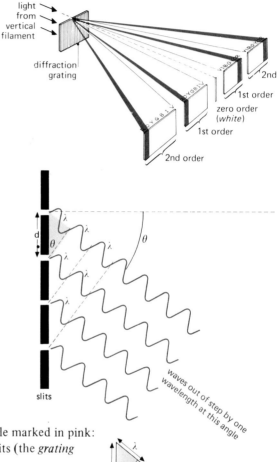

Experiment 29.7   Diffraction grating
Repeat experiment 29.5 with the same vertical filament but looking through a *diffraction grating*. This is a sheet of plastic divided into about 300 slits *per millimetre*!

What do you see?
What happens if you rotate the grating?
How many *spectra* can you see?

Which colour is deviated through the largest angle – red or violet? Notice that this is the opposite to the spectrum formed by a prism (page 222).

Each slit is very narrow so the light waves diffract and spread out from each slit. Consider red light of a particular wavelength (λ). At a certain angle of deviation (θ), the wave from each slit will be out of step with its neighbours by exactly *one wavelength*, as shown here:

If these waves are collected by your eye lens then they will add up to give a bright light at this particular angle (constructive interference).

Other wavelengths will not add up at this angle. However they will add up at some other angle (depending on their wavelength).

Looking more closely at the small right-angled triangle marked in pink:
– one side is d, the distance between two adjacent slits (the *grating space*)
– another side is λ, because these waves are out of step by one wavelength
– one of the angles is θ, the angle of deviation of this *first order spectrum*

Therefore: $\sin \theta = \dfrac{\lambda}{d}$ or $\lambda = d\, sin\, \theta$

For a *second order* spectrum, the waves would be out of step by 2λ and so the formula would be $2\lambda = d \sin \theta$.

Experiment 29.8   Measuring the wavelength of light
We can use the formula to find λ if we know *d* and θ.

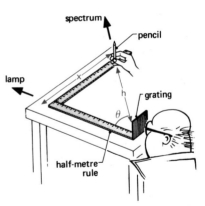

Arrange the apparatus as shown and look at a distant lamp with a
vertical filament.
Move the pencil along the rule until it appears to be in the middle
of the first order spectrum.
Measure *x* and *h* and calculate $\sin \theta = \dfrac{x}{h}$

If the diffraction grating has 300 lines per mm,
then $d = \frac{1}{300}$ mm $= \frac{1}{300\,000}$ m.

Now calculate the average wavelength $\lambda = d \sin \theta$.
In you have time, do the experiment for red light and blue light separately.

Summary

Waves can be diffracted through a narrow opening.

Two or more waves can interfere constructively or
destructively.

For Young's double slits: $\lambda = \dfrac{yd}{D}$

For a multiple slit grating: $\lambda = d \sin \theta$ (1st order)

Questions

1. a) When waves travel through a narrow gap in
       a wall, they are . . . . . This effect is most
       obvious when the . . . . . of the gap is
       similar to the . . . . .
    b) If two waves overlap and are in step
       (in . . . . .) then they interfere to give a
       . . . . . wave (. . . . . interference)
    c) If two equal waves overlap and are
       exactly out of step (in . . . . .) then they
       interfere to give . . . . . wave (. . . . . inter-
       ference).
    d) In Young's . . . . . slit experiment, inter-
       ference . . . . . are seen. The formula is:
    e) When white light passes through a . . . . .
       -slit . . . . . grating, ... are seen. Red
       light is deviated . . . . . than blue light.
       The formula is:

2. The diagram shows a double slit and a screen.
    C is the central fringe, D is the first dark
    fringe and B is the first bright fringe.
    At which points are the waves
    a) in phase b) in antiphase?

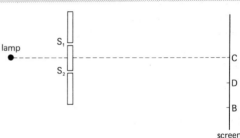

If λ is the wavelength, give the value of the
difference between
c) S_1C and S_2C d) S_1D and S_2D e) S_1B and S_2B.

3. Referring to the same diagram, describe the
    effect of a) using red light instead of green light
    b) increasing the distance from the slits to C
    c) increasing the distance S_1S_2
    d) covering up *one* of the slits.

4. In an experiment like 29.6, the slits are
    0·4 mm apart with the screen 2 m away. Four
    fringes take up a distance of 1·2 cm.
    Find the wavelength and colour of the light.

5. In an experiment like 29.8, a grating with
    300 lines per mm gives a spectrum such that
    x = 6 cm and h = 50 cm. Find the wave-
    length and colour of the light.

Further questions on page 256.

chapter 30

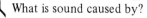

What is sound caused by?

Experiment 30.1
Hold down one end of a ruler firmly to the bench. Flick the other end to make a sound.

Look at the end of the ruler. What is it doing? Does it make any sound when it has stopped vibrating?

Experiment 30.2
Place your fingertips against the front of your throat. What can you feel when you make a noise?

Experiment 30.3
Get a tuning fork and bang it on a cork to make it vibrate (it vibrates like two rulers fastened together).

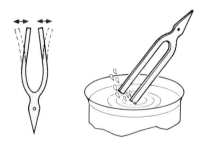

What can you hear?

Can you see that the ends are vibrating? If not, touch the ends to the surface of some water in a beaker. What do you see?

These experiments show that *sound is caused by vibrations* and is a form of kinetic energy (see page 128).
How are these vibrations caused in
a) a drum   b) a guitar?

How are the vibrations caused by a gramophone record?

Experiment 30.4
Look at an old gramophone record through a magnifying glass. Can you see the wobbly grooves?

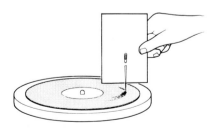

Push a needle through a piece of card and then put the point of the needle on to the record as it revolves on a turntable.

What can you hear?
What are the wobbly grooves doing to the needle and card?

How do the vibrations travel to your ear?

Experiment 30.5
Stretch a long 'slinky' spring
along a smooth bench and
vibrate one end to and fro along
the length of the spring, to send
a longitudinal wave down the
spring (see page 174 again).

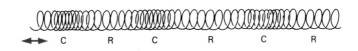

If you look closely at the spring you can see that,
at any instant, some parts of the spring are pushed
closer together (*compression*) and some parts are
pulled farther apart (*rarefaction*)

It is the same with a sound wave in air. In some
places the *molecules* of air are pushed together
at a slightly higher pressure (compression) and in
some places the molecules are farther apart at
a slightly lower pressure (rarefaction).

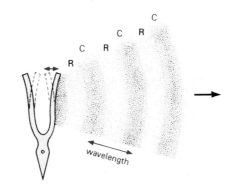

These compressions and rarefactions shoot out
across the room to your ear, travelling at the
speed of sound.

The *wavelength* of the sound is the distance between two successive compressions (or rarefactions).

For sound waves, like all other waves,

velocity = frequency x wavelength (see page 175)
  (m/s)   (Hz)   (m)

What happens if there are no molecules?

Experiment 30.6
Hang an electric bell inside a jar connected to a
vacuum pump.
Switch on the bell. Can you hear it?

Start the pump to take the air molecules out of
the jar. What happens to the sound of the bell?

Can sound travel through a vacuum? Why can't
we hear the sound of the explosions on the Sun?

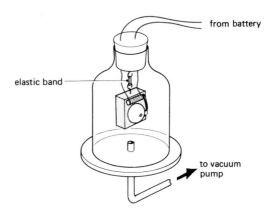

Reflection of sound

Have you ever heard an *echo*?
Can sound be reflected by walls?
Why do noises sound louder in a room than in the open air?

You may have seen that some of the compressions on the slinky spring (in experiment 30.5) were reflected back down the spring. In the same way, compressions and rarefactions are reflected back from the walls of this room.
If the distance is long enough you may hear a clear echo.

Experiment 30.7
Put a watch with a loud tick inside a cardboard tube on the bench. Put another tube nearby as shown and listen through it (in a very quiet room).
Can you hear the watch?

Now place a sheet of glass or hardboard near the ends of the tubes as shown.
Can you hear the watch now?

Turn the sheet until the watch sounds loudest.
What do you notice about the angles?

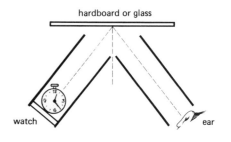

Echo-sounding.

Ships can use echoes to find the depth of the sea.

Example
A ship sends out a sound wave and receives an echo after 1 second. If the speed of sound in water is 1500 m/s, how deep is the water?

Time for sound to reach the bottom = ½ second (and ½ second to return)

∴ Depth of water = ½ x 1500
= 750 metres

Echo sounding is also used by ships to detect submarines and by bats and dolphins who make noises and listen to the echoes to 'see' their surroundings (see the examples on page 250).

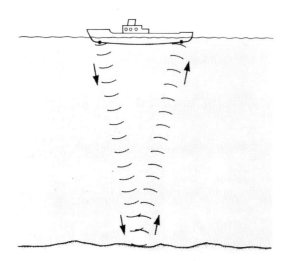

Measuring the speed of sound

Experiment 30.8   Outdoors

Stand a measured 50 metres from a large wall.
Bang two metal rods together and listen to the
echo. Then try to bang in an even rhythm of
bang-echo-bang-echo-bang . . . . . while a friend
times 100 of your bangs with a stopwatch.
During the time from one bang to the next bang,
the sound would have time to go to the wall and
back, **twice** — that is a distance of 200 metres. In
the time of 100 bangs, the sound could travel 200
x 100 = 20 000 metres.

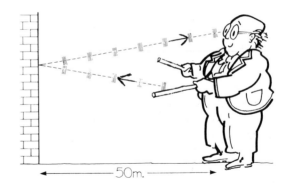

∴ Speed of sound = $\dfrac{\text{distance travelled}}{\text{time taken}} = \dfrac{20\,000 \text{ metres}}{\text{time in seconds.}}$

Experiment 30.9   Indoors

An alternative method, for indoors, is to use sound switches with an electronic timer.

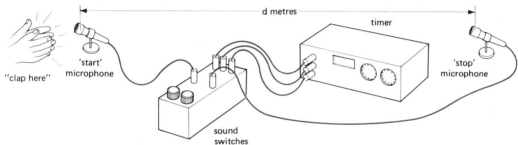

The sound arriving at the first microphone switches *on* the timer and
the same sound arriving at the second microphone switches it *off*.
The timer shows the time for the sound to travel the distance d.

∴ Speed of Sound = $\dfrac{\text{distance travelled (d)}}{\text{time taken}}$

At 0°C the speed of sound in air is 331 m/s (1200 km/hr or 740 m.p.h.)
At room temperature, its speed in air is faster, about 340m/s.
If sound travels 1 kilometre in 3 seconds, how can you use lightning
and thunder to find your distance from a storm?

Experiment 30.10

Measure the speed of sound in wood by
clamping the microphones (of experiment 30.9)
face down to a wooden bench. What do you find?

Sound travels faster through solids than through
gases (or liquids) as you can see in the diagram.

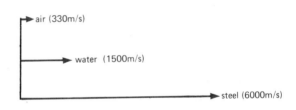

The ear

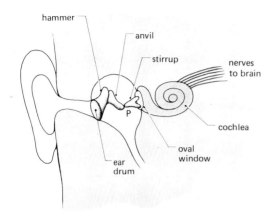

Sound waves are collected by the outer ear and passed in to the *eardrum* which is made to vibrate by the compressions and rarefactions.

These vibrations are passed to the *oval window* by three bones (called the *hammer, anvil* and *stirrup*) which act as a lever (with the pivot at point P). This means that they magnify the force of the vibrations (see page 140).
Also, the oval window has a smaller area than the eardrum, so this increases the pressure on the oval window and on the liquid in the *cochlea* (see page 97).

The vibrations of the liquid in the cochlea affect thousands of *nerves* which send messages to your brain.
What range of *frequencies* can you hear?
(See page 175).

Experiment 30.11
Connect a loudspeaker to a signal generator (which produces different frequencies as the pointer is moved to different positions).

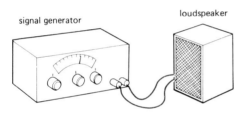

a) Turn the pointer to lower and lower frequencies. What happens to the *pitch* of the note from the loudspeaker?
What is the lowest frequency you can hear?
b) Turn the pointer to higher and higher frequencies.

What happens to the pitch of the note?
What is the highest frequency you can hear?
Does this vary from one person to another?

Young children can hear sounds as low as 20 hertz (20 cycles per second) and as high as 20 000 hertz but as you get older, this range becomes less.

Frequencies higher than 20 000 Hz are called *ultrasonic* frequencies. Many animals (including dogs and bats) can hear ultrasonic frequencies.

Doctor: Have your ears been checked lately?
Professor Messer: No, they've always been a pinkish colour.

Resonance

A swing with someone sitting on it has a certain *natural frequency* of vibration. If you are asked to push the swing to make it go higher, you will obviously push it each time it comes near you. That is, the frequency of your pushes will be the *same* as the natural frequency of the swing. Then the swing vibrates with a large amplitude.

This is an example of *resonance*. We say the swing is *resonating*. If you push at a different frequency, it will not swing as high (and you might hurt your hand).

Resonance occurs when

> the applied frequency of the pushes $=$ the natural frequency of the object

A short-sighted singer called Groat, Could do wonderful things with his throat.

At his specs he aimed sound, And with resonance found, That they cracked when he sang the right note.

Experiment 30.12

Hang a weight from a length of string (like a pendulum) and then use a straw to make it swing with a large amplitude

a) by giving it one strong tap.
b) by giving it lots of little taps at just the right frequency.

Which way gives the bigger swings?

In a similar way, if soldiers march in step over a bridge, they can make the bridge vibrate so much at its natural frequency that it may break.

A large bridge in America once fell down because it resonated with vibrations caused by the wind.

Have you heard windows or seats on a bus start to rattle (as they resonate) when the engine speeds up to a certain frequency?

If a singer sings near a wine glass with a frequency equal to the natural frequency of the glass, it may resonate so strongly that the glass breaks.

Pitch, loudness and quality of musical notes

The notes from a musical instrument can vary in three ways:

1. in *pitch*
2. in *loudness*
3. in *quality* or *tone*.

What do these three things depend on?

1.   Pitch

We have already seen (in experiment 30.11) that
the pitch of a note depends on the frequency.

Experiment 30.13
An oscilloscope (a kind of television set, see page 342) is very useful for showing vibrations.
Connect a microphone to an oscilloscope and hold a tuning fork nearby.
When the oscilloscope is correctly adjusted, it shows on the screen a *transverse* wave of the *same frequency* as the *longitudinal* sound wave from the tuning fork.

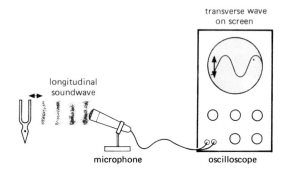

longitudinal soundwave

transverse wave on screen

microphone   oscilloscope

Experiment 30.14
Look at the waves on the screen caused by a low-pitched tuning fork and a high-pitched tuning fork. What do you notice?

2.   Loudness

Experiment 30.15
Whistle a soft note into the microphone and then the same note but louder. Look at the waves on the screen. What do you notice?

The loudness depends upon the amplitude of the wave.
A wave with a larger amplitude contains more energy.

The loudness of a sound also depends on how much air is made to vibrate.

Experiment 30.16
Repeat experiment 30.4 with the gramophone record using a small piece of card and then a large piece of card. Which sounds louder?

low pitch
(low frequency)
long wavelength

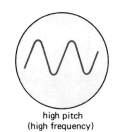

high pitch
(high frequency)
short wavelength

soft note

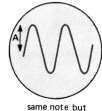

same note but
LOUDER

3. Quality (or tone)

If a violin and a piano play the same note (at the same pitch and the same loudness), you can still tell them apart because the notes have a different *tone* or *quality*.

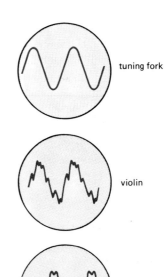
tuning fork

violin

piano

Experiment 30.17
Play different musical instruments in front of a microphone connected to an oscilloscope.
Play the same note on each instrument and sketch the *waveform* that you see.

What do you notice?

The quality of a sound depends upon the waveform.

What causes these different waveforms?

A tuning fork produces a pure note with only one frequency.
Other instruments usually produce many frequencies or *harmonics* at the same time.
All the harmonics add together to give a complicated waveform.

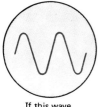 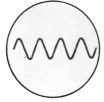 

If this wave (the first harmonic)	is added to	this wave of **twice** the frequency (the second harmonic)

the result is this waveform

If more harmonics are added in, different waveforms are produced.
Different instruments give different harmonics and waveforms depending on the shape and size of the instrument, and so they give different sounds.

Experiment 30.18
Look at the oscilloscope while you sing or hum into the microphone.
Try to keep the same pitch and loudness but change the shape of your mouth with your lips and your tongue. What do you notice?

Musical instruments

1. Pipes

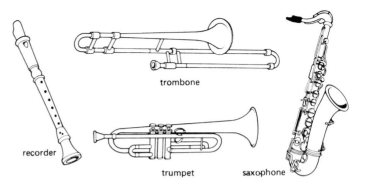

trombone

recorder

trumpet   saxophone

The diagrams show some pipe
instruments.
These make a sound because
the air inside the pipe is made to
vibrate and resonate at its
natural frequency.

**How does the natural frequency (pitch) depend
upon the length of the pipe?**

Experiment 30.19
Flatten the end of a drinking straw and then trim
the corners with scissors as in the diagram. Put
the trimmed end in your mouth and make a steady
note by blowing.

Now cut bits off the other end to shorten the
pipe and listen to the note. What happens to the
pitch (frequency) of the note?

A shorter pipe produces a higher pitch (frequency)
The sound is produced by the trimmed end acting as a *reed*. Which
of the instruments at the top of the page uses a reed?

Experiment 30.20
Set up a row of test-tubes and fill them with
different amounts of water, so that the length of
air is different in each tube.
Blow gently across the top edge of each tube to
make different notes. See if you can adjust the
levels to make a musical scale.

Which tube produces the highest frequency (pitch)?
Which of the instruments at the top of the page produces sound
by blowing across an edge like this?

How are the vibrations produced in a trumpet and a trombone?

Why do different pipes have a different tone (quality)?

2. Strings

The diagrams show some string
instruments. They make sounds
because each string vibrates at
its own natural frequency (pitch).

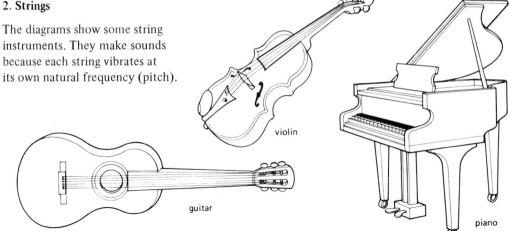

violin

guitar

piano

What does the natural (resonant) frequency of a string depend on?

Experiment 30.21
Use a guitar (or a sonometer, a kind of one-string guitar).

a) Pluck a string and listen to the note. Then use
your finger to shorten the length and pluck it again.
What happens to the pitch (frequency)?
b) Use a screw at the end to increase the tension.
What happens to the pitch (frequency)?
c) Pluck a light thin string and then a thick heavy
string (with about the same tension and length).
What do you find?

The natural frequency (or pitch) of a string can be
increased by
– *shortening* the length
– *increasing* the tension
– using a *lighter* string.

Of the three instruments at the top of the page,
which makes the strings start to vibrate by
– plucking?
– scraping?
– hitting?

Why do different stringed instruments have a
different quality (tone)?

increasing the frequency:

2. increase
the tension

1. shorten
the length

3. use a lighter
string

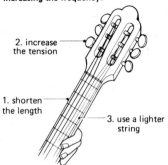

Professor Messer: If you drop a piano down a mineshaft, what key does it play in?
Mrs Messer: A flat minor?

Something to do—a string telephone

Get two clean tin-cans and use
a nail to punch a hole in the
bottom of each can.
Connect the cans together by a
long piece of string (by thread-
ing it through the holes and tying
knots).

Pull the string tight and listen carefully to one can while someone
speaks into the other can. How is the sound passing from one can
to the other? What happens if the string is slack? Why?

Summary

Sound is caused by vibrations and cannot travel
through a vacuum.
A sound wave consists of compressions and
rarefactions of the air.
Speed of sound = frequency x wavelength.

Echoes are caused by the reflection of sound.
Resonance occurs when the applied frequency
equals the natural frequency of the object.

The pitch of a note depends on the frequency.
Loudness depends on the amplitude.
Quality depends on the waveform.

The natural frequency of a pipe is increased as
the pipe is shortened.
The natural frequency of a string is increased as
the length is shortened; as the tension is increased;
as the mass of the string is decreased.

Questions

1. Copy out and complete:
    a) Sound is caused by . . . . .
    b) A sound wave consists of places at higher
       pressure (called . . . . .) and places of
       . . . . . pressure (called . . . . .)
    c) Velocity (in metres per second) equals
       frequency (in . . . . .) multiplied by . . . . .
       (in . . . . .).
    d) Sound cannot travel through a . . . . .
    e) Echoes are caused by the . . . . . of sound.
    f) The speed of sound in a solid is . . . . .
       than the speed of sound in air.

    g) Resonance occurs when the applied . . . . .
       of the pushes equals the natural . . . . . of
       the object.
    h) Pitch depends on . . . . .
       Loudness depends on . . . . .
       Quality depends on . . . . .
    i) If the length of a pipe is decreased, its
       natural frequency is . . . . .
    j) The natural frequency of a string can be
       increased by . . . . . the length, ... the
       weight of the string or . . . . . the tension.

2. The speed of sound is 340m/s. If thunder is heard 20 seconds after a lightning flash, how far away is the storm? What must you assume?

3. If the speed of sound is 340m/s, what is the wavelength of a sound wave of frequency 100 Hz?

4. a) What are the highest and lowest frequencies that the human ear can detect?
b) What are the shortest and longest wavelengths that the human ear can detect?
(Speed of sound = 340m/s).

5. a) Describe an experiment to show that sound cannot travel through a vacuum.
b) Why is the moon a silent world? If their radios broke down, how could two astronauts talk to each other using only a piece of string?

6. A man fires a gun and hears the echo from a cliff after 4 seconds. How far away is the cliff? (Speed of sound = 340m/s).

7. A sound wave sent out by a boat arrives back after 3 seconds. If the speed of sound in water is 1500 m/s, how deep is the water?

8. Explain how echoes are used
a) by a ship to find the depth of the water
b) by a ship to find shoals of fish or a submarine
c) by geologists to find likely places to drill for oil
d) by bats and dolphins
e) by blind people fitted with special equipment.

9. A boy stands 90 metres from a wall and claps his hands to hear clap-echo-clap-echo at an even steady rate of 1 clap per second. What result does he get for the speed of sound?

10. Draw a large labelled diagram of a human ear, explaining how the energy of the sound wave is transmitted to the nerves in the cochlea.

11. Explain, with diagrams, what is meant by
a) amplitude   b) wavelength   c) frequency
d) quality of a note   e) resonance.

12. The diagram shows a graph of a sound wave given by a tuning fork.

Copy this diagram and then draw graphs to show:
a) a sound wave of higher pitch but the same loudness
b) a sound wave of the same pitch but louder
c) a sound wave of the same pitch but given by a different instrument.

13. Explain three methods by which the note from a guitar may be lowered in pitch.
How could the note be changed in   a) loudness
b) quality (tone)?

14. A man is kidnapped, blindfolded and imprisoned in a room. How could he tell if he was in   a) a bare room   b) a furnished room
c) a town   d) the country?

Further questions on page 257.

Physics at work: Ultra-sonic echoes

Humans cannot hear sounds which have frequencies above 20 000 Hz, but many animals can. Dogs will respond to an ultrasonic whistle even though it seems silent to us.

Bats and dolphins use ultrasonic frequencies of about 150 000 Hz to 'see' by listening to the echoes. Echo-sounders on ships (see page 240) also use a very high frequency.

There is a reason for using a very high frequency: a high frequency means a *short* wavelength. To see this we can use the formula: **velocity = frequency x wavelength** (see page 239) to calculate the wavelength. The speed of sound in water is 1500 m/s, so an *audible* sound of frequency 1500 Hz has a wavelength of 1 metre. With this long wavelength it is difficult to detect small objects (because the wave diffracts and bends round them, see page 232). However if the sound has an ultrasonic frequency of 150 000 Hz, the wavelength is only 0.01m (= 1cm). With this shorter wavelength the sound is reflected back from smaller objects and so the echo can be timed (and the distance to the object calculated).

The ultrasonic beam is also much more directional and can be aimed rather like a torch. This has been used in ultrasonic spectacles for blind people. The spectacles have a transmitter and a receiver. The receiver produces a high or low sound in the person's ear depending on whether the object causing the echo is near or far.

Echo sounding is also used to detect flaws inside pieces of metal. A transmitter sends out pulses of ultrasound and a receiver picks up the echoes from different parts of the metal and shows the results on a CRO (cathode ray oscilloscope, page 342): Pulse A is the transmitted pulse; pulse B has been reflected by the flaw (b) in the metal; pulse C is the echo from the end (c) of the metal.

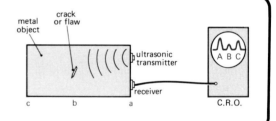

Ultrasonic echoes are also used in medicine (instead of X-rays, which can be dangerous).
Unborn babies can be seen by moving an ultrasonic transmitter/receiver across the mother's stomach. Different tissues (skin, muscle, bone) reflect the sound waves to produce a lot of echoes. The machine uses these echoes to build up a picture on the screen of a CRO:

A similar machine allows Physicists to 'see' what is happening to the components inside a nuclear reactor.

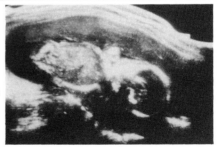

An ultrasound picture of an unborn baby (19 weeks after conception)

Physics at work: Electromagnetic Waves

Electromagnetic waves can have very short or very long wavelengths (see page 224). Different wavelengths have different properties and different uses. Different wavelengths give us different information about a hand:

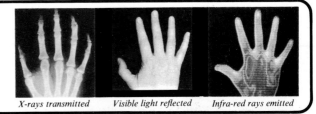

X-rays transmitted   *Visible light reflected*   *Infra-red rays emitted*

X-rays

X-rays can be used in medicine (see page 344).

They can also be used in detective work. This old painting has been photographed in visible light and then using X-rays:

You can see that the artist changed his mind and then painted the head in a new position.

Ultra-violet

Ultra-violet waves can also be used in detective work, in *forensic science*. The photograph shows a cheque photographed in ultra-violet — you can see it is a forgery.

Some chemicals glow or *fluoresce* in ultra-violet rays. Washing powders include these chemicals to make clothes look brighter in sunlight.

In visible light, and in ultra-violet:

Infra-red

Some uses of infra-red waves were shown on page 58 and 59. Satellite photos using infra-red can tell the difference between fields of healthy and diseased crops.

Missiles can follow the infra-red from the hot engine of an aircraft. A remote control for your TV sends commands in infra-red rays.

Microwaves

A microwave oven uses radio waves to cook food very quickly. The microwaves are produced by a 'magnetron' and guided to a metal stirrer which reflects the waves into different parts of the oven. Microwaves are reflected by metal but absorbed by food.
The waves have a wavelength of 12cm and a frequency of 2500 Hz. At this frequency the electromagnetic waves are absorbed by water molecules in the food, heating them up and so cooking the food. The food turns on a turntable so it is cooked evenly.
The door has a wire mesh over the window (to reflect the microwaves back inside). The door must have a safety switch to turn off the microwaves if you open the door — otherwise you could cook your fingers!
Microwaves are also used for radar and for communicating with satellites.

A microwave oven

rotating 'stirrer'

metal wave-guide

magnetron produces microwaves

food on turntable

metal box to keep microwaves inside

window has fine metal mesh over it

Water waves

1. Describe a demonstration to illustrate the difference between transverse and longitudinal waves. Give an example of each, other than those you have used in your demonstration. (L)

2. Illustrate the reflection of a circular wavefront by a plane reflecting surface. (AEB)

3. A spherical wave, after reflection from a concave mirror, converges so as to pass through its point of origin. Explain, with the help of a wave-front diagram, where the point of origin must be situated for this to happen. (O&C)

4.

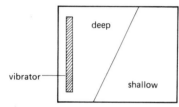

The diagram represents a plan view of a horizontal ripple tank, set up so that the water in part of the tank was shallower than that in the remainder. The vibrator was set to produce plane waves.

a) Ignoring reflections from the sides of the tank, draw **one** diagram to show the wave pattern you would expect to see (i) in the deep water, (ii) as the waves pass from deep to shallow water, (iii) in the shallow water.
Explain why the waves take the form suggested in your diagram.
If the vibrator is set to vibrate at 5 Hz and produces ripples 2 cm apart, calculate the speed at which the energy is carried over the water surface.

b) The waves produced in a ripple tank are transverse waves whereas those produced by a tuning fork in air are longitudinal waves. Explain how the two types of wave differ. (JMB)

There are more questions on waves on page 256.

Light rays

5. An object is placed in front of a pinhole camera. An image is seen at the centre of the translucent screen when viewed from behind. If the object is now moved slightly further away and to the observer's left, the image becomes

   A larger and moves to the observer's left.
   B larger and moves to the observer's right.
   C smaller and moves to the observer's left.
   D smaller and moves to the observer's right
   E slightly blurred and remains central.
   (O&C)

6. An illuminated arrow is 6 cm high and is placed 24 cm in front of a very small hole in a pinhole camera. The distance from the hole to the screen is 8 cm.
   The size of the image on the screen, in cm, will be
   A 12   **B** 6   **C** 3   **D** 2   **E** 1
   (O&C)

7. Describe experiments, one for each case, to show the formation of shadows produced by (a) a point (small) source of light, (b) a source of light approximately half the size of the object casting the shadow. Draw diagrams to show how the shadows are formed.
   (JMB)

8. Explain, with the aid of diagrams, how total, partial and annual eclipses of the sun are formed. (WAEC)

9. Explain, with the aid of a ray diagram, how the image is formed in a pinhole camera. Discuss how the size, sharpness and brightness of the image are affected by the diameter of the pinhole. (S)

10. Describe the essential features of a pinhole camera, and state one property of light that it demonstrates. What would happen to the image if
    a) the camera were made longer,
    b) the pinhole were triangular but still very small,
    c) the pinhole were increased in size? (L)

Plane mirrors

11. On a diagram, draw **two** rays from the point A to illustrate how the eye sees the image of the top of the candle.
    Also draw the image of the whole candle and mark with B′the image of the point B.

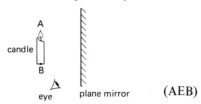

(AEB)

12. a) Describe an experiment which you could use to locate the position of the image of a pin in a plane mirror.
    b) Name **four** properties of this image.
    c) A plane mirror is frequently seen below the scale of pointer instruments such as ammeters. With the aid of digrams, explain why this is an advantage and describe how the mirror must be used to obtain accurate pointer readings.

(JMB)

13. The diagram represents the right eye R, and the left eye L, of a man standing in a room which has mirrors M_1 and M_2 on two walls at right angles. Make a diagram, preferably on graph paper, showing the positions of the images of R and L in the mirror M_1. Label these R_1 and L_1. Show also the positions of the images of R_1 and L_1 in the mirror M_2.

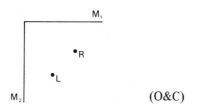

(O&C)

14. Draw a clear diagram to show how two mirrors may be used to make a periscope. You should show the paths of two rays of light through the instrument.

(EA)

Curved mirrors

15. An object 4·0 cm tall is placed 18·0 cm from a concave mirror of focal length 6·0 cm so that it is perpendicular to and has one end on, the principal axis of the mirror. **Either** by calculation **or** by *full-scale* ray drawing, determine
    a) the distance of the image from the mirror,
    b) the height of the image,
    c) whether the image is erect or inverted,
    d) whether the image is real or virtual.

(AEB)

16. An object is placed 15 cm in front of (i) a plane mirror, (ii) a concave mirror of radius 60 cm, and (iii) a convex mirror of radius 60 cm.
    a) Describe how you would locate the image in case (i) by experiment. State clearly the nature and relative size of this image, as well as its position.
    b) Draw scale diagrams for both (ii) and (iii) to show where the images are produced. State the nature and position of each image. State a practical application of each case.

(S)

17. a) Rays of light parallel and close to the principal axis are incident on a convex spherical mirror. Show on a diagram what happens to the rays after reflection, and account for their subsequent paths.
    b) An object 2 cm tall situated 10 cm from a concave mirror produces an erect image 6 cm tall and 40 cm *from the object*. Find, graphically, or by calculation, the radius of curvature of the mirror. State a practical use of this arrangement.

(L)

18. A concave mirror of radius of curvature 1·0 metre is used to light a piece of paper by focusing the sun's rays. If the diameter of the Sun is 1·2 million km, and its distance from the Earth is 150 million km, what is the diameter of the Sun's image on the paper?

(L)

Refraction

19.   The velocity of light in air = 3×10^8 m s^{-1}.
The refractive index of glass = 1·5. Calculate
a) the velocity of light in glass, and (b) the
apparent thickness of a glass block which is
actually 6 cm thick. What is assumed in your
answer for (b)?

(S)

20.   The diagram illustrates the refraction of a
beam of light by a semi-circular glass block.

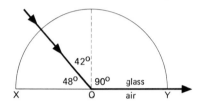

What is the angle of incidence at O?
What is the refractive index of the glass?

(AEB)

21.   A ray of light travels from glass to air
making an angle with the normal inside the
glass of 30°. The refractive index of the glass
is 1·60.
a) What angle does the ray make with the
normal as it enters the air? Show your
working.
b) What happens to this angle as the angle
inside the glass is gradually increased?
c) Calculate the critical angle at the glass/
air boundary.   (S16+)

22.   Draw a diagram showing how a glass prism
can deviate a narrow beam of light through a
right angle. Show how this forms the basis
for the design of a simple periscope.

(JMB)

23.   Explain the terms (a) *refractive index*
(b) *critical angle.*
Under what conditions does total internal
reflection take place?
On a hot summer's day the surface of a
tarmac road may appear to be wet and shiny
some distance ahead. Explain this with the
aid of a diagram.   (W)

Lenses

24.

The diagram illustrates an approximate
method for finding the focal length of a converging (convex) lens. The distance of the
lens from the screen was 20 cm, the height
of the clear image of the *distant* house was
2·0 cm. Given that the height of the house
was actually 8·0 m, find
a) the approximate focal length of the lens.
b) the distance of the house from the lens.

(AEB)

25.   What is the linear magnification produced by
a magnifying glass of focal length 15 cm
placed 10 cm from an object?   (O&C)

26.   An object 1·5 cm high stands upright on the
axis of a converging lens of focal length 4·0
cm. The object is placed in turn at distances
8 cm, 7 cm, and 2 cm from the lens. For
each of these positions determine (**either** by
careful drawing **or** by calculation) the
position, nature and size of the image. If you
use a calculation method state the sign convention which you use. Use graph paper if a
ray diagram is used.   (N16+)

27.   a) Describe two methods by which you
could determine the focal length of a converging (convex) lens,the first a quick but
approximate method, the second a more
accurate method.
b) An object 3·0 cm tall is placed 10·0 cm
from a diverging (concave) lens of focal
length 6·0 cm so that it is perpendicular
to, and has one end on, the principal axis
of the lens. **Either** by a ray drawing on
graph paper **or** by calculation using a
clearly stated sign convention, determine
the height, position and character of the
image.   (AEB)

Optical instruments

28. A lens camera has a lens of focal length 50 mm. The lens must be moved 2 mm (from its position for the photography of distant objects) if a close object is to be photographed.
    a) What is the distance between lens and film for the distant object?
    b) What is the distance between lens and film for the close-up shot?
    c) If the image on the film of the close object is 1/25th life-size, what is the distance of the object from the lens?
    (N16+)

29. a) Draw clearly labelled diagrams of
    (i) a single lens camera,
    (ii) the optical system of the human eye.
    List the points of similarity and difference between the two, in relation to focusing, image formation and image brightness.
    b) Explain with the aid of ray diagrams
    (i) short-sightedness,
    (ii) how this defect may be overcome by use of spectacles. (AEB)

30. Explain what is meant by a long-sighted eye and show how a suitable lens can correct this defect. (L)

31. A simple projector is used to project an image of a slide on to a screen. The slide is 20 mm high and is placed in a carrier 40 mm in front of a lens of focal length 30 mm. State, with a reason, the type of lens used. Draw a ray diagram, to a stated scale, to show how the lens projects an image of the slide on the screen.
    Use the diagram to find
    a) the height of the image on the screen,
    b) the distance of the screen from the lens,
    c) the magnification produced.
    State **two** changes which would have to be made to produce a larger magnification using the same lens.
    Describe how you would measure the magnification produced by this lens for a stated object distance. (C)

32. Draw a labelled diagram showing the arrangement of the optical components of a slide projector, showing where the slide is placed, and explain the role of each component. (O&C)

33. Draw a labelled ray diagram of a compound microscope, indicating the nature of both the intermediate and final images. Distinguish clearly between actual light rays and construction lines. (JMB)

34. Describe, using a ray diagram, a telescope which uses two converging (convex) lenses. State clearly the nature of the intermediate and final images and distinguish between rays of light and construction lines on your diagram. (JMB)

35. a) Draw a clear diagram to show the construction of one half of a prismatic binocular. Show the position of the lenses, prisms and the eye and the path of one **ray** from a distant object to the eye.
    b) What are the principal reasons for the use of prisms in a binocular?
    c) What are the main differences between the results obtained by using binoculars marked
    (i) 8 x 30.
    (ii) 8 x 50.
    (iii) 10 x 50.
    (S16+)

36. a) Name three types of astronomical telescope.
    b) Herschel discovered the planet Uranus with a 6 inch reflector telescope. The focal length of his telescope was 2 m.
    (i) How far from the objective mirror was the image of Uranus formed?
    (ii) What kind of mirror was used for the objective of the telescope?
    c) Draw a fully labelled diagram of a reflecting telescope. (NW)

The spectrum and colour

37. State **four** ways in which *light, heat radiation* and *radio waves* are similar and **two** ways in which those three types of radiation differ.

    (AEB)

38. a) State two forms of electromagnetic radiation, other than light.
    b) How may the forms of radiation given in (a) be detected? Outline one method for each case.
    c) Name **two** properties that all electromagnetic radiations have in common.
    d) Draw a labelled diagram of the electromagnetic spectrum, indicating the high frequency end.   (C)

39. Assuming that 'white light' consists of three primary colours, complete the following table relating to the appearance of different objects in different coloured lights.

Appearance in white light	red light	green light	yellow light
Magenta			
Yellow			
Green			
Black			

    (AEB)

40. a) What colour would an object which is white in daylight appear to be when illuminated by overlapping beams of red light and green light?
    b) What colour would an object which is red in daylight appear to be when illuminated by green light?
    c) What colour would an object which is magenta in daylight appear to be when illuminated by red light?
    d) If a beam of green light is incident on a cyan (peacock-blue) filter, what is the colour of the light transmitted?
    e) A beam of yellow light passes through a filter and emerges as a beam of red light. What colour could the filter be? Either
    . . . . . or . . . . .   (AEB)

Light waves

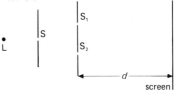

41. The diagram shows the arrangement of a monochromatic light source L, a single slit S and two parallel slits S_1 and S_2 set up to produce interference fringes on the screen. The distance between the screen and the plane of the slits S_1 and S_2 is d.
    a) Describe the pattern of fringes seen on the screen. Explain how the pattern arises.
    b) The following changes are now made separately. (i) the separation of S_1 and S_2 is reduced, (ii) the distance d is doubled, (iii) the monochromatic light source is replaced by a monochromatic source emitting light of greater wavelength. How will the pattern seen, in each case, differ from that described in (a)?

    (O&C)

42. The diagram indicates a beam of white light falling normally on a diffraction grating G, which has 600 lines per mm. Green light in the first order spectrum leaves the grating in the direction of the arrow and meets a distant screen at Y.

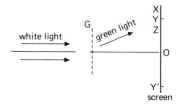

    a) What is the sequence of colours observed in the visible spectrum from X to Z?
    b) What can be deduced from this sequence about the relative wavelengths of different parts of the visible spectrum?
    c) What is observed in the region of Y', where $OY' = OY$?
    d) How would the pattern alter if a grating with 300 lines per mm were used in place of G?   (O)

Sound

43. When longitudinal waves pass through a medium the particles of the medium oscillate

A   up and down.

B   parallel to the direction of the wave.

C   to and fro.

D   perpendicular to the direction of the wave.

E   at 45° to the direction of the wave.

(NW)

44. a) Sound is produced by vibrating objects. Sound travels as a longitudinal wave at a speed of 330 m/s in air at normal pressure.

 (i)   What is meant by a 'longitudinal wave'?

 (ii) What is meant by 'frequency' of a sound wave?

 (iii) What is the wavelength in air of a sound wave of frequency 165 Hz?

 b) The quality of a note produced by a vibrating string depends on the overtones (or harmonics) present in the note.

 (i)   What does 'quality' mean in this sentence?

 (ii) What are overtones (harmonics)?

 (iii) Suggest **two** ways of producing a *higher* note from a stretched string.

 (iv) Two musical notes can differ in quality. In what other **two** ways can they differ?

 c) In a violin the strings are mounted on a hollow box. The air in the box vibrates as the strings vibrate, and so does the box. Explain the effect this has on the sound produced. (WM)

45. The sound produced by a musical instrument is picked up by a microphone and the resulting waveform displayed on the screen of a suitably-adjusted cathode ray oscilloscope. Sketch the waveforms you would observe when

 a) the instrument sounds a steady tone,

 b) the instrument sounds the same steady tone but louder,

 c) the instrument sounds a steady tone of higher pitch. (O&C)

46. A boy can hear sounds which range in frequency from 30 Hz to 20 000 Hz. Calculate the longest wavelength the boy can hear if the velocity of sound in air is 330 m/s. (W)

47. Describe an experiment to measure the speed of sound in air. State clearly (i) the observations you would make and the units used, (ii) how the observations would be used to calculate the final result. (JMB)

48.

Shoal of fish

The diagram above shows a fishing boat using sonar to detect a shoal of fish. A short pulse of sound waves is emitted from the boat, and the echo from the shoal is detected 1/10 s later. The sound waves travel through sea-water at 1500 m/s.

 a) How far has the pulse travelled in 1/10s? [2 marks]

 b) How far below the boat is the shoal of fish? [2]

 c) The reflected pulse lasts longer than the emitted pulse. Suggest a reason for this. [2]

(MEG)

Further Reading

Science and Crime Detection – Ellen (Wheaton)

The Discovery of the Universe – Osman (Usborne)

Stars and Planets – Kerrod (Kingfisher)

Lasers – Myring (Usborne)

Physics Plus: *Applications of Ultrasonics, Optical Fibres, Light and the Motor Car* (CRAC)

ELECTRICITY
and magnetism

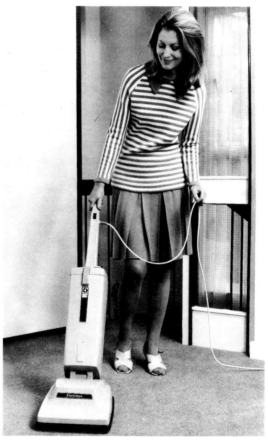

The photographs show some of the many uses of electricity and magnetism.

What difference would it make to your life if electricity and magnetism had not been discovered?

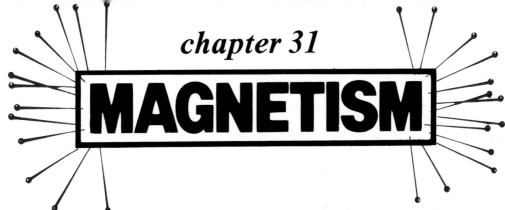

chapter 31

MAGNETISM

What is the quickest way of picking up some pins that have been spilt on the floor?

If a magnet is used, the pins stick to the ends or *poles* of the magnet. Can you think of any other uses for magnets?

Experiment 31.1
Place a bar magnet in a paper stirrup as shown and hang it from a wooden table or a wooden retort stand.

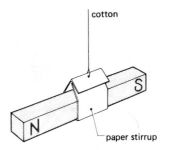

Leave it hanging until it stops moving. From a map or from the position of the sun, find out which direction is north.

Which direction does the magnet point to?

You have made a *compass*. A freely moving magnet comes to rest pointing roughly north-south (see also page 265).
The end pointing north is called the *north-seeking pole* or *north pole* (*N-pole*) of the magnet. The other is the *south-seeking pole* (*S-pole*).

Experiment 31.2
Mark the N-pole of the magnet in experiment 31.1. Then take that magnet well away and repeat the experiment with another magnet.

Now bring the N-pole of the first magnet close to the N-pole of the hanging magnet.

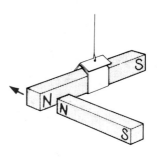

What happens?
What happens if you bring the S-poles close together?
What happens if you bring a N-pole near a S-pole?

We find that: **Like poles repel each other.**
                 Unlike poles attract each other.

All Physics students should learn well
This scientific fact:
Like magnetic poles repel
And unlike poles attract.

Making a magnet

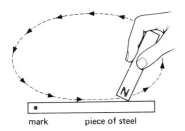

mark   piece of steel

A. Stroking method
Experiment 31.3
Mark one end of a piece of unmagnetised steel. Check that it is not magnetised by dipping it into a dish of small iron nails.

Stroke the piece of steel with the N-pole of a magnet. Start at the marked end of the steel and lift the magnet clear at the end of each stroke. Do this about 10 times in the same direction.

How many nails can the steel pick up now?
Is the marked end a N-pole or a S-pole?
(Use a small compass to test this, remembering that like poles repel.)

B. Electrical method
Experiment 31.4   An electromagnet
Wind about 1 metre of insulated copper wire round and round a large nail. Then connect the ends of the wire to a battery, as shown.

How many nails can it pick up?
This is an *electromagnet* (see also page 315).

Disconnect the battery and test the large nail again.
Is it magnetised as strongly as before?

Why are cranes in steelworks often fitted with electromagnets?

Destroying magnetism

A. Hammering
Experiment 31.5
Magnetise the piece of steel (as in experiment 31.3) and then hammer it. Test it with nails.
Is it magnetised as strongly as before?

B. Heating
Experiment 31.6
Magnetise the piece of steel as before and then heat it in a bunsen flame until it is red hot.
When it cools down, test it. Is it demagnetised?

C. Alternating current method
Experiment 31.7
This can be done only if your teacher can provide you with a safe supply of alternating current.
DO NOT use mains voltage – it is dangerous!
Pass alternating current (see page 331) through the coil you used in experiment 31.4 and then slowly reduce the amount of current in the coil (or gradually pull the nail out of the coil).
Then test the nail. Is it completely demagnetised?

Which of these three methods would you use to demagnetise a watch?

Theory of magnetism

Magnetise a thin piece of steel, as in experiment 31.3.
Then break it in half and use a compass to test each piece.
Is each half a magnet?

What happens if you break each piece again?

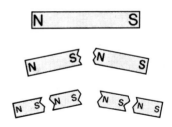

If we repeated this again and again, then we can imagine ending up
with a very very small "molecular magnet"

It seems that all pieces of iron and steel are made of millions of
these tiny molecular magnets.
In the following diagrams, the tiny molecular magnets are shown
by arrows, with the arrowhead as the N-pole.

Magnetised and unmagnetised objects
In an unmagnetised piece of iron, the molecular magnets are
pointing in all directions and so cancel out each other.

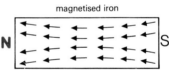

unmagnetised iron

If the molecular magnets can be turned round to point the same way
then the piece of iron becomes magnetised because all the tiny N-
poles add up at one end and all the S-poles add up at the other end.

magnetised iron

Experiment 31.9
Fill a small test-tube with iron filings and test it to check that it is
unmagnetised.

Stroke it with the N-pole of a magnet, while looking carefully inside
the test-tube.
Can you see the iron filings being pulled round by the magnet?

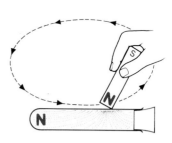

Bring a small compass near each end of the test-tube. Is the test-tube
now a magnet?
Which end is its N-pole? Can you explain this (remembering that like
poles repel)?

Shake the test-tube (as though you are hammering it) and test it
again.
Is it demagnetised? Why?

When an object is heated, do the molecules vibrate slower or faster?
When an object is heated, why is it demagnetised?

Magnetic induction

Experiment 31.10
Hold an unmagnetised piece of iron near some
nails. If it is unmagnetised, it will not attract the
nails.

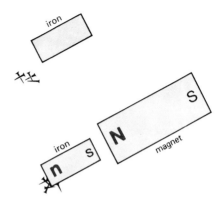

Now bring near a strong magnet as shown in the
diagram.
What happens to the nails?
This means that without being touched or stroked,
the piece of iron has magnetism *induced* into it.

What do you think has happened to the molecular
magnets in the piece of iron?

Which end of the iron becomes a S-pole? Why?
Why is the iron now attracted towards the magnet?

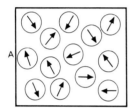

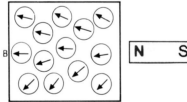

Experiment 31.11
To understand *magnetic induction* more clearly,
place some small 'plotting' compasses on a table
(or on an overhead projector if the compasses are
clear-sided).

Count how many compasses (or 'molecular
magnets') are pointing in each direction.
Does this represent a slightly magnetised or a
strongly magnetised specimen?

Now bring a magnet nearby. What happens?

At the side of the specimen nearest the magnet's
N-pole, do you get a N-pole or a S-pole?
Why is there attraction now between the end of
the specimen and the magnet?

Because this magnetic induction causes attraction,
the only sure way to test if an object is magnetised,
is to see if it *repels* a compass.

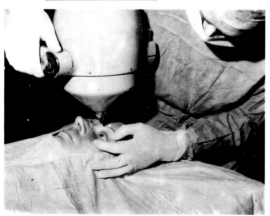

Objects like steel tools and girders in bridges are
often found to be slightly magnetised by magnetic
induction from the Earth's magnetism (particularly
if the object is shaken by vibrations).

In a hospital, a strong magnet may be used to
induce magnetism in a steel splinter and attract
it from someone's eye.

Magnetic materials

Experiment 31.12
Get some specimens of many different substances e.g. iron, steel, nickel, plastic, wood, copper, etc.
Use a magnet to test which ones are attracted (by magnetic induction).
Draw up lists of magnetic and non-magnetic substances.

magnetic substances	non-magnetic substances
iron	

Iron and steel are both magnetic but do not behave in exactly the same way.

Experiment 31.13
a) Hold an *iron* bar above some iron nails (or iron filings), with a magnet over the iron bar to magnetise it (by induction)

   How many nails will it hold up?

   Now take away the magnet? What happens to the nails?

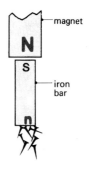

b) Now repeat the experiment with a *steel* bar of the same size.

   Do all the nails fall when you remove the magnet?

Iron is called a 'soft' magnetic material because it loses its magnetism easily. Iron is used in electromagnets.

Steel is called a 'hard' magnetic material because it does not lose its magnetism so easily. It is used to make permanent magnets.

Permanent magnets are used to keep shut the doors of cupboards and refrigerators.
A permanent magnet is placed in the oil tank of a car engine
so that it collects any small bits of steel that are worn off the engine.

Magnetic shielding
Experiment 31.14
Set up an 'Indian rope trick' as shown.
Using the same materials as in experiment 31.12, put them one at a time into the gap between the magnet and the paper clip.

Which materials shield the paper clip from the magnet?

Iron is often used to shield sensitive instruments from strong magnets or the Earth's magnetism.

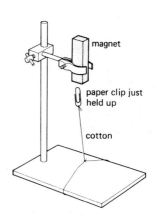

Magnetic fields

The region round a magnet where it has a magnetic effect is called its *magnetic field*.

Experiment 31.15
Place several small plotting compasses near a weak bar magnet (preferably on an overhead projector).

The compasses point along curved lines as shown in the diagram. These lines are called *lines of flux* (or lines of force) and they show the direction of the magnetic field at each point.

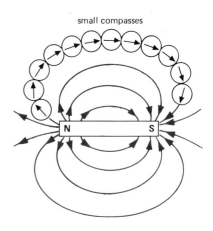

small compasses

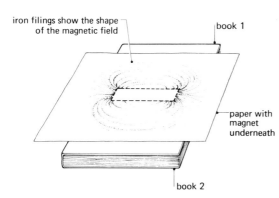

iron filings show the shape of the magnetic field

book 1

paper with magnet underneath

book 2

Experiment 31.16
Place a sheet of paper over a bar magnet (supporting the paper with books on either side to keep it level).

Sprinkle a thin layer of iron filings over the paper and then tap the paper gently. The iron filings act like thousands of tiny compasses and point themselves along the lines of flux.

Look carefully at the pattern that appears. Can you see that it is the same shape as in the diagram at the top of the page?

Sketch the pattern or, if you have used waxed paper (e.g. from the inside of a cornflakes packet), warm it very gently well above a low bunsen flame to fix the pattern in the wax.

Experiment 31.17
Use the iron filings method of experiment 31.16 to find the shape of the magnetic field round two magnets placed so that the N-pole of one is near the S-pole of the other.

Where do the iron filings gather most thickly? Where is the field strongest?

Experiment 31.18
Use the same method to find the shape of the field when a N-pole is placed near another N-pole.

At point P in the diagram, there is no magnetic field. P is called a *neutral point*.

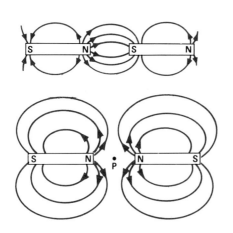

The magnetic field of the earth

How does a compass show you that the Earth has a magnetic field?

In fact, the Earth behaves as if there is a bar magnet inside it (although such a magnet cannot really exist because the centre of the Earth is too hot).

Notice in the diagram that the *S-pole* of the imaginary magnet is in the northern hemisphere so as to attract the N-pole of a compass.

Angle of declination

The diagram shows that the axis of the imaginary magnet does not point to the geographical north pole, but to a point in northern Canada called the magnetic north pole.
Because of this, a compass does not point towards the true north. The angle between the true north and magnetic north is called the *angle of declination* (or the angle of variation).

It varies from place to place. If you look at a detailed map of your area, you should find the angle of declination marked in the margin.
What is the angle of declination in your area?

Angle of dip

Experiment 31.19   Using a dip needle
A 'dip needle' is a compass turned on its side so that the needle can move vertically (instead of horizontally) while it is lined up north-south.

Where does the compass needle point to?
In the northern hemisphere, a compass points down into the ground. This is because the Earth's lines of flux are pointing into the ground (towards the imaginary magnet, see the diagram at the top of the page).

Where does a compass needle point in Australia?

The angle between the horizontal and the compass needle is called the *angle of dip* (or angle of inclination).

What is the angle of dip at your school?

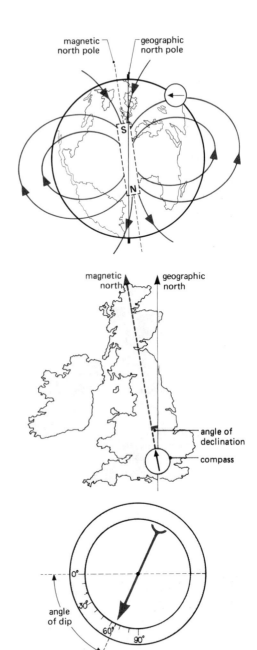

Summary

Like magnetic poles repel. Unlike poles attract.
Magnets can be made by
a) the stroking method (the end where a N-pole starts stroking is also a N-pole) or
b) the electrical method (more details on p. 315).

Strong magnets can be demagnetised by
a) hammering
b) heating
c) using a coil with alternating current.
Iron (unlike steel) is easily magnetised and easily de-magnetised.
Lines of flux show the direction of the field round a magnet.

Questions

1. Copy out and complete:
   a) Like poles . . . . . ; unlike poles . . . . . .
   b) If a N-pole is used in the stroking method of magnetisation, the end where the stroking begins is a . . . . . pole.
   c) Strong magnets can be demagnetised by (i) a coil carrying . . . . . current, (ii) . . . . . or (iii) . . . . .
   d) If the N-pole of a magnet is brought near an unmagnetised nail, then magnetism is . . . . . into the nail, with a . . . . . pole at the end of the nail nearest the N-pole of the magnet. The nail is then . . . . . to the magnet.
   e) The Earth's magnetic field is rather like that of a . . . . . magnet with its . . . . . pole in the northern hemisphere. The angle between true north and magnetic north is called the angle of . . . . . The angle between the horizontal and the Earth's magnetic field is called the angle of . . . . .

2. Explain the following:
   a) Magnets are often fitted to the doors of refrigerators and cupboards.
   b) A crane in a steelworks is fitted with a large electromagnet (see page 258).
   c) Flour is usually passed near a magnet before being packed.
   d) A steel ship becomes magnetised as it is constructed.
   e) The sump plug on a car is usually magnetised.

3. Use the idea of molecular magnets to explain:
   a) If a magnet is broken in half, each half is a magnet.
   b) Heating or hammering destroys magnetism.
   c) Magnetism is felt strongly at the poles but not at the centre of a magnet.
   d) There is a limit to how strongly an iron bar can be magnetised (*saturation*).
   e) Magnetism is induced into an iron bar placed near a magnet.

4. Choose a material, giving a reason, for each of the following: a) the core of an electromagnet b) a compass needle c) the case of a compass.

5. Draw diagrams showing the magnetic fields round a) a single bar magnet b) two magnets with their unlike poles close together c) two magnets with their N-poles close together d) the Earth.

6. Professor Messer says that 'like poles must attract' because the north pole of the Earth attracts the N-pole of a compass. Explain why he is wrong.

7. Look, here's a new compass the Prof. has designed. But how many silly mistakes can you find? [8?]

Professor Messer's new compass

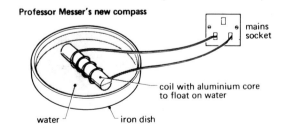

mains socket

coil with aluminium core to float on water

water

iron dish

Further questions on page 300.

chapter 32

STATIC ELECTRICITY

Experiment 32.1
Rub a plastic pen or comb on your sleeve and then hold it near some tiny pieces of paper.
What happens?

Experiment 32.2
Rub a plastic object on your sleeve and then hold it near (but not touching) a thin stream of water from a tap.
What happens?

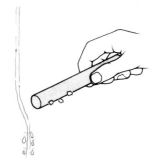

Because of friction with your sleeve, the plastic object has been *charged* with stationary or *static electricity*.

Have you ever heard a crackling noise as nylon clothes rub together when you undress?
In the dark you can see the tiny sparks of electricity (like tiny flashes of lightning).

Experiment 32.3
Rub a strip of polythene (a grey plastic) on a dry woollen cloth and hang it in a paper stirrup as shown.
We say the electricity on polythene is a *negative* charge.

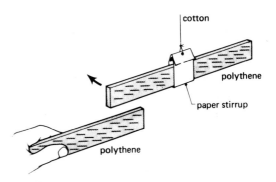

cotton

polythene

paper stirrup

polythene

Rub another strip of polythene (so it is also negatively charged) and bring it close.
What happens?

Repeat the experiment with two strips of cellulose acetate (a clear plastic).
What happens?

Now repeat the experiment with one strip of charged polythene and one strip of charged cellulose acetate.
What happens?

Because it behaves differently, we say that cellulose acetate is charged *positively*.

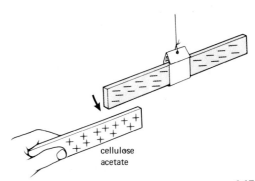

cellulose
acetate

We find that: **Like electric charges repel.**
                Unlike electric charges attract.

The electron theory

Objects become charged because the atoms of all substances contain both negative and positive charges (see also page 376).

The positive charges (called *protons*) are in the central core or *nucleus* of the atom.
The negative charges (called *electrons*) are spread in orbits round the outer part of the atom.

In an uncharged or *neutral* atom, the number of protons (+) is equal to the number of electrons (−).

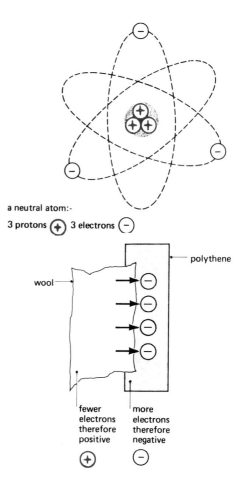

a neutral atom:-
3 protons (+) 3 electrons (−)

We believe that when a polythene strip (or an ebonite rod) is rubbed on wool, some of the outer electrons are scraped off the wool and move on to the polythene. This means that the polythene has an extra number of electrons (so it is negatively charged) and the wool then has fewer electrons than protons (so it is positively charged).

Notice that only the negative electrons can move – the positive protons remain fixed.

When cellulose acetate (or glass) is rubbed, it becomes positively charged. Draw a diagram to show the movement of electrons in this case.

wool — polythene

fewer electrons therefore positive (+)

more electrons therefore negative (−)

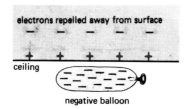

electrons repelled away from surface

ceiling

negative balloon

Electrostatic Induction

Experiment 32.4
Rub a balloon on your clothes and then put it on the wall or ceiling so that it stays there!

This happens because the negative charge on the rubbed balloon repels some of the electrons in the ceiling away from the surface. This leaves the surface positively charged and so the negative balloon is attracted to the ceiling.

The separated charges in the ceiling are called *induced* charges.

This also explains why the paper and the water were attracted in experiments 32.1 and 32.2

Why do gramophone records tend to gather a lot of dust?

The gold leaf electroscope

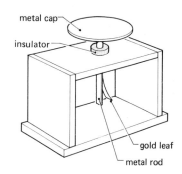

metal cap

insulator

gold leaf

metal rod

This instrument can be used to tell if an object is charged.
The metal cap is connected to a metal rod and a strip of very thin
metal (the gold leaf). This gold leaf is hinged at the top so that it
can move.

Experiment 32.5
Rub a polythene strip and bring it close to, but not touching, the
cap of an electroscope.
What happens to the leaf?

The negative polythene induces charges in the electroscope by
repelling negative electrons down to the metal rod and the leaf.
Because like charges repel each other, the light leaf is repelled and
lifted.

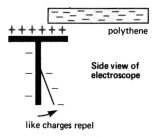

polythene

Side view of
electroscope

like charges repel

What happens when you remove the polythene? Why?

What happens if you bring up a positively-charged strip? Draw a
diagram to show how the electrons move in this case.

Experiment 32.6
Charge a polythene strip and then slide the strip firmly across the
edge of the metal cap of the electroscope.

What happens? Does the leaf stay up even if you remove the
polythene?

In this case, some of the negative electrons moved from the
polythene to the electroscope, where they spread out over all the
metal parts and made the leaf rise by repulsion.

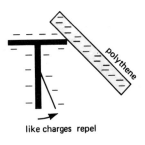

polythene

like charges repel

What happens if you now bring up a positively-charged strip?

Experiment 32.7   Conductors and insulators
Charge an electroscope as in experiment 32.6, then touch the cap
with your finger.

What happens to the leaf?
What happens to the electrons on it?

Your skin is a *conductor* of electricity because it allows electrons to
move through it.

Charge the electroscope again and then touch it with a plastic pen.
Does anything happen this time?
Plastic is an *insulator* because it does not allow electrons to pass
through it.
What happens with other materials?

conductors	insulators
skin	plastic

Electrostatic induction

In experiment 32.4 on page 384, we saw that a charged object can *induce* charges in another object. This is *electrostatic induction*. It happens because electrons can be moved to different parts of a conductor by attraction or repulsion.

Experiment 32.8   Charging two objects by induction
Follow the diagrams to charge two metal spheres by induction.

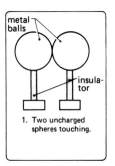

1. Two uncharged spheres touching.

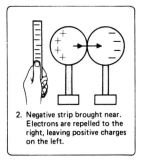

2. Negative strip brought near. Electrons are repelled to the right, leaving positive charges on the left.

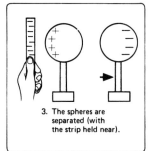

3. The spheres are separated (with the strip held near).

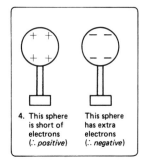

4. This sphere is short of electrons (∴ *positive*)   This sphere has extra electrons (∴ *negative*)

How can you use a leaf electroscope to test these spheres?
Repeat this with a *positive* strip and draw diagrams to show what happens.
(Remember: only the electrons can move, not the positive charges).

Experiment 32.9   Charging one object by induction

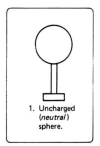

1. Uncharged (*neutral*) sphere.

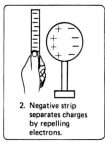

2. Negative strip separates charges by repelling electrons.

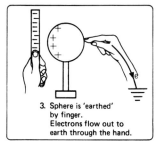

3. Sphere is 'earthed' by finger. Electrons flow out to earth through the hand.

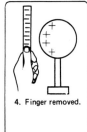

4. Finger removed.

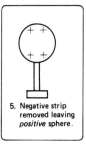

5. Negative strip removed leaving *positive* sphere.

Notice that a *negative* strip gives a *positive* sphere.
Now repeat this with a positive strip and draw the correct diagrams.

Experiment 32.10   Charging an electroscope by induction
This is a much better method than the one used in experiment 32.6. This time, begin with a *positive* strip (and end with a negative charge). Repeat this with a *negative* strip and draw the correct diagrams.

1. Uncharged (*neutral*) electroscope.

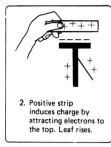

2. Positive strip induces charge by attracting electrons to the top. Leaf rises.

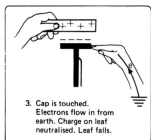

3. Cap is touched. Electrons flow in from earth. Charge on leaf neutralised. Leaf falls.

4. Finger removed.

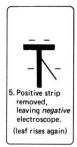

5. Positive strip removed, leaving *negative* electroscope. (leaf rises again)

Distribution of charge on a conductor

We can investigate where the charges go to on a conductor by using a *proof plane* and a gold-leaf electroscope.
A proof plane is a small metal disc on an insulating handle. It is used to take a sample of charge and carry it to an electroscope for testing.

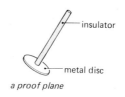

insulator

metal disc

a proof plane

Experiment 32.11   *Charge on the inside of a conductor*

Use a Van de Graaff generator (see the next page) to charge a hollow metal sphere.
Then use a proof plane to test:
a) the inside of the sphere, and
b) the outside of the sphere.
What do you find?

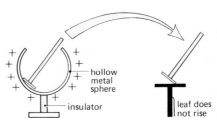

hollow metal sphere

insulator

leaf does not rise

There is *no* charge on the inside of a conductor; *all* the charge is on the *outside*.
This is because all the charges are repelling each other and so move as far apart as they can – to the outside.
On a sphere they are *evenly* spread on the outside.

One of the safest places in a thunderstorm is *inside* a metal car. Any charge that leaks from a thunder-cloud will stay on the outside of the metal. Concrete buildings with metal frames are safe for the same reason.

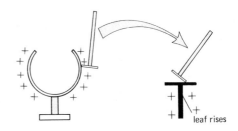

leaf rises

Experiment 32.12   *Charge on a pear-shaped conductor*

Use a Van de Graaff generator to charge a pear-shaped conductor.
Then use a proof plane to test different parts of the conductor.
What do you find?

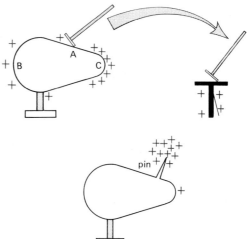

B   A   C

pin

At **A**, where the surface is almost flat, there is very little charge.
At **B**, where the surface is curved, there is more charge.
At **C**, where the surface is strongly curved, there is a lot of charge (the greatest *charge density*).

If a pin was attached to the conductor, most of the charge would move to the very sharply curved point of the pin (see also experiment 32.17).

271

Experiments with a Van de Graaff generator

This is a machine used to charge objects easily. (A Whimshurst machine can be used instead).
These experiments should be done only under the supervision of a teacher — be careful you do not get an unexpected shock from the machine (it is safer to approach the dome while holding a pin in front of you — see experiment 32.19).

Experiment 32.13
Place a piece of fur on the dome and start the machine running.

What happens to the hairs on the fur?
Why is this? (Remember all the hairs have the same kind of charge).

Experiment 32.14   A hair-raising experience
If someone in the class has soft dry hair, ask them to stand on a thick sheet of polythene (to insulate them from the Earth) while they touch the dome.

Explain what happens when the machine starts.

Experiment 32.15
While standing on an insulator and touching the dome as in experiment 32.14, blow some soap bubbles over the dome.

What happens? Explain this. (See also page 275).

Experiment 32.16
Without touching the machine, blow some bubbles near it.

What happens? Why?
How is this used in electrostatic paint spraying?

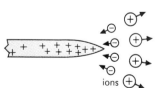

Experiment 32.17   An electric wind
While standing on an insulator and touching the dome as in experiment 32.14, aim the point of a pin at a candle flame.

Which way does the flame move?

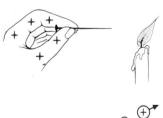

This happens because, if the point is positively charged, it attracts to it some of the electrons off atoms in the air. This leaves the atoms positively-charged.
We say the air has been *ionised*.
The positively-charged atoms (called *ions*) are repelled away from the point, as an 'electric wind'.

272

Experiment 32.18   An electric windmill

Connect a Van de Graaff machine to a metal windmill on an insulated stand.

What happens when you start the machine?

As the ions are repelled from the point, so also is the point repelled by the ions (Newton's 3rd Law; see page 114).

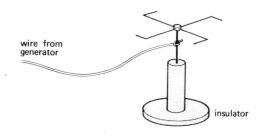

wire from generator

insulator

Experiment 32.19   A lightning conductor

Place a metal ball near a Van de Graaff generator so that a spark jumps between them.
This is a tiny lightning flash (some flashes of lightning may be up to 6 km long).

The spark makes a sharp cracking sound because it is heating and expanding the air very rapidly. What is thunder?

Now approach the dome with a pointed metal pin in your hand.
Does the machine stop sparking?

This is how a lightning conductor works — if a negatively-charged cloud passes overhead, it induces a positive charge on the point at the top of the lightning conductor (as in experiment 32.4). The point then repels positive ions to the cloud (as in experiment 32.17) to neutralise it, so that it is less likely to produce a lightning flash.
The electrons that are attracted to the conductor travel down it to the Earth.

Why might it be dangerous to use an umbrella in a violent storm?

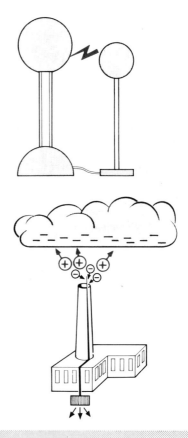

Summary

When rubbed, polythene (or ebonite) becomes charged negatively (it has an excess number of electrons).
Cellulose acetate (or glass) becomes charged positively (it has a shortage of electrons).
Like charges repel; unlike charges attract.

A gold leaf electroscope can be used to detect charges.

Charged objects can separate (induce) charges in other objects (by attracting or repelling electrons).

A conductor allows electrons to move through it; an insulator does not.

Charges go to the *out*side of a conductor.

Sharp points lose their charge easily. This is used in a lightning conductor.

Questions

1. Copy out and complete:
   a) Like charges . . . . . ; unlike charges . . . . .
   b) When polythene is rubbed with wool, some . . . . . leave the . . . . . and move on to the . . . . . so that the polythene becomes . . . . . charged and the wool is left . . . . . charged.
   c) When rubbed, cellulose acetate becomes . . . . . charged because it . . . . . some electrons.
   d) A negatively-charged object attracts a piece of paper because it . . . . . electrons away from the surface of the paper. This leaves the surface of the paper . . . . .-charged so that it is . . . . . to the object. These separated charges are called . . . . . charges.
   e) In a gold leaf electroscope, the leaf rises because like charges . . . . .
   f) Conductors (unlike . . . . .) allow . . . . . to travel through them.
   g) Sharp points . . . . . their charge easily by . . . . . the air.
   The . . . . . at the top of a lightning conductor loses . . . . . to the clouds to neutralise them so that they are less likely to produce a flash of . . . . .

2. Explain the following:
   a) Nylon clothing crackles as you undress.
   b) In dry weather, people walking on nylon carpets may get a shock if they touch a radiator.
   c) Passengers sliding off a car seat on to the ground sometimes get a shock.
   d) Petrol road-tankers usually have a length of metal chain hanging down to touch the ground.
   e) Because some anaesthetics are explosive, the floor tiles in an operating theatre are made of a conducting material.

3. Explain the following:
   a) A rubbed balloon will stick to the wall for some time.
   b) A mirror or window polished by a dry cloth on a dry day soon becomes dusty.
   c) Gramophone records and other plastic objects soon become dusty.
   d) As sellotape is pulled off a roll it is attracted to your hand.
   e) When spraying an object with paint, less paint is wasted if the object is charged.
   f) TV screens soon become very dusty.
   g) The amount of smoke leaving a factory can be reduced by placing a charged object inside the chimney.

4. Explain the following:
   a) It might be dangerous to raise an umbrella in a storm.
   b) The dome of a Van de Graaff machine is spherical with no sharp edges on the outside.
   c) It is safer to approach it while holding a pin in front of you.
   d) On high voltage electrical equipment, the blobs of solder should be rounded and smooth.

5. State and explain what happens during each of the following steps:
   a) A polythene rod is rubbed with wool.
   b) This polythene rod is brought close to, but not touching, an electroscope.
   c) The polythene is scraped across the edge of the cap of the electroscope and then removed.
   d) The polythene is brought close to the electroscope and then removed.
   e) A charged cellulose acetate strip is brought close to the electroscope and then removed.
   f) The cap is touched with a finger.

6. Explain how a lightning conductor works. Sound travels 1 km in 3 seconds (1 mile in 5 seconds). If you hear thunder 15 seconds after you see the flash, how far away is the lightning? What have you assumed in calculating this?

Have you started revising for your examinations? See page 390.

274

chapter 33 **more about**

ELECTRO STATICS

Electric fields

The region round a charged object where it has an electric effect is called its *electric field*. Any charge placed in this field will have a force exerted on it.

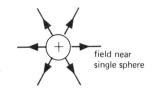

field near
single sphere

On page 264, we showed magnetic fields by lines of flux. In the same way, we can show electric fields by *lines of electric flux* (or lines of force).
Electric charges tend to move along these lines of flux.
Positive charges would move in the direction of the arrows (from + to –).
Negative charges (electrons) move in the opposite direction.

In experiment 32.15, the charged soap bubbles tended to move along the lines of flux of the electric field (but they were also pulled down by the Earth's gravitational field).

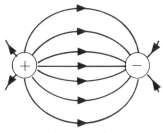

field near two spheres

Electric potential

If a charged soap bubble is in an electric field (experiment 32.15), it moves and gains kinetic energy. While doing this, it is losing *electric potential energy* (just as an object falling in a gravitational field loses P.E. and gains K.E., see page 129).

Positive charges (e.g. positive ions or soap bubbles) move from a place of *high* electric potential energy (near a positive charge) to a place of *low* electric potential (near a negative charge).
Electrons move the other way, from a low potential to a high potential.

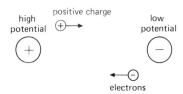

high
potential

positive charge

low
potential

electrons

In a gravitational field, heights or depths are often measured relative to sea-level. Similarly, in an electric field, the electric potential is usually measured relative to the potential of the Earth.
The Earth and any earthed object is said to be at *zero potential*.

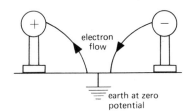

electron
flow

earth at zero
potential

Potential difference (p.d.)

When a charge moves from one place to another, the change in energy depends on the *potential difference (p.d.)* between its starting and finishing points. This potential difference is measured in *volts* and is often called *voltage*.

The greater the voltage between two points, the greater the change in energy when a charge moves between the two points.

One way to measure p.d. is to use a gold-leaf electroscope.
It is a *voltmeter*. The deflection of the leaf depends on the voltage between the leaf and the case.

Charge, potential difference and capacitance

Charge (in coulombs)

We measure charges in units called *coulombs*, usually shortened to C, (see page 325). A coulomb is a very large amount of charge (about 6 million million million electrons), so we often use micro-coulombs ($1\mu C = 10^{-6} C$) or nano-coulombs ($1nC = 10^{-9} C$).

Potential difference (in volts)

The diagram shows a charge of 1 coulomb (1C) being moved from point **A** (earthed) to point **B** (on a charged sphere).
If 1 joule of energy is needed to move the 1 coulomb of charge from A to B (against the repulsive force), then the potential difference is 1 joule per coulomb and this is called **1 volt (1V)**.
That is, 1 volt = 1 joule/coulomb (see also page 297).

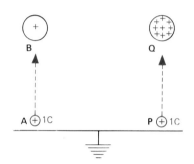

To move 1 coulomb from **P** to **Q** would need more energy (see diagram). If 10 joule of energy is needed to move the 1 coulomb from P to Q, then the p.d. is 10 joule/coulomb = 10 volt.

Capacitance (in farads)

We have seen that objects (such as metal spheres and electroscopes) have the capacity to store charges.
Which do you think could store more charge – a small sphere or a large sphere?

We say a larger sphere has a larger *capacitance*.

In the diagram higher up the page, we saw that a sphere at a higher voltage stores more charge than one at a lower voltage.
We find that doubling the voltage of an object doubles the amount of charge on it.

In fact: the charge stored, Q $\propto$ the potential difference, V.

To change this proportionality into an equation (see page 402), we introduce a constant **C**, called the **capacitance** (measured in *farads*).

> **charge stored, Q = capacitance, C × potential difference, V** or $Q = CV$ or $C = \dfrac{Q}{V}$
> (in coulombs) (in farads) (in volts)

If 1 coulomb of charge changes the potential difference by 1 volt, then the capacitance is **1 farad (1F)**.
Because the coulomb is a large unit, so also is the farad.
In practice we often use micro-farads ($1\mu F = 10^{-6} F$) or nano-farads ($1nF = 10^{-9} F$) or even pico-farads ($1pF = 10^{-12} F$).

Capacitors

An object that stores charge is called a *capacitor*.
A common kind of capacitor is a *parallel-plate capacitor*. It consists of two metal plates separated by an insulator.

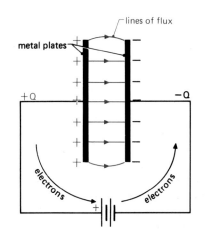

It can be charged by connecting it to a Van de Graaff generator or a battery, so that one plate becomes short of electrons (so it is positive) and the other plate gains electrons (negative).
The diagram shows the electric field between the plates.

Experiment 33.1   Factors affecting capacitance
Charge an electroscope by induction (see experiment 32.10 on page 270).

a) Hold a metal plate in your hand. Move it parallel to the cap of the electroscope so that more and more of the area of the plate overlaps the cap. What happens to the leaf?

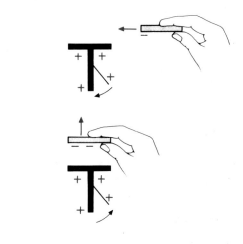

If the leaf *falls*, this means that the potential of the electroscope falls (remember: it is a volt-meter). Since the charge is constant, this means that the capacitance *increases*, because

capacitance $C = \dfrac{\text{charge } Q}{\text{p.d. } V}$ (see opposite page).

b) Now move the plate upwards, away from the cap
Does the leaf fall or rise?
Has the capacitance increased or decreased?

c) Now hold the plate still, about 1 cm above the cap. Then move a strip of polythene (called a 'dielectric') into the gap.
What happens to the capacitance?

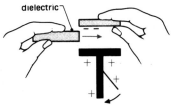

The capacitance *increases* if:
a) the area of overlap of the plates *increases*,
b) the distance between the plates *decreases*,
c) the insulator (dielectric) is a solid rather than air.

Example
A capacitor of capacitance $10\mu F$ is charged to a p.d. of 5 volt. How much charge is stored on each plate?
formula first: charge Q = capacitance (in F) x p.d. (in V)
then numbers:
$$= 10 \times 10^{-6} \text{ farad} \times 5 \text{ volt}$$
$$= 50 \times 10^{-6} \text{ coulomb}$$
$$= 50\mu C \text{ (microcoulomb)}$$

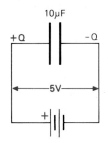

Practical capacitors (used in radios, TV's and computers)

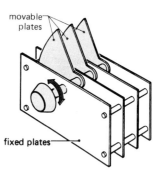

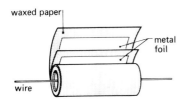

1. A *waxed paper capacitor* consists of two long strips of metal foil (the plates) separated by strips of waxed paper (the dielectric). It is rolled up to form a compact cylinder. Typical value = 0·1μF.

2. An *electrolytic capacitor* has a very thin dielectric which is formed chemically. It can store a larger charge but still be compact. However the plate marked + must never be charged negatively. Typical value = 100 μF.

3. The tuning knob on your radio is a *variable capacitor*. This has two sets of parallel plates: one fixed, one movable. Turning the movable plates varies the area of overlap and so varies the capacitance. Typical value = 0·001μF (= 1nF).

Summary

Lines of flux show the direction of an electric field.

1 volt is the potential difference between two points when 1 joule of energy is needed to move 1 coulomb of charge between the two points. 1 volt = 1 joule per coulomb.

Charge Q = capacitance C × potential difference V
(coulombs)   (farads)   (volts)

The capacitance can be increased by (i) increasing the area of the plates (ii) decreasing the separation of the plates (iii) adding a dielectric.

Questions

1. a) The arrows on a line of . . . . . show the direction in which a . . . . . charge would move.
   b) Positive charges move from a place of . . . . . potential to one of . . . . . potential. Electrons move the . . . . . way.
   c) 1 volt = 1 . . . . ./. . . . .
   d) charge = capacitance × . . . .
      (in . . . .)   (in . . . .)   (in . . . .)
   e) 1 farad is the . . . . . of an object if one: . . . . . of charge changes the . . . . by one . . . . .

2. A capacitor is marked 10 μF.
   a) What is its capacitance in farad?
   b) It is connected across a 10V battery. What charge is stored?
   c) What p.d. is needed to store 200μC?

3. The diagram shows two identical electroscopes with two cans of different size.

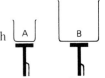

   a) Which can has the greater capacitance? (Hint: see page 276).
   b) Each can is given exactly the same positive charge. Which can has the higher potential? (Hint: $Q = CV$).
   c) Which leaf then shows the greater divergence? (Hint: an electroscope is a voltmeter).
   d) Draw a diagram of A and B showing the charge and the divergence of each leaf.
   e) The cans are connected by a wire. Which way do electrons flow? Why?
   f) When the electrons stop flowing, what can you say about (i) the potential and (ii) the charge of A compared with B.

Further questions on page 301.

chapter 34

Experiment 34.1   The simple cell

Place a copper strip and a zinc strip (called *poles* or *electrodes*) in a beaker containing dilute sulphuric acid (called the *electrolyte*). Take care with the acid.

Use wires to connect the poles to a small lamp, as shown.

What happens when you first connect the lamp? What happens a short time later? What happens if you scrape the bubbles off the copper?

The disadvantages of this **simple cell** are:

a) it soon becomes *polarised* (when hydrogen bubbles on the copper stop it working).

b) *local action* eats away the zinc even when it is not producing electricity.

This cell is an electron pump which puts extra electrons on the zinc (so that it is negatively charged) and takes electrons from the copper (so that it is the positive pole).

The electrical pressure of this pump is about one *volt*.

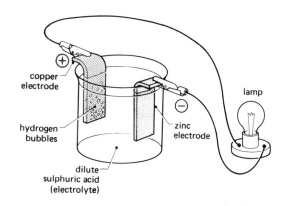

The dry (Leclanché) cell

This is the common cell used in torches. Study the diagram to see which substances are used for the poles and the electrolyte.

A *depolariser* (powdered manganese dioxide and carbon) prevents the formation of hydrogen bubbles.

The electrical pressure of this electron pump is about 1.5 volt.

As in the simple cell, the zinc is slowly eaten away as the chemical energy of the zinc is converted to electrical energy. These cells are called *primary* cells.

A series of cells connected together is called a *battery*.

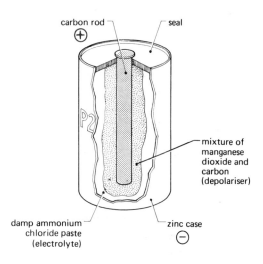

279

The lead-acid accumulator

Experiment 34.2
Place two strips of clean lead in a beaker containing dilute sulphuric acid.
Connect the strips to a battery of dry cells (or a similar supply) with an electrical pressure of about 4 volt.

Look at the lead plates. Which one changes colour?

After 5 minutes, disconnect the wires from the supply and, without letting them touch each other, connect them to a lamp as shown.
What happens?

Re-charge the accumulator for 10 minutes.
Does it light the lamp for a longer time?

The lead-acid accumulator can be re-charged again and again. It is called a *secondary* cell.

The electrical pressure of this electron pump is about 2 volt.
The strip connected to the positive of the charging supply becomes coloured with lead dioxide and is the positive pole of the accumulator.
The electrical energy is stored as chemical energy in the lead dioxide.

A battery of accumulators is used in a motor-car, because it can be recharged easily and it can produce large currents for the starter motor, headlights etc.

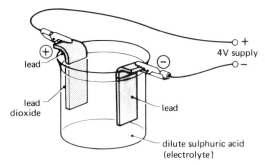

charging an accumulator

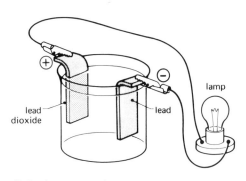

discharging an accumulator

Care of accumulators
1. Accumulators must not be left discharged for a long time. They are discharged if the density of the acid falls to 1.15 g/cm^3 (1150 kg/m^3 (see page 92).
The density is usually measured using a hydrometer (see page 110).
2. Accumulators should be charged slowly (with the positive of the supply connected to the positive pole of the accumulator) until the density of the acid rises to 1.25 g/cm^3 (1250 kg/m^3).
3. They should be topped up regularly with distilled water.
4. The terminals should never be connected directly together in a 'short circuit'.

Power packs

In the laboratory we often use 'power packs' instead of dry cells or accumulators.

Power packs are connected to the mains electricity supply which they 'transform' and 'rectify' to supply electricity (see p. 334 and 349).

Experiment 11.3

Your teacher may be able to provide a high voltage battery (or a high voltage power pack) to demonstrate this experiment. Taking care, a high voltage battery or power pack is connected to a gold leaf electroscope.

What happens?

What can you say about the voltage of the electricity produced by a battery (or power pack) compared with the voltage of the static electricity produced by friction.

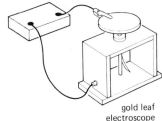

high voltage battery
(or power pack)

gold leaf
electroscope

Something to do

You can make your own battery using some clean copper coins and aluminium milk-bottle caps, with pieces of blotting paper soaked in salt water, as shown.

Connect your battery to the earphone from a portable radio. What happens as the wires make contact?

Leave it connected for a few hours and then dismantle the battery. What has happened to the aluminium? Where does the electrical energy come from?

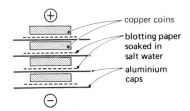

copper coins
blotting paper
soaked in
salt water
aluminium
caps

a battery of 4 cells

Questions

1. Copy out and complete:
   a) A simple cell has . . . . . as the positive pole and . . . . . as the negative . . . . . with dilute . . . . . as the electrolyte.
      Its electrical pressure is about . . . volt.
   b) Its defects are (i) . . . . . and (ii) . . . . .
   c) The dry (. . . . .) cell has . . . . . as the positive pole, ... as the negative pole, ... paste as the electrolyte and a de-polariser consisting of powdered . . . . . and . . . . .
      Its electrical pressure is about . . . volt.

   d) The accumulator is a . . . . . cell with an electrical pressure of about . . . volt.
      The positive pole is made of . . . . . , the negative pole is . . . . . , and the electrolyte is . . . . .
   e) The state of an accumulator can be checked by measuring the density of the . . . . . using a . . . . .

2. Make a list, as long as possible, of objects that use electric cells.

   Q. What did the police do with the man *charged* with assault and *battery*?
   A. Put him in a *dry cell*.

Physics at work: Magnets

Paint thickness gauge

If paint is to protect steel from rusting – for example on ships or cars – it has to cover all the surface. The paint is usually about $\frac{1}{5}$ mm thick and this has to be checked with a paint thickness gauge:

The gauge uses a magnet and a spring balance.

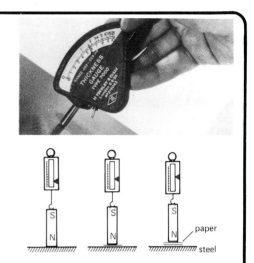

Experiment

Hang a bar magnet from a spring balance (using sellotape). Touch the magnet to a piece of steel and measure the force (in newtons) needed to pull it away. Then measure the force with one or more pieces of paper ("the paint") between the magnet and the steel. What do you find?

This is how a paint thickness gauge works. (It can also tell you if the steel has rusted away or been replaced by fibre-glass.)

Magnetic inks and paints

It is possible to make magnetic inks by mixing very small particles of a magnetic substance with a liquid. The mixture (a dark brown colour) can be used as a paint to make magnetic tape for tape-recorders (see page 347), and floppy discs for computers.

Banks use magnetic ink on cheques so that the cheques can be sorted automatically by a machine which detects the magnetic field round each number:
Automatic cash-cards also use magnetic ink to store information on the card.

Cheque number   bank code   customer code

Magnetic liquids

The magnetic ink can be made with oil so that it does not dry out. This magnetic liquid can be used to detect very small cracks in the surface of steel pipelines. These cracks can be very dangerous if left untreated.

The pipe is magnetised by a simple coil and painted with magnetic liquid:
If there is a tiny crack, some of the magnetic lines of force leak out of the pipe at the crack. Then the magnetic liquid shows up the shape of the crack.

Magnetic oil is also used to lubricate the shafts of motors and the coils of loudspeakers. A magnet is used to keep the magnetic oil in the right place.

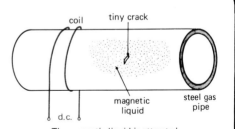

The magnetic liquid is attracted to the edge of the crack.

Physics at work: Static Electricity

An Electrostatic Precipitator

Power stations and factories produce enormous amounts of smoke pollution. This smoke is a cloud of small dust particles or ash. It can be removed by using static electricity.

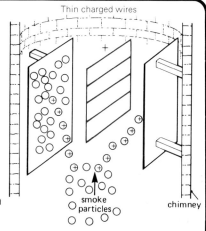

Thin charged wires

Some thin wires are stretched across the centre of the chimney: These wires are charged positively to about 50 000 V and they cause the gas around them to be ionized (see also experiment 32.17 on page 272).
Positive ions are repelled from the wire and attach themselves to some of the dust particles in the smoke. These dust particles are then repelled towards earthed metal plates, where the dust sticks.

A mechanical hammer hits the plates every few minutes and the ash falls down into a bin.
Later, the ash is used to make bricks for houses.

smoke particles   chimney

Fingerprinting

A similar method can be used to show up fingerprints on paper: The paper is placed near a charged wire like one of the plates in the chimney and a fine black powder is used instead of smoke. The powder sticks to the fingerprint but not to the clean paper.

Photocopiers are based on a similar method.

Top Secret   Top Secret

before   after

Paintspraying

Bicycles and cars are painted using an electrostatic paintspray. The nozzle is given a charge and this makes a better spray: the droplets all have the same charge and repel each other so that the paint spreads out to form a large even cloud.
Less paint is needed because the charged droplets are all attracted to the object (even the back of it) because it has an opposite charge (see page 267).
The same idea is used to make crop-spraying more efficient for farmers.

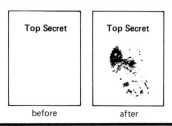

nozzle is charged positively

object is negative

The droplets of paint repel each other but are attracted to the object.

Preventing fires

A liquid flowing through a pipe can become charged in the same way as two objects rubbed together. This can be dangerous if it causes a spark and the liquid is inflammable.
For this reason, whenever an aeroplane is being re-fuelled by a tanker lorry, they are always connected together by a copper wire.
For the same reason, spare petrol for cars should always be carried in metal cans, never plastic.

Large ships have been known to explode because their tanks were being cleaned out by a high speed water-jet which became charged and caused a spark. Nowadays the tanks are filled with an inert gas before cleaning starts.

Walking on nylon carpets can give you a shock when you touch the door handle or a radiator. Modern carpets are made conducting to prevent this.

chapter 35

CIRCUITS

Experiment 35.1

Connect a cell to a lamp. This can be done by loose wires as in experiment 34.1 or on a *circuit board* as shown here.

When the lamp lights, we say an electric *current* is flowing round the *circuit*.

In fact, negative electrons are being pushed out of the negative pole of the cell and are drifting slowly round the circuit, from atom to atom in the wire, to the positive pole of the cell.
A current flowing in one direction like this is called a direct current (d.c.).

When drawing diagrams of circuits we use symbols for each component as shown here.

Now make a gap in your circuit.
What happens?
How can you switch the lamp on and off?
Does it matter where you make the gap?

For an electric current to flow, there must be a *complete* circuit, with no gaps.

The diagram shows the circuit with a switch added. Look at some switches of different kinds and include one of them in your circuit.

Here are some other symbols used in electric circuit diagrams:

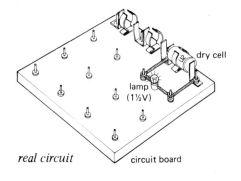

real circuit   circuit board

circuit diagram

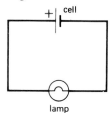

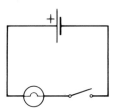

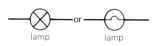

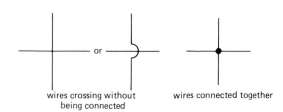

284

Experiment 35.2   Conductors and insulators
Connect a circuit as shown (very like the last experiment but with a gap in it).

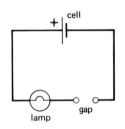

Place a piece of copper across the gap.
What happens?

Place a piece of plastic across the gap.
What happens?

Some materials resist the flow of electricity more than others. Repeat the experiment with other materials and draw up a table of conductors and insulators. Compare this with the list from experiment 32.7 on page 269 and explain any differences.

Conductors	Insulators
copper	plastic

Is air an insulator or a conductor?
Why are the handles of electrician's screwdrivers made of plastic?
Why are copper wires usually covered with plastic or rubber?

Experiment 35.3
Connect a dry cell to a lamp as shown.
Use a length of thick copper wire to connect one side of the lamp to the other side for just a second. (Do not do this experiment with anything but a dry cell, unless your teacher gives permission).

What happens? If the current is not flowing through the lamp, where is it going?

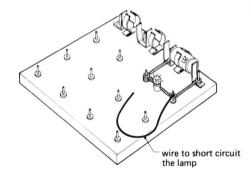

wire to short circuit the lamp

We say we have *shorted* the lamp or caused a 'short circuit'. Your wire has *less resistance* to the flow of electrons than the thin wire inside the lamp. *Electricity travels by the easiest path*, not necessarily the shortest path.

Conventional current and electron flow
We know now that an electric current is a flow of electrons from negative to positive.
However when cells were first invented scientists guessed, *wrongly*, that something was moving from positive to negative and they marked arrows this way on their diagrams.
Unfortunately we have kept to this convention when drawing circuit diagrams, but you should remember that really electrons are moving the opposite way.

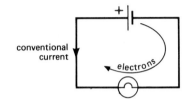

Series circuits

Experiment 35.4

Connect a cell (or a power pack) to two lamps as shown.

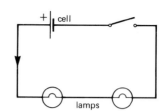

All the electrons that go through one lamp must also go through the other. We say the lamps are connected *in series*.

The electric current flowing in a circuit is measured in *amperes* or *amps* (usually shortened to A). 1 ampere is a flow of about 6 million million million electrons per second past each point!

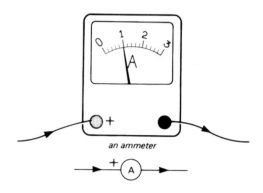

an ammeter

The size of a current can be measured using an *ammeter* (see page 322).

In experiments, it is important to connect an ammeter so that its red (+) terminal is always nearer to the positive (+) pole of the cell than to the negative pole. Otherwise the pointer will be moved the wrong way.

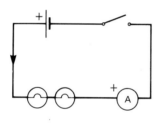

Experiment 35.5   Using an ammeter

Add a suitable ammeter to the circuit of experiment 35.4.

Read the scale. (On some ammeters, the numbers may have to be scaled up or down. Your teacher will explain if this is necessary).

How much current is flowing in your circuit?

Experiment 35.6

Move the ammeter to different positions in the series circuit.

What do you find?

The same current flows through each part of a series circuit.

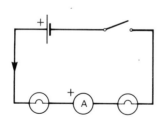

Experiment 35.7

If you unscrew one lamp in your series circuit, what happens to the other lamp?

Christmas tree lamps are usually wired in series. What happens if one lamp breaks?

Are the lights in your house wired in series?

Parallel circuits

Experiment 35.8
Connect a cell (or a power pack) to two lamps in the way shown here.

We say the two lamps are connected *in parallel*.

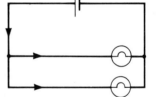

Experiment 35.9
Add switches to your circuit as shown (or loosen the wires instead).

Can you switch the lamps separately?

Are the lamps in your home wired in series or in parallel?

Where in this circuit would you put a switch to control both lamps together? Try it.

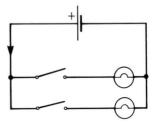

Experiment 35.10
If you have 3 ammeters, connect the circuit shown here. (With just one ammeter, place it in each position in turn).

Read the ammeters. What do you notice about the reading on A_1 compared with the total of the readings on A_2 and A_3?

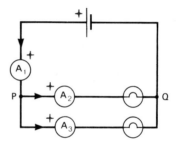

The current in the main circuit is the sum of the currents in the separate branches (sometimes called Kirchhoff's first law).

As the electrons (from the negative pole of the cell) reach point Q, they divide between the two branches of the circuit until they reach P where they join together again.

Experiment 35.11   Different parallel circuits
There are many different ways of drawing the same **parallel** circuit. Because bending the connecting wires does not change the move-ment of the electrons, all the circuits shown below are *electrically the same* as each other. (In each case the current divides to go through the lamps and then rejoins).
Try connecting some of these circuits yourself.

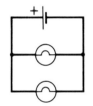
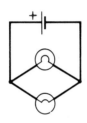
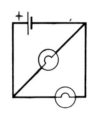
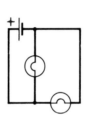
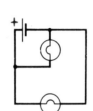
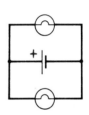

E.M.F. and potential difference

We have already seen (page 279) that cells are electron pumps and
that these pumps have different electrical pressures.
The scientific name for this electrical pressure is *electromotive force
(e.m.f.)* and it is measured in *volt* (V). (See also page 276.)

The e.m.f. of a dry cell is $1\frac{1}{2}$ V; the e.m.f. of a Van de Graaff
generator may be 100 000 V.

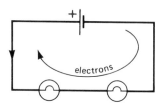

When a cell is connected to some lamps as in the
diagram, the e.m.f. of the cell pushes electrons
round the circuit. The energy to do this comes
from the chemical energy of the cell.
The electrons give up this energy to the thin wires
inside the lamps, so that they get hot and glow.

Across each lamp there is an electrical pressure
difference, called a *potential difference (p.d.)* or
"voltage". It is measured in *volt* (V) using a
voltmeter.

Experiment 35.12   Using a voltmeter
Connect a cell to two lamps in series as shown in
the diagram above.

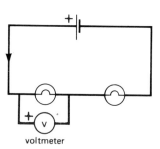
voltmeter

Then connect a suitable voltmeter *across* one of
the lamps as shown here.
As with an ammeter, the voltmeter must be connected the correct way round (+ to +).

What is the reading on your voltmeter?

Reconnect the voltmeter (always in *parallel, across*
a component) to find the potential difference
across the other lamp.

Then use your voltmeter to find the potential
difference across the cell.

What do you notice about your three readings?

Repeat the experiment with a battery of two cells
in the circuit.
What do you notice about the readings now?

Prefixes
These are used to describe very small or large
currents or p.d.'s.
e.g. 1 kilovolt (1kV) = 1000 volt
     1 milliamp (1mA) = $\frac{1}{1000}$ amp
     100 milliamp (100mA) = $\frac{100}{1000}$ amp = $\frac{1}{10}$ amp.

Prefix	Value
micro (μ)	one-millionth = $\frac{1}{1\,000\,000}$ = 10^{-6}
milli (m)	one-thousandth = $\frac{1}{1\,000}$ = 10^{-3}
kilo (k)	thousand = 1000 = 10^{3}
mega (M)	million = 1 000 000 = 10^{6}

Ohm's law and resistance

Experiment 35.13
Connect a single cell (or power pack) in series with
a 6V lamp and a suitable ammeter.
Note the ammeter reading.

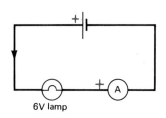
6V lamp

Connect another cell in series to assist the first cell
(or you can double the p.d. of the power pack).
What happens to the ammeter reading?

Use a third cell to increase the p.d. again.
What do you notice about the current as you
increase the potential difference?

In 1826, George Ohm discovered that **the current
flowing through a metal wire is** *proportional* **to
the potential difference across it (providing the
temperature remains constant).**

"Proportional' means that if you double the p.d.,
the current is doubled (see page 402).
Ohm's law does not apply precisely to your lamp
because its temperature changes (see page 295).

Resistance
The thin wire in a lamp tends to resist the movement of electrons in
it. We say that the wire has a certain *resistance* to the current. The
greater the resistance, the more p.d. is needed to push a current
through the wire. The resistance of a wire is calculated by:

$$\text{Resistance, R} = \frac{\text{p.d. across the wire (V)}}{\text{current through the wire (I)}}$$

or, in symbols: $R = \dfrac{V}{I}$ where V = p.d. (in V)
                                                    I = current (in A)
or, rearranging: $I = \dfrac{V}{R}$ R = resistance in a unit
                                                    called *ohm* (Ω).

or, easiest to
remember: $\boxed{V = I \times R}$

This fact you should know evermore,
(the formula comes from Ohm's law)
Overcome all resistance,
Learn by dogged persistence
That V = I x R

Example
A 2V accumulator is connected to a wire of
resistance 20 ohm.
What current flows in the circuit?

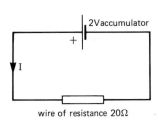
2V accumulator

wire of resistance 20Ω

write down the
formula first: $V = I \times R$

then put in
the numbers: $2 = I \times 20$

∴ The current, $I = \frac{2}{20} = \frac{1}{10}$ amp = <u>0·1 A</u>

What resistance would give you twice the current with the same accumulator?

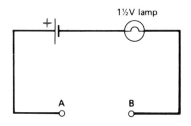

1½V lamp

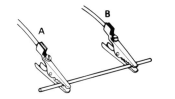

4 Factors affecting resistance

Experiment 35.14   Varying length

Connect the circuit shown , with crocodile clips at **A** and **B**.
To complete the circuit you need about 30cm of 30-gauge nichrome wire.
Put the clips on the wire close together and note the brightness of the lamp.

Now move the clips farther apart.
Is the lamp brighter or dimmer?
Is the current greater or less?
Has the resistance decreased or increased?

1. As the length increases, the resistance increases.
This fact is used in a rheostat (see opposite).

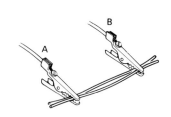

Experiment 35.15   Varying cross-sectional area

Using the same circuit, place your clips about 20cm apart on the nichrome wire. Note the brightness of the lamp.

Now add another (identical) length of nichrome wire between the clips so that the cross-sectional area is doubled.
Is the lamp brighter or dimmer?
Is the current greater or less?
Has the resistance decreased or increased?

2. As the cross-sectional area increases, the resistance decreases.

Experiment 35.16   Changing the substance

Using the same circuit, place your clips 20cm apart on a nichrome wire. Note the brightness of the lamp.

Now replace the nichrome wire by a *copper* wire of the same length and the same area (30-gauge).
Which substance has the greater resistance?

3. Copper is a good conductor and is used for connecting wires.
Nichrome has more resistance and is used in the heating elements of electric fires.

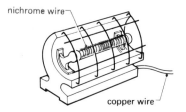

nichrome wire

copper wire

The resistance of a wire also changes as the temperature changes (you can investigate this in experiment 35.23).

4. The resistance of a wire increases as the temperature increases.
This is used in a resistance thermometer (see page 35).

Which has the greater resistance – a long, thin, hot nichrome wire or a short, thick, cool, copper wire?

Resistors

Resistors (sometimes made of a length of nichrome wire) can be used to reduce the current in a circuit.

fixed resistors

circuit symbol

A variable resistor or *rheostat* is used to vary the current in a circuit. As the sliding contact moves, it varies the length of wire in the circuit.

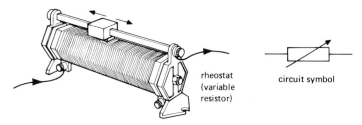

rheostat (variable resistor)

circuit symbol

Experiment 35.17 *Using a rheostat*

Connect a rheostat in a circuit as shown and use it to control the brightness of the lamp.

How are spotlights dimmed in a theatre?

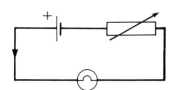

Experiment 35.18 *Measuring resistance*

To find the resistance of a resistor, use the circuit shown here. Connect all the series components first (the black parts of the diagram) and then add the voltmeter (in parallel) last of all.
Make sure that the meters are the correct way round and that you understand the marks on their scales.

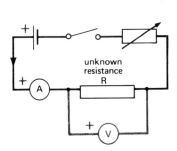

unknown resistance R

Use the rheostat to adjust the current to a convenient value and note the readings of the ammeter and voltmeter.

Move the rheostat and take more readings.

Calculate the resistance from the formula (page 289)

$$\frac{\text{Resistance}}{\text{(in } \Omega)} = \frac{\text{potential difference (in V)}}{\text{current (in A)}}$$

The average of the last column of the table should give you an accurate value of the resistance.

p.d. (in V)	current (in A)	resistance = $\dfrac{\text{p.d.}}{\text{current}}$ (in Ω)

Resistors in series

Experiment 35.19

Get two resistors whose values are marked (in Ω) and connect them in series as shown here (in black).

We know already (from page 286) that **the same current flows through each part of a series circuit.**

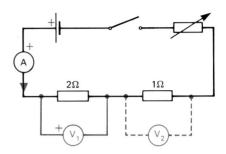

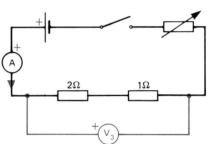

Now investigate the potential differences in a series circuit by connecting a suitable voltmeter:
a) in position V_1 across one of the resistors
b) in position V_2 across the other resistor
c) in position V_3 across both together.
What do you notice about the voltmeter readings?

For resistors in series, the total p.d. is the sum of the p.d.'s across the separate resistors ($V_3 = V_1 + V_2$).

What is the combined resistance of the two resistors in series? To find out, use the reading of V_3 and the ammeter reading to calculate the combined resistance (using $R = \dfrac{V}{I}$).

Do you find that a 2Ω resistor and a 1Ω resistor in series combine to act like a 3Ω resistor?
Try this with other resistors.

The total resistance (R) of a series circuit is equal to the sum of the separate resistances (R_1, R_2):

$$\boxed{R = R_1 + R_2}$$

Example.
A p.d. of 4V is applied to two resistors (of 6Ω and 2Ω) connected in series.
Calculate a) the combined resistance
           b) the current flowing
           c) the p.d. across the 6Ω resistor.

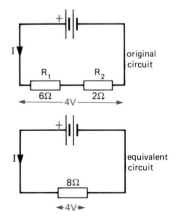

a) formula first: Combined resistance $= R_1 + R_2$
    then numbers: $= 6\Omega + 2\Omega$
                                         $= 8\Omega$

b) formula first: $V = I \times R$ for the combined resistance
    then numbers: $4 = I \times 8$
                     $\therefore \quad I = \frac{4}{8} = \frac{1}{2}$ amp $= \underline{0.5A}$

c) formula first: $V = I \times R$ for the 6Ω only
    then numbers: $V = 0.5 \times 6$
                     $V = \underline{3V}$ across the 6Ω resistor.

What is the p.d. across the 2Ω resistor?

Resistors in parallel

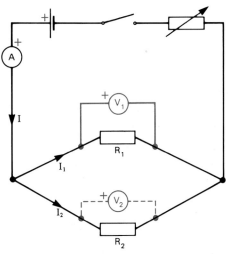

Experiment 35.20
Get two resistors whose values are marked (in Ω)
and connect them in parallel as shown here (in black).

Connect a suitable voltmeter in position V_1 across
one resistor and then in position V_2 across the
other resistor.
What do you notice about the readings?

**For resistors in parallel, the potential difference
across each is the same.**

We already know (from page 287) that
**the current in the main circuit is the sum of the
currents in the parallel branches ($I = I_1 + I_2$).**

What is the combined resistance of the two resistors in parallel? Use
your voltmeter reading with the ammeter reading to find the value
of the combined resistance (using $R = \dfrac{V}{I}$).

Notice that the **combined resistance (R) is less than either of the
separate parallel resistances** (R_1, R_2) because the electrons find it
easier to travel when they have more than one path they can take.

The combined resistance R can be found from:

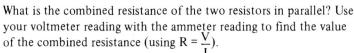

$$\frac{1}{R} = \frac{1}{R_1} + \frac{1}{R_2}$$

Do your values agree with this formula?

Example.
A p.d. of 6V is applied to two resistors (of 3Ω and 6Ω) connected
in parallel.
Calculate a) the combined resistance
          b) the current flowing in the main circuit
          c) the current in the 3Ω resistor.

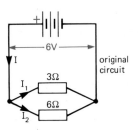

original circuit

a) formula first: $\dfrac{1}{R} = \dfrac{1}{R_1} + \dfrac{1}{R_2}$

     then numbers: $\dfrac{1}{R} = \dfrac{1}{3} + \dfrac{1}{6} = \dfrac{2+1}{6} = \dfrac{3}{6} = \dfrac{1}{2}$

           $\therefore$ combined resistance R = 2Ω

b) formula first: $V = I \times R$ for the combined resistance
     then numbers: $6 = I \times 2$
           $\therefore$ $I = \frac{6}{2} = 3$A in the main circuit.

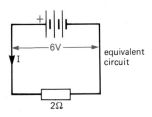

equivalent circuit

c) formula first: $V = I_1 \times R$ for the 3Ω resistor.
     then numbers: $6 = I_1 \times 3$
           $\therefore I_1 = \frac{6}{3} = 2$A in the 3$\Omega$ resistor.

What is the current, I_2, in the
6Ω resistor?

Potential divider (voltage divider)

A battery provides a fixed e.m.f. For example a 6V battery provides 6V only. However we often wish to provide a different voltage. To do this we use a *potential divider* circuit.

Experiment 35.21
Connect a 6V battery to two 100Ω resistors in series, as shown: Use a voltmeter to check that the p.d. across the 2 resistors is 6V. Then use the voltmeter to find the voltage across each resistor in turn. Do you find that the 6V is shared equally between the two resistors?

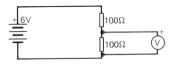

Now replace one resistor by a 200Ω resistor. The 200Ω resistor has two-thirds of the total resistance. Do you find two-thirds of the voltage is across it?

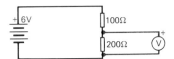

If we apply Ohm's Law to this circuit:

$$V_1 = IR_1 \text{ for resistor } R_1 \qquad \ldots (1)$$

$$V = I(R_1 + R_2) \text{ for the resistors in series} \ldots (2)$$

Dividing equation (1) by equation (2):

$$\frac{V_1}{V} = \frac{R_1}{R_1 + R_2} \quad or \quad V_1 = \frac{R_1}{R_1 + R_2} \times V$$

This is the potential divider equation

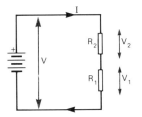

Experiment 35.22
Connect a battery to a rheostat (see page 291) so that all three of its terminals are used:
The fixed voltage of the battery is applied across the full length of the wire AB.
As the slider S is moved to different positions, the voltage across BS varies. If S is at B, the output voltage is zero. If S is at A, the output voltage is the full input voltage. If S is halfway between B and A, the output voltage is half the input voltage; and so on.

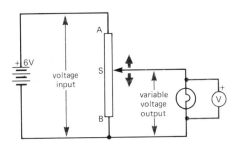

What happens to the brightness of the lamp as you move the slider S?
What happens to the voltmeter reading as you move S?

When a rheostat is used like this it is called a *potentiometer* or 'pot'.

Uses:
Potentiometers are used in radios and record players as volume controls and tone controls. They are often circular with a rotating slide or 'wiper':
Potential dividers are often used in electronic circuits.

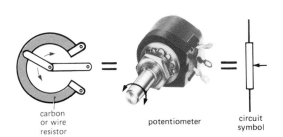

carbon or wire resistor   potentiometer   circuit symbol

Current: Voltage Graphs

In experiment 35.18 (on page 291), you measured several values of the **current** I through a resistor, and the **voltage** V across the resistor. If you plot a graph of these values, you get a *straight line through the origin:*

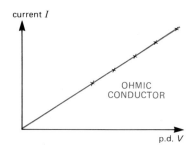

This means that the current I is *proportional* to the voltage V.
This is Ohm's Law (see page 289).
The steeper the graph, the lower the resistance.
The flatter the graph, the higher the resistance.

A substance that gives a straight graph like this is called an *ohmic conductor*. Copper wire and all other metals give this shape of graph, unless they change temperature.

Not all objects and substances give a straight graph like this. You can use the same circuit as before (experiment 35.18) to investigate what happens with other objects and materials. Alternatively, you can adapt the potentiometer circuit shown on the page opposite (experiment 35.22). Where should you include an ammeter in that circuit?

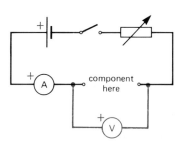

Experiment 35.23   A filament lamp
Use your circuit to investigate how the current I varies with the voltage V in a filament lamp (e.g. a torch bulb, see page 305).

You can see that the lamp is a *non-ohmic* conductor because the graph is not a straight line. It does not obey Ohm's Law.

As more current flows, the metal filament gets hotter and so its resistance *increases*. This means the graph gets flatter, as shown:

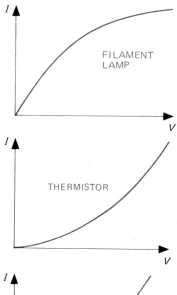

Experiment 35.24   A thermistor
A thermistor is used in electronics and is made of a semi-conductor substance (see page 351).
Look at the graph. Is a thermistor an ohmic conductor?
Why does the graph bend the opposite way to the lamp?

As more current flows, the thermistor gets hotter and so its resistance *decreases* (see page 351). The graph gets steeper.

Experiment 35.25   An ionic solution
An ionic solution (like dilute sulphuric acid) can carry an electric current (see pages 280 and 311).
In this case, no current flows until a certain voltage is reached. This voltage depends on the chemicals that are present.

Another non-ohmic conductor is a diode (see the graph on page 348).

295

Identifying Resistors

Resistors are often so small that it is difficult to label them clearly.
Special codes are used: either the British Standard Code or the older Colour Code:

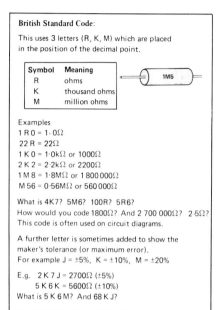

British Standard Code:

This uses 3 letters (R, K, M) which are placed in the position of the decimal point.

Symbol	Meaning
R	ohms
K	thousand ohms
M	million ohms

Examples
1 R 0 = 1·0Ω
22 R = 22Ω
1 K 0 = 1·0kΩ or 1000Ω
2 K 2 = 2·2kΩ or 2200Ω
1 M 8 = 1·8MΩ or 1 800 000Ω
M 56 = 0·56MΩ or 560 000Ω

What is 4K7? 5M6? 100R? 5R6?
How would you code 1800Ω? And 2 700 000Ω? 2·5Ω?
This code is often used on circuit diagrams.

A further letter is sometimes added to show the maker's tolerance (or maximum error).
For example J = ±5%, K = ±10%, M = ±20%

E.g. 2 K 7 J = 2700Ω (±5%)
 5 K 6 K = 5600Ω (±10%)
What is 5 K 6 M? And 68 K J?

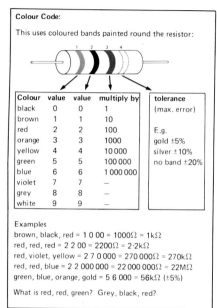

Colour Code:

This uses coloured bands painted round the resistor:

Colour	value	value	multiply by	tolerance
black	0	0	1	(max. error)
brown	1	1	10	
red	2	2	100	E.g.
orange	3	3	1000	gold ±5%
yellow	4	4	10 000	silver ±10%
green	5	5	100 000	no band ±20%
blue	6	6	1 000 000	
violet	7	7	–	
grey	8	8	–	
white	9	9	–	

Examples
brown, black, red = 1 0 00 = 1000Ω = 1kΩ
red, red, red = 2 2 00 = 2200Ω = 2·2kΩ
red, violet, yellow = 2 7 0 000 = 270 000Ω = 270kΩ
red, red, blue = 2 2 000 000 = 22 000 000Ω = 22MΩ
green, blue, orange, gold = 5 6 000 = 56kΩ (±5%)

What is red, red, green? Grey, black, red?

Resistivity

On page 290 we found that the resistance R of a wire depends on its length L and its cross-sectional area A.
In fact, R is proportional to L: $R \propto L$

and R is inversely proportional to A: $R \propto \dfrac{1}{A}$

Combining these two, we get $R \propto \dfrac{L}{A}$

We can change this into an equation (see page 402) by introducing a constant called the *resistivity*, written ρ (pronounced *ro*).
ρ is a number which depends upon the substance (see the table).

$$\therefore \quad \boxed{R = \rho \frac{L}{A}}$$

where: L is in metre, A in metre squared,
 R is in ohm, ρ is in *ohm-metre* (Ωm).

Substance	Resistivity ρ at $0°$C (in Ω m)
Insulators:	
polythene	$\sim 10^{16}$
ebonite	$\sim 10^{14}$
Semi-conductors:	
germanium	$\sim 10^{-3}$
silicon	
Conductors:	
iron	$8·9 \times 10^{-8}$
copper	$1·6 \times 10^{-8}$

Example

What is the resistance of a copper wire, diameter 2 mm, 100 m long?
what do we know: radius = 1 mm = $\frac{1}{1000}$ m = 10^{-3} m
 area = $\pi r^2 = 3·14 \times 10^{-3} \times 10^{-3} = 3·14 \times 10^{-6}$ m²

then formula: $R = \dfrac{\rho L}{A} = \dfrac{1·6 \times 10^{-8} \times 100}{3·14 \times 10^{-6}} = \dfrac{1·6 \times 10^{-6}}{3·14 \times 10^{-6}} = \underline{0·51\Omega}$

Energy changes in a circuit

When electrons flow round a circuit, they gain potential energy in the cell and then lose this energy in the resistors. The change in energy depends on the potential difference (p.d.), measured in volts.

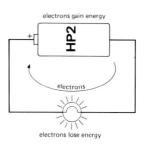

1 volt is the p.d. between two points if 1 joule of energy is needed to move 1 coulomb of charge between the two points (see also page 276). If a voltmeter connected across the lamp shows $1\frac{1}{2}$ V, this means that each coulomb of charge flowing through the lamp gives up $1\frac{1}{2}$ joule of energy (to heat the filament).

If Q coulombs pass through a p.d. of V volts, then

> the change in energy = $Q \times V$ joule

From page 325, since $Q = I$(amps) $\times t$(seconds), then

> the change in energy = $I \times t \times V$ joule

(see also page 306)

Experiment 35.26   E.m.f. and terminal p.d.
Connect a high resistance voltmeter across a dry cell on a circuit board. Note the reading of the voltmeter.

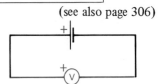

If we assume for the moment that the voltmeter is not taking any current from the cell, then this voltage is called the *electromotive force (e.m.f.)* of the cell. It is measured in volts and is the *total energy per coulomb* that is available from the cell.

Now connect a lamp, as shown, so that a current flows from the cell. What has happened to the voltmeter reading? What happens if a second lamp is connected, as shown?

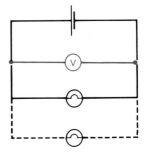

The voltmeter is now measuring the *terminal potential difference* across the terminals of the cell.
This terminal p.d. is always *less* than the e.m.f. of the cell because the cell has some *internal resistance.*
Some of the cell's energy (the 'lost volts') is used up in pushing the current through this internal resistance.

In the diagram, the e.m.f. = E volts and the internal resistance = r ohms.

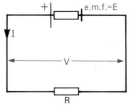

Then: energy *supplied*
      from chemical = energy *delivered* + energy *wasted*
      energy of cell   to external circuit   in internal resistance

$$\therefore \quad E = \text{terminal p.d., } V + \text{'lost volts'}$$

from page 289: $E = \quad I \times R \quad\quad + I \times r$

factorising: $\therefore \boxed{E = I(R + r)}$

Example. A battery of e.m.f. 6V and internal resistance 2Ω is connected to a resistance of 4Ω. What current flows?

formula first: $E = I(R + r)$
then numbers: $6 = I(4 + 2)$
           $\therefore I = \underline{1A}$

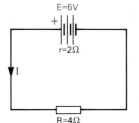

Summary

An electric current in a wire is a flow of electrons and can be measured in ampere (A) by a (low resistance) ammeter placed in series in a circuit.

The potential difference (p.d.) can be measured in volt (V) by a (high resistance) voltmeter placed in parallel, across a component.

Ohm's law: the current flowing in a wire is proportional to the p.d. across it, providing the temperature is constant.
$V = I \times R$
Changing this round gives $R = \dfrac{V}{I}$ or $I = \dfrac{V}{R}$.

To find the one you want, cover up that letter in the triangle and the remainder shows you the formula.

The resistance of a wire increases with length and temperature; it decreases as the cross-sectional area is increased. It also depends on the substance.

In a *series* circuit:
a) the same current flows through each component
b) total p.d. = sum of p.d.'s across the separate resistors.
c) combined resistance $R = R_2 + R_2 + R_3 + \ldots$

In a *parallel* circuit:
a) there is the same p.d. across each component in parallel
b) total current = sum of currents in the separate resistors.
c) $\dfrac{1}{\text{combined resistance } R} = \dfrac{1}{R_1} + \dfrac{1}{R_2} + \dfrac{1}{R_3} + \ldots$

Questions

1. Copy out and complete:
   a) A current is a flow of . . . . . For this to happen there must be a . . . . . circuit.
   b) Copper is a good . . . . . Plastic is an . . . . .
   c) Current is measured in . . . . . using an . . . . . placed in . . . . . in a circuit.
   d) Potential difference (p.d.) is measured in . . . . . using a . . . . . placed in . . . . . across a component.
   e) Ohm's law: the . . . . . flowing through a conductor is . . . . . to the . . . . . difference across it, providing the . . . . . is constant.
   f) The formula connecting V (. . . . . difference), I (. . . . .) and R (. . . . .) is . . . . .

   g) Resistance is measured in . . . . . The resistance of a wire increases as the length . . . . . ; as the temperature . . . . . ; and as the cross-sectional area . . . . .

2. Arrange the following materials into conductors and insulators: copper, iron, rubber, nylon, sulphuric acid, polythene.
   List 10 uses of electrical insulators.

3. Draw a circuit diagram to show how two 4V lamps can be lit brightly from two 2V cells.

4. The Prof. is lost, can't see the light,
   So help him get his circuits right.

298

5. Explain what happens to the identical lamps X and Y in the circuit shown, when a) switch S_1 only is closed b) S_2 only is closed c) S_1 and S_2 are closed.

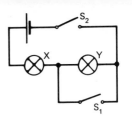

6. Draw a circuit diagram to show how 3 lamps can be lit from a battery so that 2 lamps are controlled by the same switch while the third lamp has its own switch.

7. In the diagram, the lamps are identical and ammeter A_2 reads 0.5A.
   a) What are the readings on A_1, A_3, A_4?
   b) Redraw the diagram to include a switch to control both lamps together.
   c) Redraw the diagram to include two switches to control the lamps separately.
   d) Redraw the diagram to include a rheostat to control lamp P only.

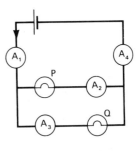

8. a) What p.d. is needed to drive a current of 2A through a 5Ω resistor?
   b) A p.d. of 6V is applied across a 2Ω resistor. What current flows?
   c) What is the value of a resistance if 20V drives 2A through it?

9. In the circuit shown, the p.d. across the 4Ω resistor is 8V.
   a) What is the current through the 4Ω resistor?
   b) What is the current through the 3Ω resistor?
   c) What is the p.d. across the 3Ω resistor?
   d) What is the p.d. applied by the battery?

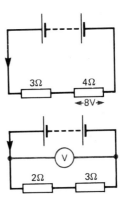

10. In the circuit shown, the voltmeter reads 10V.
    a) What is the combined resistance?
    b) What current flows?
    c) What is the p.d. across the 2Ω resistor?
    d) What is the p.d. across the 3Ω resistor?

11. In the circuit shown:
    a) What is the combined resistance?
    b) What is the p.d. across the combined resistance?
    c) What is the p.d. across the 3Ω resistor?
    d) What is the current through the 3Ω resistor?
    e) What is the current through the 6Ω resistor?

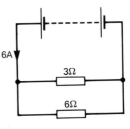

12. If the battery in question 10 has an internal resistance of 1Ω, what is its e.m.f.?

13. What is the resistance of a copper wire, diameter 4 mm, 800 m long?

14. It seems the Prof's gone out of his mind. How many errors can you find?

Magnetism

1. Describe with the help of a suitable diagram, how you would use an electrical method to
   (a) magnetise a piece of steel,
   (b) demagnetise a magnet.

2. Use the molecular theory of magnetism and labelled sketches to explain each of the following.
   a) When a bar magnet is broken in two, each piece is a magnet.
   b) Heating a bar magnet destroys its magnetism.   (EA)

3. a)

   A piece of unmagnetized steel and a piece of unmagnetized iron are both placed in contact with the N pole of a bar magnet as indicated in the diagram. Copy the diagram and on it indicate the polarity of the steel and the iron.
   b) Explain why the lower ends of the steel and the iron do not come together.
   c) Both pieces are removed from the magnet and placed in some iron filings. Which picks up most filings? Explain why.
   d) If you were given a short length of steel and a magnet, how would you turn the steel into a magnet other than by just leaving it in contact with the magnet for some time?   (SE)

4. A short soft-iron bar, free to move, is placed in line with the axis of a strong bar magnet with one end of the bar a few centimetres from the N pole of the magnet. State and explain what happens.

If the bar is surrounded by a thick soft-iron ring before the magnet is again put in position, explain any difference in the effect produced.   (JMB)

5. Describe how you would plot the magnetic field in the region around a bar magnet, and draw a diagram to show the field lines you would expect to obtain, ignoring the effect of the earth's magnetic field.

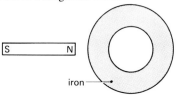

   Show how the magnetic field of the bar magnet would be modified by an iron ring placed as shown in the diagram. Describe how you would verify what you have shown for the region inside the ring.
   Describe how you would magnetise an iron bar to produce a N-pole at a marked end.   (C)

6. Using a circle of radius 5 cm to represent the Earth, draw lines to illustrate the Earth's magnetic field.
   Indicate clearly
   a) the Earth's geographic north pole,
   b) the axis of rotation of the Earth,
   c) the magnetic north pole,
   d) the direction of the magnetic lines of force.   (AEB)

7. Explain why
   a) the north seeking pole of a magnetised knitting needle, when suspended so that it can rotate in a *horizontal* plane, does not usually point to the geographic north pole, and
   b) the same needle, when suspended so that it can rotate in a *vertical* plane, usually comes to rest at an angle to the horizontal. Describe what steps you could take to demagnetise the needle completely.   (L)

Electrostatics

8. Describe and explain what happens when a charged rod is brought near to an uncharged light metal sphere which hangs on an insulating thread.

(C)

9. a) Draw and clearly label a diagram of an electroscope.
   b) Describe with the aid of a series of simple diagrams how you would charge an electroscope by the method of induction, using a negatively charged rod.

(AEB)

10. a) An electroscope is given a positive charge. Draw a diagram showing the distribution of charge.
    b) A positively charged rod is then brought near to, but not touching, the cap of the electroscope. Draw another diagram to show the new distribution of charge.

(W)

11. When an insulated metal plate is brought close to, but not touching, the cap of a positively charged gold leaf electroscope, the divergence of the leaf is reduced. What possible states of charge of the insulated plate would cause this effect?

(O&C)

12. a) A pear shaped conductor is mounted on an insulating stand and given an electrostatic charge. How would you expect the charge per unit area to vary over the surface of the conductor? Describe how you would demonstrate this by experiment. How would you expect the electrostatic potential to vary over the surface? Give a reason.
    b) Describe with a clear diagram the action of a lightning conductor.

(S)

13. a) What is meant by electrostatic induction?
    b) How would you show that there are two kinds of static electricity?
    c) Describe an experiment to show the discharging action of a needle.

(WAEC)

14. Define the *capacitance* of a capacitor. Draw a clearly labelled diagram of a gold-leaf electroscope. Describe how you would use the electroscope to investigate the effect on the capacitance of a parallel-plate capacitor of changing
    a) the area of the plates,
    b) the distance between the plates,
    c) the dielectric between the plates.
    State clearly the conclusions which you would expect to reach.

(AEB)

15. a) *A positively charged plastic rod is brought close to the cap of an uncharged gold leaf electroscope.*
    (i)   What will be seen as the rod is brought close to the cap?
    With the rod still held close, the cap is touched with a finger.
    (ii) What will then be seen?
    The rod and the finger are then removed.
    (iii) What will then be seen and what will be the charge (if any) on the electroscope?
    b) (i)   Describe, with the aid of a diagram, the construction of a paper dielectric capacitor.
    (ii) On what factors does the capacitance of such a capacitor depend?
    (iii) A $2\mu F$ capacitor is connected to a 400V d.c. supply. Calculate the charge (in μC) which will be stored.

(S16+)

Circuits

16. Calculate the resistance of the single resistor which is equivalent to two resistors, one 5 ohm and one 20 ohm, when they are connected (a) in series, (b) in parallel. (AEB)

17. Two resistors, of $2 \cdot 0\Omega$ and $3 \cdot 0\Omega$, are connected first in series, and then in parallel with a battery of e.m.f. $6 \cdot 0$ V and negligible internal resistance. For each case draw a circuit diagram and calculate the current through the battery.   (C)

18.
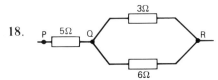

The potential difference across PR is 14 volts. Calculate
a) the equivalent resistance of the network QR,
b) the current flowing through PQ,
c) the potential difference across PQ,
d) the current flowing through the 3 ohm resistor.   (AEB)

19.
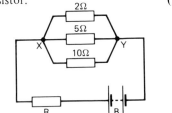

In the circuit above, R is a resistor and B an 8-volt battery of negligible internal resistance. The potential difference between X and Y is 5 V.
a) Determine the effective resistance of the set of three parallel resistors between X and Y.
b) What is the potential difference across the 2 Ω resistor?
c) Determine the current flowing through the 10 Ω resistor.
d) Determine the current flowing through the resistor R.
e) Determine the magnitude of the resistance of R.   (NI)

20. A pupil made an electric circuit like the one below in order to measure the current through the lamps.

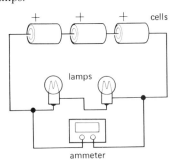

a) Are the two lamps in *series* or in *parallel*?   [1 mark]
b) The pupil has made a mistake in this circuit. What is the mistake?   [2]
c) Draw a diagram showing the *correct* way to connect this circuit. Use the proper circuit symbols in your diagram.   [4]
                    (NEA)

21. a) A length of wire has a resistance of about 8 ohm. How would you measure this resistance, using an ammeter and a voltmeter? Draw a labelled circuit diagram. Explain how you would calculate the resistance from your readings.
b) For a particular specimen of wire, a series of readings of the current through the wire for different potential differences across it is taken and plotted as shown. Describe how the resistance is changing. Suggest the most likely explanation.   [4 marks]

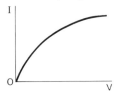

c) How would the resistance of a piece of wire change if
i) the length were doubled,   [2]
ii) the diameter were doubled?   [2]
The growth of cracks in a brick wall can be detected by measuring the change in resistance of a fine piece of wire. The wire

is stretched tightly and fixed firmly to the wall on each side of the crack:

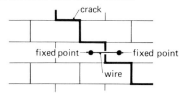

d) If the crack opens slightly, what happens to
   i) the length of wire, [1]
   ii) the thickness of wire, [1]
   iii) the resistance of the wire? [2]
   (NEA)

23. a) An experiment was performed to measure the direct current passing through a 12 V 24 W tungsten filament lamp when various potential differences were applied to the lamp.
   (i) Draw a labelled diagram of the circuit you would use if you were asked to carry out this experiment. (Do **not** describe the experiment.)
   (ii) State suitable ranges for the various components in the circuit.
   (iii) What is the normal operating current for such a lamp?
   b) The following set of results was obtained from the experiment referred to in (a).

p.d. (V)	2	4	6	8	10
current (A)	0·74	1·10	1·38	1·60	1·80

   (i) Plot a graph of p.d. on the y-axis against current on the x-axis.
   (ii) Explain why the graph is not a straight line.
   (iii) Use the graph to obtain a value for the resistance of the filament when the p.d. applied to the lamp is 3·0 V.
   (JMB)

24. X and Y are reels of wire of the same metal. A 1 m length from reel X has a resistance of 5 Ω. The wire from reel Y has a cross-sectional area three times that of the wire from reel X. Calculate the resistance of:
   a) 4 m of wire from reel X;
   b) 1 m of wire from reel Y.
   c) How long would a piece of wire cut from the reel Y have to be for its resistance to be 10 Ω?
   (S16+)

25. A six-volt battery is used with a potential divider to provide a variable voltage across a torch bulb rated at 3 V 0.3 A.
   a) Sketch a circuit diagram showing
      (i) the arrangement of the potential divider and bulb,
      (ii) how the voltage across the bulb could be measured,
      (iii) how the current through the bulb could be measured. [4 marks]
   b) Sketch a graph showing how the current through the bulb varies with the voltage across it. [4]
   c) Describe and explain how the resistance of the bulb filament changes with increasing current. [4]
   (NEA)

26. An experiment was set up to investigate the heating effect of an electric current. A constant current I was passed through a coil of constant resistance immersed in water, T, the temperature rise per minute, was measured for various values of current I. The values obtained are given in the table.

I in A	0	1·00	2·00	2·50	3·00	3·50	4·00
T in K/min	0	0·06	0·26	0·40	0·57	0·74	0·95

   a) Draw up a table of values I^2 and T. [4]
   b) Draw a graph of T (y-axis) against I^2 (x-axis) [6]
   c) Determine the slope of the straight line portion of the graph. [4]
   d) Determine the slope of the graph at a rate of temperature rise of 0·8 K/min. [4]
   e) Suggest a reason for the change in the slope of the graph. [2]
   (SEG)

Further Reading

Physics Plus: *Electrostatics outside the Laboratory* (CRAC)
Over 1000 Physics formulae – Moore (Hamlyn)
Mr Tompkins in Wonderland – Gamow (C.U.P.)

chapter 36

Heating effect of a current

You know that when electrons pass through a wire, they can give some of their energy to the atoms in the wire and make them vibrate more, so that the wire gets hotter.

The greater the resistance, the hotter it becomes – this is why an electric fire glows red-hot while the connecting wires stay cool (see also page 290).

Electric fires

Radiant electric fires glow red-hot about 900°C. Sometimes the coil of nichrome wire is inside a silica tube.

Why is this safer?

What is the concave mirror for?

Convector electric fires are cooler at about 450°C, and are used to warm a room by convection currents (see page 52).

Alec Trishan once plugged in his fire,
To raise the room temperature higher,
But he soon got too hot,
And he thought he should not,
So he tied a large knot in the wire.

    (Find *two* things wrong with this.)

Immersion heaters

The diagram shows an *immersion heater* fitted into a hot water tank. The heater contains resistance wire, like an electric fire but insulated from the water.

Where is the tank hottest? Why?

Through which pipe is hot water delivered?

What is the other pipe for?

Why should the tank be thickly lagged?

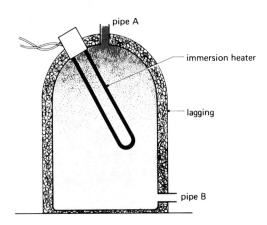

Irons

An electric *laundry iron* contains a heating element of resistance wire wound on to a piece of mica (an insulator) and held between sheets of mica so that it cannot touch the metal parts of the iron.

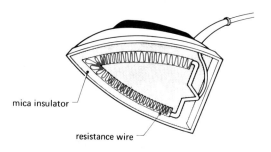

Make a list of all the electrical heaters in your home – remember to include electric kettles and cookers, hair-driers, toasters, electric blankets etc.

Electric filament lamp

Look closely at a light bulb — can you see that the filament is a coil of wire which has been coiled again to give a hotter "coiled coil"? The filament (heated to 2500°C) is made of tungsten which has a high melting point.

An inert gas is usually included to reduce the evaporation of the tungsten.

These lamps are not very efficient. The sodium discharge lamps used for street-lighting and the fluorescent tubes used for room lighting are about 4 times as efficient and so cheaper to run.

Which kind of lamp would you use to reduce the shadows in a classroom?

How do you know that the lamps in your home are wired in parallel and not in series?

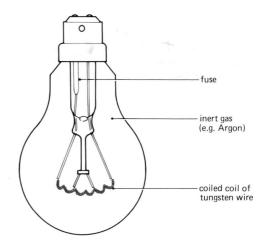

The circuit diagram shows how a lamp is wired to a ceiling rose and a switch (L and N refer to the 'live' and 'neutral' wires from the a.c. mains supply — see also page 308).

The switch *must* be in the live wire for safety when you are changing lamps.
N.B. *Never* inspect any mains wiring without first switching it off at the mains fuse box.

Why should wall switches not be fitted in bathrooms? Why are string pull switches safer?

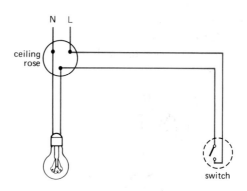

Two-way switches

With a two-way switch as shown in the diagram, wire C can be connected to either A or B. These switches can be used to switch a lamp from two positions (e.g. the top and bottom of the stairs).

Experiment 36.1

If you have 2 two-way switches, connect them as shown.

Can you control the lamp with either switch? In the diagram, is the lamp on or off?

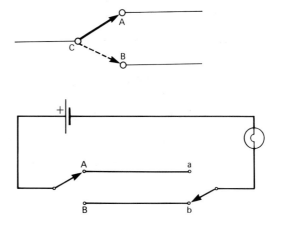

Electrical power

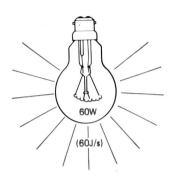

60W

(60J/s)

In Physics, we always measure energy in joules (see page 127). Therefore the rate of using energy, called *power*, is measured in joules per second, called *watts* (W), (see page 132).

If a lamp is marked 60W, it means it is converting electricity to heat and light at the rate of 60 joules per second.

From page 297,
the change in electrical energy (in joules) = current, I x time x p.d., V

$\therefore$

| **Power, P = current, I x potential difference, V** |
| (in watt)   (in amp)   (in volt) |

or

$$P = I \times V$$

For resistors, we can combine $P = I \times V$ with $V = I \times R$ (from Ohm's Law) so that we get alternative equations:

$$P = I \times V = I^2 \times R = \frac{V^2}{R} \text{ watt}$$

If these equations are multiplied by the time t, then they become equations for the electrical energy converted to heat:

$$\text{Electrical energy converted} = IVt = I^2Rt = \frac{V^2 t}{R} \text{ joule}$$

Example.

An electric kettle connected to the 240V mains supply draws a current of 10A.
Calculate a) the power of the kettle
 b) the energy produced in 20 seconds.
 c) the rise in temperature if *all* this energy is given to 2 kg of water (see page 45).
 (specific heat capacity of water = 4200 J/kg°C)

a) formula first: Power, P = current, I x p.d., V
   then numbers: = 10A x 240V
                       = <u>2400W</u> (= 2·4kW)
                    (where 1kW = 1 kilowatt = 1000W).

b) A power of 2400W means that the energy produced is
   a) 2400 joules per second.
   $\therefore$ In 20 seconds, the energy produced = 2400 x 20 joules
                                = <u>48 000J</u> (= 48kJ)

c) from page 45: $\dfrac{\text{energy}}{\text{given}} = \dfrac{\text{specific heat}}{\text{capacity}} \times \text{mass} \times \dfrac{\text{change in}}{\text{temperature}}$

   then numbers: 48 000 = 4200 x 2 x $\dfrac{\text{rise in}}{\text{temperature}}$

   $\therefore \dfrac{\text{rise in}}{\text{temperature}} = \dfrac{48\,000}{4200 \times 2} = \underline{5\cdot7°C}$

Power of appliances in the home

This can vary widely, from 3 watt for an electric clock to 5 kilowatt for an electric cooker. The power is usually marked on a label on the back of the appliance. Look carefully at some in your home and at the values shown in the table.

Appliance	Power
lamp	100 W = 0.1 kW
T.V.	200 W = 0.2 kW
vacuum cleaner	500 W = 0.5 kW
one-bar fire	1000 W = 1.0 kW
kettle	2500 W = 2.5 kW

The cost of electricity

If you look at your electricity meter at home you will see it is marked in *kilowatt-hour* (kWh).

The *kilowatt-hour* is a unit of *energy* (*not* power) and is calculated by:

> **Energy consumed = power x time**
> (in kWh)   (in kW) (in hours)

Since 1 kW = 1000 W and 1 hour = 3600 seconds, it follows that 1 kWh = 3 600 000 joules.

Example
A 2000W electric fire is used for 10 hours. What is the cost, at 8p per kWh?

formula   Energy consumed = power x time
first: (in kWh)   (in kW) (in hours)
then numbers: = 2kW x 10 hours
                                  = 20 kWh

∴ Cost at 8p per kWh   = 20 x 8p
                           = 160p

Example
The diagram shows the dials of an older type of meter at an interval of a month.
What is the electricity bill for the month at 8p per unit (kWh)?

Look carefully at the dials and the pointers.
The first reading is 1386.
Do you agree that the second reading is 2586?

∴ Number of units used = 2586 − 1386 kWh
                        = 1200 kWh

∴ Cost at 8p per unit   = 1200 x 8p
                        = 9600p = £96.00

The Prof. got a bill for his heater,
The shock turned him into a cheater.
He tried to reverse
The effect on his purse
By reversing the wires to the meter.
(*Can you explain why this would not work*).

first reading

second reading

307

Fuses

Experiment 36.2

Connect the circuit shown in black in the diagram, with crocodile clips at **A** and **B**.

Complete the circuit with a short length (about 3cm) of 36-gauge tin wire, placed between **A** and **B**. Does the lamp light?

Now imitate a fault in the circuit by using a piece of copper wire to 'short' out the lamp (as shown in red).
What happens to the thin tin wire?

The thin tin wire melted or *fused* and so stopped the current flowing.
Fuses are included in circuits as deliberate weak links for safety, so that if a fault occurs and too much current flows, the fuse wire melts before anything else is damaged or starts a fire.

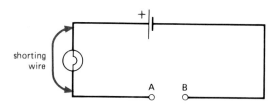

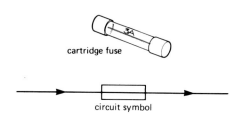

cartridge fuse

circuit symbol

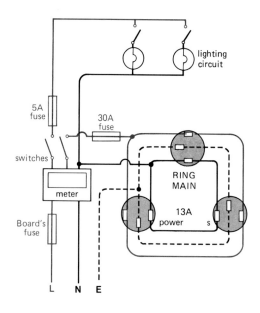

A young electrician called Hyde,
Learned nothing because of his pride.
He got his wires wrong,
And before very long,
He cried — then he sighed — then he died.

Household wiring

Here is a simplified diagram of a modern household circuit. It is in two parts: — a lighting circuit protected by a 5 amp fuse and a *ring main* power circuit (for electric fires etc) protected by a 30 amp fuse.
Each power point is also protected by a fuse inside the plug (see opposite page).

The current from the mains supply is a.c. (alternating current, see page 331).

L and **N** in the diagram refer to the Live and Neutral wires coming from the power station. The third wire (**E**) is connected to earth.
The *Live* wire is the most dangerous — the mains voltage may easily kill you.
The *Neutral* wire also carries the current, but because it is earthed back at the power station, its voltage is usually not as high as the live wire.
The *Earth* wire is for safety and only carries a current when there is a fault, so that a fuse melts (see opposite page).

Never touch any part of a mains circuit without first switching off at the main switch near the meter.

The earth wire

If a fault develops in an appliance and a live wire touches the metal case, a large current flows from the live wire to the earth wire and the fuse breaks. This makes the appliance safe until the fault is mended and the fuse replaced.

Notice that the fuse (like a switch) *must* be placed in the *live* wire so that it can isolate the appliance from the live wire.

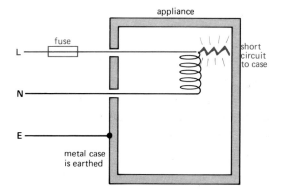

Fused plugs

It is absolutely essential that a plug is wired with the coloured wires going to the correct places:
Live – brown wire (old colour was red)
Neutral – blue wire (old colour was black)
Earth – green and yellow (old colour green)

Practise wiring a plug, taking care not to damage the insulation or the wires by cutting too deeply. Do not make the bare ends longer than necessary and do not over-tighten the cord-grip screws. Make sure that you fit a fuse of the correct value.

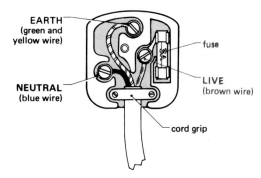

Fuse rating

The fuse rating is the maximum current that the fuse can carry without melting.
Only certain values are available: e.g. 3A, 5A, 10A, 13A.

Example.
A vacuum cleaner has a rating of 720W on the 240V mains. What fuse should be fitted in the plug?

formula first: Power = current x p.d. (from page 306)
720W = current x 240V
∴ Current $= \dfrac{720}{240} = \underline{3A.}$

A 3A fuse would be likely to melt so the next size (5A) should be chosen.
A 13A fuse should *not* be fitted – it would allow a large and damaging current to flow before melting.

What fuse rating would you choose for
a) a table lamp of 100W
b) an electric fire of 2400W?

A careless young student called Hughes,
Once fitted the wrong size of fuse.
This fault caused a wire
In his house to catch fire.
– His death made the 6 o'clock news!

Summary

Power = IV watt (or I^2R or $\dfrac{V^2}{R}$) where I in ampere, V in volt, R in ohm.

Energy (in joule) = power (in W) x time (in seconds).

Energy (in kWh) = power (in kW) x time (in hours).

A fuse is used to protect a circuit and is always placed in the live wire.

Mains wiring colours: brown (live), blue (neutral), green and yellow (earth).

Questions

1. Copy out and complete:
   a) The power (in watts) in an electric circuit is equal to the current (in . . . . .) multiplied by the . . . . . (in . . . . .).
   b) To calculate the energy (in joule), the power (in . . . . .) is multiplied by the . . . . . (in . . . . .).
   c) To calculate the energy (in kilowatt-hour), the . . . . . (in . . . . .) is multiplied by the . . . . . (in . . . . .).
   d) A fuse (or a light switch) should always be placed in the . . . . . wire of a mains circuit.
   e) When wiring a fused plug, the brown wire should go to the . . . . . terminal, the blue wire to the . . . . . terminal and the green and yellow wire to the . . . . . terminal.
   f) The earth wire should be connected to the . . . . . of an appliance.

2. A vacuum cleaner is labelled 250V 500W. When connected to a 250V supply, how much current does it take?

3. Copy out and complete the table:

4. The Prof. has got a mental block – Now tell me why he gets a shock.

5. A 60W lamp is connected to the 240V mains supply.
   a) What current does it take?
   b) What is the resistance of the filament?

6. A 2kW fire is switched on for 6 hours. What is the cost if one unit (kWh) costs 8p?

7. A 2kW fire, a 200W T.V. and three 100W lamps are all switched on from 6 p.m. to 10 p.m. What is the cost at 8p per kWh?

8. The diagram shows the inside of a 3-pin plug.
   a) What is the colour of each of the wires attached to A, B, C?
   b) Which of these is (i) the earth wire (ii) the live wire?
   c) What is placed in the gap CD? d) Why is this component necessary? e) How is its value chosen? f) Why is the direction of the wire round screw A not as satisfactory as those round B and C?

Appliance	Power (W)	p.d. (V)	current (A)	correct fuse (from 3A,5A,10A,13A)
car headlamp	48W	12V		
T.V.	200W	240V		
hairdrier		240V	2A	
iron	960W	240V		
kettle		240V	10A	

Further questions on page 366.

chapter 37
CHEMICAL effect of a current

What happens if you try to pass electricity through a liquid?

Experiment 37.1
Connect the circuit, as shown, to two electrodes in a beaker.
Pour a liquid into the beaker and use the lamp to see if the liquid can conduct electricity. Try this with distilled water, salt water, vinegar, paraffin, acids and copper sulphate solution. (Remember to wash out the apparatus each time you change the liquid.)

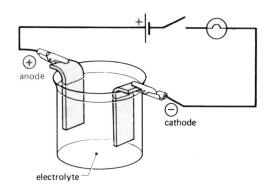

Liquids which conduct electricity are called *electrolytes* (see page 279). Show your results in a table of electrolytes and non-electrolytes.

When electricity is passed through a liquid, it may change the liquid – this is called *electrolysis*. The apparatus used for electrolysis is called a *vol**t**ameter* (not a voltmeter).

Experiment 37.2   Electrolysis of water
Pass electricity through water, using a water voltameter with two platinum electrodes connected to a battery as shown. Water is a poor conductor of electricity, so some sulphuric acid must be added to make an electrolyte.

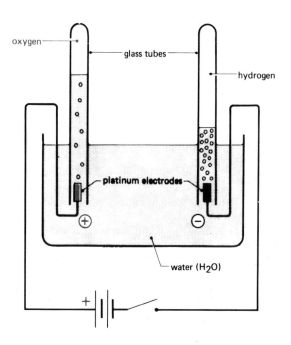

What happens?

When a current is flowing, bubbles appear at both the anode (+) and the cathode (–).
If your teacher can collect and test these gases, you will find that it is oxygen at the anode and hydrogen at the cathode.

What do you think has happened?

The electricity has a chemical effect on the water and has split it into oxygen and hydrogen. (The volume of hydrogen is twice that of oxygen and so the formula for water is H_2O).

Electroplating

Experiment 37.3   Electrolysis of copper sulphate
Connect the circuit shown here to a copper voltameter consisting of a beaker of copper sulphate
solution and two clean copper electrodes.
Label and find the mass of each electrode (to
0·1 gram) before putting them in the electrolyte.

Pass a current of about $\frac{1}{2}$ A for about $\frac{1}{2}$ hour.

Then carefully dry the electrodes and find the
new mass of each one. What do you find?

The cathode (–) has gained mass and is covered
with a bright new coating of copper.
We say it has been *electroplated* with copper. The
anode (+) has lost an equal mass of copper.

Experiments (first done by Michael Faraday) show
that the *mass* of metal deposited in electrolysis
is proportional to the *current* and proportional to
the *time* for which it flows.

Experiments also show that:
> **the metal (or hydrogen if it is present) is
> always deposited at the cathode.**

Experiment 37.4   Copperplating an object
Replace the copper *cathode* of experiment 37.3
by a clean metal object (e.g. a key or a badge).

Pass a very small current for several minutes
until your object becomes plated with copper.

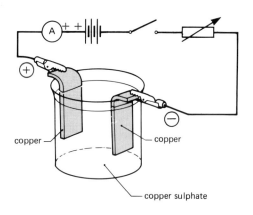

copper — copper

copper sulphate

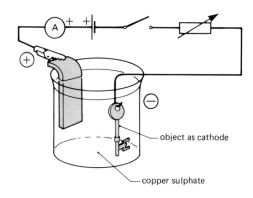

object as cathode

copper sulphate

The ionic theory
A molecule of copper sulphate has the chemical
formula $CuSO_4$ (= 1 atom of copper Cu, 1 atom
of sulphur S, 4 atoms of oxygen O_4).
When copper sulphate is dissolved in water, we
believe that it splits into two charged parts (Cu^{++}
and SO_4^{--}. These are called *ions* (see page 272).

The copper atom Cu, loses 2 electrons to become
a positive ion Cu^{++}.
As unlike charges attract, this Cu^{++} ion drifts
toward the negative cathode where it gains 2
electrons to become an atom of copper plating on
the cathode. The SO_4^{--} ion drifts to the positive
anode.

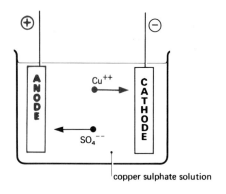

copper sulphate solution

Uses of electrolysis

1. Electroplating

In a similar way to your copper-plating in experiment 37.4, spoons can be silver-plated if they are used as the cathode of a voltameter containing silver.

Chromium plating for cars and bicycles is achieved by a similar method.

2. Refining copper.

In experiment 37.3, the copper transferred to the cathode was very pure even though the anode may have been made of impure copper. This process is used to produce pure copper for electric cables.

3. Manufacture of sodium and aluminium

These metals are obtained by electrolysis (of common salt for sodium, and of aluminium oxide for aluminium).

Plating cycle wheels with nickel

Something to do Electric writing

Soak some blotting paper in a solution of potassium iodide (from a chemist) which has had some starch added. Place the damp paper on a metal tray and connect the circuit shown. With a 12V battery connected, the positive wire leaves a mark (where iodine is liberated by electrolysis).

What happens if you use a *safe* 12V <u>a.c.</u> supply (e.g. a model railway transformer)? Why?

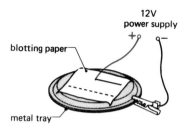

Questions

1. Copy out and complete:
   a) A liquid which conducts electricity is called an . . . . .
   b) The positive electrode of a voltameter is called the . . . . . and the negative electrode is called the . . . . .
   c) In the electrolysis of water, hydrogen forms at the . . . . . and . . . . . forms at the . . . . .
   d) In the electrolysis of copper sulphate, ... is deposited on the . . . . .
   e) A metal or hydrogen is always deposited at the . . . . .
   f) Faraday discovered that the . . . . . of metal deposited is proportional to the . . . . . and the . . . . .
   g) The current in the electrolyte is carried by charged atoms called . . . . . The positive . . . . . move towards the . . . . . In copper sulphate solution, the atoms of copper have lost . . . . . and so have a . . . . . charge which causes them to move towards the . . . . .

2. The diagram shows some apparatus used to copper-plate an object at B.

   a) What is the substance at A?
   b) What is the liquid C?
   c) What would be connected to the positive of the battery?
   d) What factors determine the mass of copper deposited?

3. In a particular voltameter, 1 gram of copper is deposited. The experiment is repeated at three times the current for twice the time. What mass of copper is deposited?

4. Explain why Professor Messer had difficulty in electroplating a plastic toy.

How much revision have you done? See page 390.

chapter 38

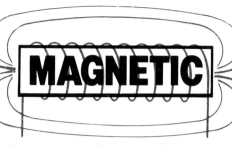

MAGNETIC effect of a current

We have seen in the last two chapters that an electric current has a *heating* and a *chemical* effect.

In 1819, Hans Christian Oersted discovered that electricity also has a *magnetic* effect.

Experiment 38.1   *Oersted's experiment*
Let a compass settle in a north-south direction. Then, with the circuit shown, hold the wire over the compass and press the switch (for a few seconds only).

What happens to the compass needle when the current flows?
What happens to the deflection of the compass if you reverse the direction of the current?

It seems that a wire carrying a current must have a magnetic field round it.

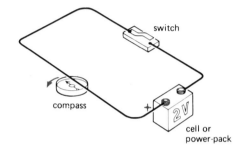

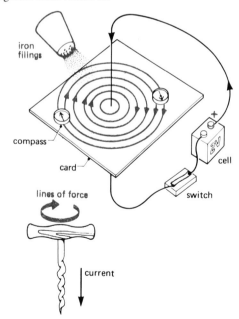

Experiment 38.2   *Finding the shape of the field*
Using the apparatus shown, sprinkle some iron filings on to the horizontal card.
Pass a large current through the vertical wire (briefly), while tapping the card so that the iron filings show the shape of the magnetic field.

Can you see that *the magnetic field round the wire is in the shape of circles*?

Place a small compass on the card to find the direction of the lines of force.

Does your result agree with the diagram?
What happens if you reverse the current?

The diagram shows how to remember this result using *Maxwell's screw rule*: if the corkscrew (or a screwdriver) moves in the direction of the current, the hand turns in the direction of the lines of force.

314

The magnetic field near a coil

Experiment 38.3   A short coil
Sprinkle iron filings round a coil as shown in the diagram.
Pass a large current and tap the card.

Study the shape of the magnetic field.
Use a small compass to find the direction of the field.
Does the direction of the field round each wire agree with the result of experiment 38.2?

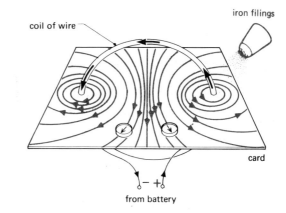

Experiment 38.4   A long coil (or 'solenoid')
Repeat experiment 38.3 but with a long coil or *solenoid*.

Study the shape of the field. Does it look familiar? (Look at page 264 if you are not sure.)

The magnetic field round a solenoid has the same shape as the field round a bar magnet.

Notice that the field *inside* the solenoid is very strong and uniform. It can be used to magnetise objects (see also page 260).

Use a small compass to find which end of the solenoid behaves like a N-pole.
Now look at this N-pole end of the solenoid. Which way is the current flowing – clockwise or anti-clockwise?

Now do the same for the S-pole end of the solenoid. What do you find?

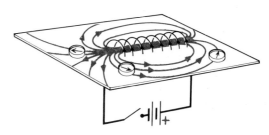

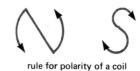

rule for polarity of a coil

The diagram shows an easy way of remembering this result. Use this rule to check the two diagrams above.

Experiment 38.5   Adding an iron core
Using the same apparatus as in the last experiment, move a small compass away from the coil until its direction shows that it is only just affected by the magnetic field.

Now slide an iron bar into the solenoid.
What is the effect on the compass?

Why do you think an iron core makes a stronger electromagnet?

Name two other ways of making a stronger electromagnet.

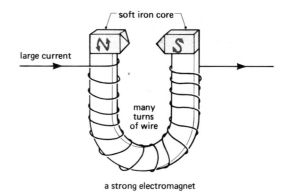

a strong electromagnet

315

Uses of electromagnets

Compared to a permanent magnet, an electromagnet has the disadvantage that it must be supplied with energy continuously. However, it has the advantage that it can be switched on and off.

1. Electromagnets are used on cranes in steel-works and scrapyards (see also page 258).

2. In hospitals dealing with eye injuries, electromagnets are used to remove splinters of iron or steel (see page 262).

3. **The electric bell**
   Study the diagram.
   a) When the bell-push completes the circuit, a current flows through the electromagnet.

   b) The soft-iron armature is attracted toward the electromagnet and the hammer hits the gong.

   c) This movement breaks the circuit at point X, so that the current stops flowing and switches off the electromagnet.

   d) The spring pulls the armature back so that contact is made and the sequence begins again.

   The 'make-and-break' contact screw at X is adjustable to give the best sound. If the gong is removed, the bell becomes a buzzer.

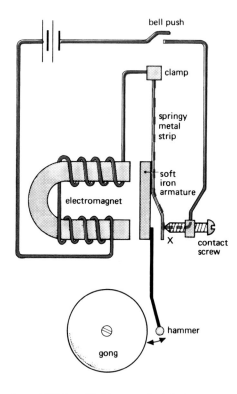

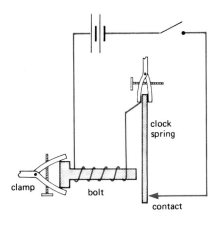

Experiment 38.6 Making an electric buzzer
Wind 100 turns of thin enamelled copper wire on to a large iron nail or bolt.
Clamp a length of steel clock-spring nearby as shown.

Complete the circuit and carefully adjust the position of the contact wire until the buzzer starts to work.

What do you see at the make-and-break contact? Suggest a good material for the contacts.

4. The electromagnetic relay

A relay is a magnetic switch, used so that a small current can switch on a large current.

For example, in a car, the starter motor needs a very large current (100A or more), which should be carried in thick wires kept as short as possible. A relay is used so that this large current is switched on by a smaller current in the dashboard switch.

In the diagram, a small current in the long thin wires to the key switch on the dashboard is used to energise the electromagnet.
The electromagnet attracts a soft-iron armature, which moves into the coil and so completes the motor circuit. What is the spring for?

A similar system can be used in lifts, so that a small current through the lift button can control the powerful lift motor.

Relays are used in electronic circuits (p 353).

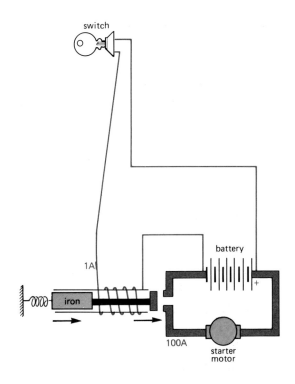

5. The telephone earpiece (or radio earphone)

This converts varying electric currents into sound waves. The varying electric currents pass round the coils of an electromagnet, which attracts an iron disc.

As the currents vary, the movement of the disc varies and makes a sound wave in the air (see also page 238).

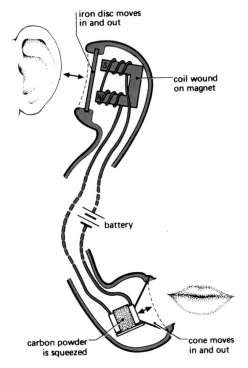

The varying electric currents are produced by a **carbon microphone** in the mouthpiece of a telephone.
The sound waves force a cone in and out so that it squeezes some carbon powder. When the carbon powder is squeezed, its resistance decreases and so the current from the battery varies as the sound wave varies.

317

The motor effect

Experiment 38.7

Hang a flexible wire between the poles of a powerful magnet.
Connect the wire to a cell and a switch as shown.

What happens when you press the switch briefly?

Does the movement reverse if you reverse the direction of the current?
What happens if you turn over the magnet?

This movement is put to practical use in electric motors.

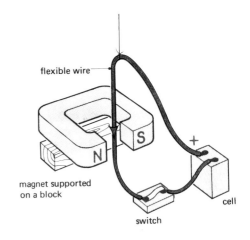

flexible wire

magnet supported on a block

S

N

+

cell

switch

The wire moves because the magnetic field of the permanent magnet reacts with the magnetic field of the current in the wire.

The diagram shows the combined field of the magnet and the wire. The wire tends to be 'catapulted' outwards.

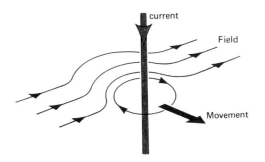

current

Field

Movement

To remember the direction of movement we use **Fleming's left hand rule**:
Hold your *left* hand in a fist and then spread out the thumb, first finger and second finger so that they are at right-angles to each other.

—Point your **First** finger in the direction of the magnetic **Field** (from N to S).

—Rotate your hand about that finger until your se**C**ond finger points in the direction of the **C**urrent (conventional current, from + to −)

—Then your thu**M**b points in the direction of the **M**ovement of the wire.

Check this rule with the movement in your experiment.

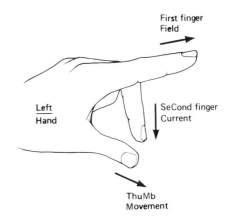

First finger
Field

Left
Hand

SeCond finger
Current

ThuMb
Movement

The moving-coil loudspeaker

The loudspeaker in a radio is used to convert electrical energy to sound energy.

It contains a movable coil attached to a large cone. The coil fits loosely over the centre of a cylindrical permanent magnet so that the coil is in a strong magnetic field.

If a current flows in the direction shown, the coil will move to the right (use Fleming's left hand rule to check this). If the current reverses, the coil moves the opposite way.

As the current (from an amplifier) varies rapidly, the coil and the cone vibrate rapidly.
As the cone vibrates out and in, it produces the compressions and rarefactions of a sound wave (see page 239).
If the current varies at a frequency of 1000 hertz, then you will hear a note of frequency 1000 hertz.

The same apparatus can be used in reverse as a moving-coil microphone (see page 329).

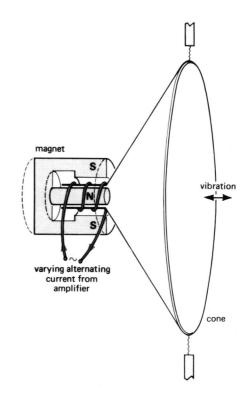

A coil pivoted in a magnetic field

Experiment 38.8
Wind several turns of wire round a cork and suspend it, from a clamp, between the poles of a magnet as shown.
Pass a current briefly through the coil.

What happens?

Does the coil turn the opposite way if you reverse the current?

This twisting movement is used in electric motors and in moving-coil galvanometers.

There are two ways to find the direction of rotation in the diagram:
a) Use Fleming's left hand rule for each side of the coil.
b) Find which end of the coil becomes a N-pole. This end will be attracted to the S-pole of the magnet.

Apply both these methods to the diagram and check that they give the same result.

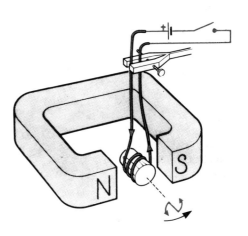

The simple electric motor

The coil in the last experiment turns so that its N-pole moves to the S-pole of the magnet but then stops.

To keep the coil turning round toward the other pole as in a motor, we have to reverse the current in the coil at just the right moment.

In a direct current motor, the current is reversed every half turn by a *commutator* which looks like a copper ring cut into two halves.

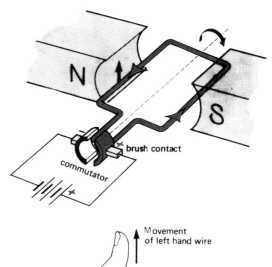

The diagram shows the current from the battery travelling through a brush contact to one half of the commutator. After travelling round the coil, the current passes through the other half of the commutator and back to the battery.

Use Fleming's left hand rule for the wire that is near the N-pole of the magnet.

Do you find that this wire moves upwards, so that the coil turns clockwise?

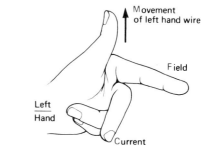

When the coil turns through 90° (so that the coil is vertical), the current stops flowing because the gaps in the commutator break the circuit. However the coil keeps turning because of its own momentum.

When the brushes make contact again, both the commutator and the coil have turned over so that the wire which is *now* nearest the N-pole has current coming out towards us.

This means that the force on it is upwards and so the coil still turns clockwise.

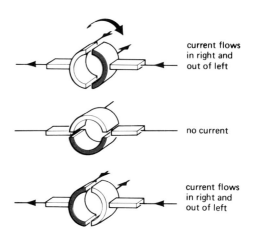

current flows in right and out of left

no current

current flows in right and out of left

That is, the commutator makes sure that whichever wire in the diagram is nearest the N-pole, it always has current moving out toward us and so the wire keeps turning clockwise.

What would happen if you reversed the battery?

Of course, in practice the coil would have more than one turn of wire.

The brushes are usually made of carbon and held against the commutator by springs.

A designer of lifts, L.E. Vator,
Built a motor with no commutator,
It turned, with a wheeze,
Through just 90 degrees,
And trapped, between floors, its creator.

320

Experiment 38.9   Building an electric motor

a) Push a short knitting needle through a large cork.

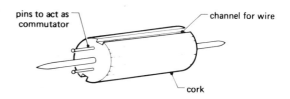

b) Push 2 pins in one end of the cork as shown, to make a simple commutator.

c) Now wind round the cork about 30 turns of thin insulated copper wire, starting at one pin and finishing at the other.
   It helps if you can cut channels in the cork to take the wire.

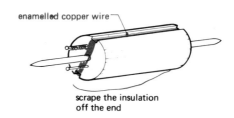

d) Scrape the insulation off the ends of the wire (using emery paper or a knife).
   Wrap each end of the wire round a pin, making good electrical contact.

e) Push 2 pairs of long pins into a baseboard to support the axle at each end.

f) Strip the insulation from the ends of 2 wires and use drawing pins to hold them in position so that they just touch the commutator.

g) Use plasticine to support magnets on each side of the coil (with opposite poles facing).

h) Connect the wires to a 2V or 4V supply and give the coil a flick to start it.

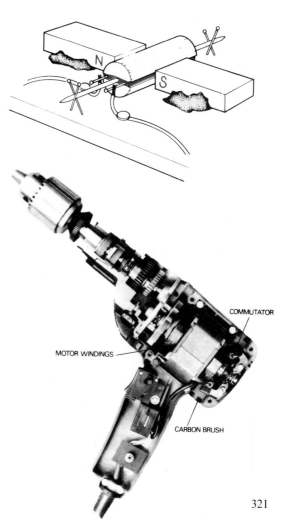

Commercial electric motors

These can be made more powerful in 3 ways:

1. A large number of turns are wound on to a soft-iron core (instead of cork) so that the magnetic field is stronger.

2. The permanent magnets can be replaced by electromagnets which give a stronger field. (This also allows the motor to work from an a.c. supply as well as a d.c. supply.)

3. The motor will be more powerful (and run more smoothly) if extra coils are added round the core. This means that the commutator must be split into more than two parts: 4 coils would need a commutator with 8 segments.

321

Galvanometers

Galvanometers are used to detect electric currents.
The simplest galvanometer is a wire and a compass,
as in Oersted's experiment (page 314).

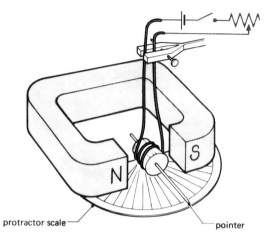

1. The moving-coil galvanometer

This is built like a motor but without a commutator.

Experiment 38.10

Use the apparatus of experiment 38.8, with a
pointer and a scale added.

What happens if you use a rheostat to vary the
current?
What happens if you reverse the current?

protractor scale — — pointer

Experiment 38.11

Connect a manufactured moving-coil voltmeter in
series with a cell and a rheostat.
Look closely at the coil's movement as you vary
the rheostat.

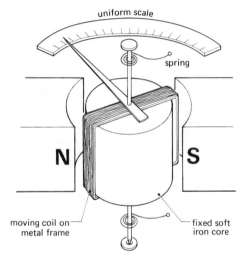

uniform scale

spring

A manufactured moving-coil meter has:

a) Two coil springs to return the pointer to zero
when the current stops.
These springs also conduct the current in and
out of the coil.

b) The magnet has curved pole faces and there is a
soft-iron core fixed at the centre of the moving
coil.
This means that the coil is always in line with a
strong uniform magnetic field. This ensures that
the meter has the advantage of a *linear*, evenly-
divided scale.

N   S

moving coil on metal frame — — fixed soft iron core

Some moving-coil galvanometers do not have a pointer: a mirror is
fixed to the coil and this reflects a beam of light on to a scale.

Moving-coil galvanometers can be made so sensitive that they can
measure a micro-amp or even less.

Disadvantages of this meter: a moving-coil meter must be connected
the correct way round in a circuit (so that the pointer moves the right
way). It cannot be used with alternating current (unless a *rectifier* is
used – see page 349).

2. The moving-iron galvanometer (repulsion type)

This galvanometer has a coil with 2 pieces of iron in it — one piece fixed and one movable, with a pointer.

If a current flows clockwise as shown in the diagram, then the nearest side of the coil is a *S*-pole.

Both pieces of iron become magnetised the same way and so repel each other (because like poles repel). This moves the pointer over the scale.

This repulsion occurs even if the current is reversed (to give two *N*-poles) and even if alternating current is supplied. This is the advantage of this instrument.

A disadvantage is that the scale is not uniform.

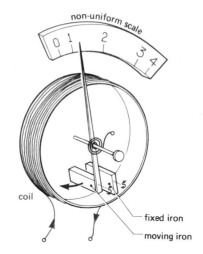

Converting a galvanometer to a voltmeter

Suppose you have a galvanometer which gives a full scale reading for a current of only 1 mA. To convert it to a voltmeter, a *high resistance* is connected in *series* with the galvanometer. This high resistance (called a 'multiplier') limits the current through the galvanometer to a safe value, so that it is not damaged.

To calculate the resistance of the multiplier, we use the Ohm's Law formula:

Example

A galvanometer has a resistance of 100Ω and gives a full-scale-deflection (f.s.d.) for 1mA (= 0·001A). What series resistance M is needed to convert it to a voltmeter with a f.s.d. of 1V?

formula first: $V = I \times R$

then numbers: $1V = 0{\cdot}001A \times (M + 100\Omega)$

$$\therefore (M+100) = \frac{1}{0{\cdot}001} = 1000\Omega$$

$$M = \underline{900 \text{ ohm}}$$

A voltmeter is always placed *across* a component, in *parallel*. It should have a *high* resistance. (In this example, it was 1000Ω). Ideally a voltmeter should have an *infinite* resistance so that it does not alter the circuit it is connected across.

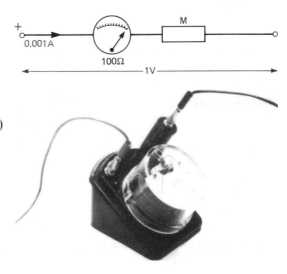

Converting a galvanometer to an ammeter (to measure larger currents)

Suppose you have a galvanometer which has a resistance of 100Ω and gives a full-scale reading for a current of only 1mA.

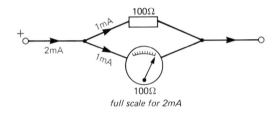

This can be converted to give a full-scale reading for 2mA by connecting an equal 100Ω resistance in *parallel*. In this circuit half the current passes through the meter and the other half passes through the 'shunt' resistor.

full scale for 2mA

If we use a shunt resistor of 25Ω (only ¼ of the meter's resistance), then 4mA passes through the shunt while 1mA passes through the meter.

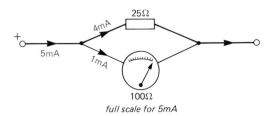

A total of 5 mA is passing through the apparatus now. This is 5 times the current through the galvanometer. This means that the reading on the scale should be multiplied by 5.

full scale for 5mA

To calculate the size of a shunt resistor, we use the Ohm's Law formula:

Example
A galvanometer has a resistance of 100Ω and gives a full-scale-deflection (f.s.d.) with 1mA.
What shunt is needed to convert it to an ammeter with a f.s.d. of 5A?

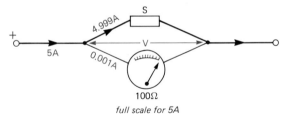

full scale for 5A

Max. current through this galvo = 1 mA = 0·001 A

Of the 5 A entering the apparatus, only 0·001 A can be allowed through the galvanometer.
The other 4·999 A must by-pass through the shunt S.
We can calculate S, in two steps:

a) For the galvanometer only: $V = I \times R$
$$= 0\cdot001\text{A} \times 100\Omega$$
$$= 0\cdot1\text{V}$$
But this is also the p.d. across the shunt S.

b) For the shunt S only: $V = I \times R$
$$0\cdot1\text{V} = 4\cdot999\text{A} \times S$$
$$\therefore S = \frac{0.1}{4\cdot999} = \underline{0\cdot02 \text{ ohm}}$$

An ammeter is always placed in *series* in a circuit. It should have a *low* resistance. (In this example it was 0·02Ω). Ideally the ammeter should have *zero* resistance so that it would not alter the current it is intended to measure.

324

Force between two parallel currents

The diagram shows two parallel wires carrying currents in the same direction. The black dotted lines show the field that is given by the left wire alone.
The right-hand wire is in this field and therefore has a force on it. By Fleming's left-hand rule, this force is towards the other wire.

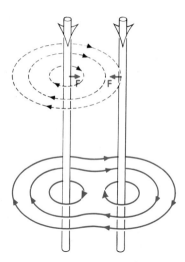

Applying the same argument to the other wire shows that the two wires are *attracted* to each other.
What happens if one of the currents is reversed?

The ampere is defined by the force between the two wires:

1 ampere is the current which, if flowing in each of two infinitely long, straight, parallel wires placed 1 metre apart in a vacuum, will produce a force on each of them of 2×10^{-7} newton per metre of wire.

The field round both the wires is shown in red.

1 coulomb (1C) is the *charge* passing any point in a circuit when a steady current of 1 ampere flows for 1 second (see also page 276).

If I amps flow for t seconds, then:

Quantity of charge Q = current, I x time, t	or $Q = I \times t$
(in coulomb) (in amp) (in seconds)	

Summary

An electric current always produces a magnetic field.
For a straight wire, the lines of force are circular.
Maxwell's screw rule: if a corkscrew moves in the direction of the current, the hand turns in the direction of the lines of force.

For a coil or a solenoid, the lines of force are like those of a bar magnet. The poles can be found by the N–S rule.
Electromagnets are used in electric bells, relays, telephone receivers.

If a wire carrying a current is placed across a magnetic field, there is a force on the wire.
Fleming's left hand rule:
First finger for Field (N to S)
seCond finger for Current (+ to −)
thuMb then gives Movement.

A coil carrying a current in a magnetic field tends to twist. This is used in an electric motor where a commutator reverses the current every half revolution so that the coil keeps turning.

A moving-coil galvanometer is similar but with springs instead of a commutator. It has a uniform scale and can be very sensitive.
A moving-iron galvanometer has a solenoid containing two iron rods that repel each other. It can be used with d.c. or a.c. but does not have a uniform scale.

A galvanometer can be converted to an ammeter with a different range by adding a low shunt resistance in parallel. It can be converted to a voltmeter by adding a high resistance in series.

Questions

1. a) The lines of . . . . . round a straight current-carrying conductor are in the shape of . . . . .
   b) Maxwell's . . . . . rule: if the . . . . . moves in the direction of the current, the hand turns in the direction of the lines of . . . . .
   c) For a solenoid, the magnetic field is like that of a . . . . .
   d) If a coil is viewed from one end and the current flows in a anti-clockwise direction, then this end is a . . . . . pole.
   e) The magnetic effect of a coil can be increased by increasing the number of . . . . . , increasing the . . . . . , or inserting a . . . . . core.

2. a) Fleming's rule for the motor effect uses the . . . . . hand.
   b) A motor contains a kind of switch called a . . . . . which reverses the current every half . . . . .
   c) A moving-coil galvanometer has the advantage that it has a . . . . . scale and can be made very . . . . . to small currents. Its disadvantage is that it must be connected the . . . . . way round in a circuit and it cannot measure . . . . . current.
   d) A moving-iron galvanometer can be used with both . . . . . and . . . . . currents. Its disadvantage is that the scale is not . . . . .
   e) A galvanometer can be converted to an ammeter with a different range by connecting a . . . . . resistance in . . . . . with the meter.
   f) A galvanometer can be converted to a voltmeter by connecting a . . . . . resistance in . . . . . with the meter.

3. For the coil in the diagram, when the switch is pressed:
   a) What is the polarity of end A?
   b) Which way will the compass point then?

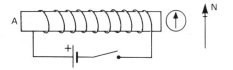

4. a) Draw a large circuit diagram of an electric bell and explain, step by step, how it works.
   b) Draw a wiring diagram to show how one electric bell can be rung by switches at both the front and back doors, using only one battery.

5. Draw a circuit diagram to show how a small current through a relay can be used to switch a large current.
   How is this used in cars?

6. The diagram shows a relay circuit for a burglar alarm. Explain, step by step, how it works. In what way is it 'fail-safe'?

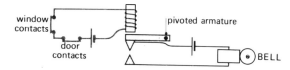

7. The diagram shows a design for an electrically operated model signal.
   a) Explain, step by step, how it works.
   b) Explain how, by adding a scale and a spring, you could use it as an ammeter.

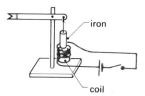

8. Explain, with diagrams, how the system shown in question 7 could be used:
   a) To unlock a door by remote control.
   b) To make a 'circuit breaker', which breaks the circuit if the current exceeds a certain value (like a fuse) but can be quickly re-set.

9. Which way does the wire in the diagram tend to move?

10. A motor has two parts, you'll note,
    A rotor and a stator.
    So now please state (and learn by rote)
    Which has the commutator.

11. a) Explain, with the aid of a diagram, how an electric motor works.
    b) Show on your diagram which way your motor would rotate.
    c) In what ways may a motor be made more powerful?

12. Draw a large diagram of a moving-coil galvanometer and explain how it works. What are its advantages and disadvantages?
    In what ways can it be made more sensitive?

13. A galvanometer has a full-scale reading of 1mA and a resistance of 100Ω.
    a) What is the full-scale reading (in mA) if a shunt resistance of 1Ω is placed in parallel?
    b) What is the full-scale reading (in V) if a resistance of 9900Ω is placed in series?

14. A galvanometer of resistance 10Ω gives a full scale deflection for 10 mA. Calculate:
    a) the p.d. needed for full scale deflection
    b) the shunt needed to measure up to 1 A
    c) the series resistance needed to measure up to 100 V.

15. a) 1 ampere (1A) is that current which, if flowing in each of . . . . . infinitely long, straight, ... wires placed . . . . . metre apart in a . . . . ., will produce a . . . . . on each of them of . . . . . newton per . . . . . of wire.
    b) 1 coulomb (1C) is the . . . . . passing any point in a . . . . . when a steady . . . . . of one . . . . . flows for one . . . . .
    c) The formula connecting charge (in . . . . .), current (in . . . . .) and time (in . . . . .) is:

17. A current of 4A flows for 10 seconds.
    a) What charge passes?
    b) If this charge moves through a p.d. of 2V, what is the change in energy? (See page 297.)

18. A motor draws a current of 2A at 10V and lifts a load of 10N through a distance of 3m in 2 seconds.
    a) What is the work done on the load? (See page 127.)
    b) What is the rate of working on the load (power output)?
    c) What is the electrical power input?
    d) What is the efficiency?

19. Comment on each of these statements made by Professor Messer:
    a) The N-pole of an electromagnet is where the current leaves.
    b) The N-pole of a coil is the end where the current flows clockwise.
    c) The coil in an electric bell should have a core of hard-steel.
    d) The armature of a relay should be a permanent magnet.
    e) A motor with 5 coils needs a carbon commutator with 5 segments.
    f) An ammeter can be made more sensitive by removing the springs.

Further questions on page 367.

How much revision have you done? See page 390.

16. Professor Messer's not too bright, So help him get his circuits right:

As soon as Oersted discovered that electricity produced magnetism, many people began to look for the reverse effect.

Eventually, in 1831, Michael Faraday discovered how to make electricity using magnetism.

Experiment 39.1
Connect a straight piece of wire to a sensitive centre-zero galvanometer. Then move the wire across a strong magnetic field.

What does the galvanometer show?

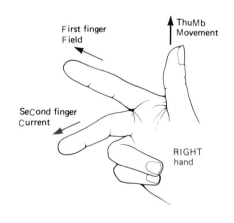

We call this current an *induced* current.
Does the induced current reverse if you reverse the movement of the wire?

Is there a current if you hold the wire still?
Is there a current if you move the wire only along a line of force and not across it?

Faraday found that *if a wire is moved to cut across lines of force, then a current is induced in the wire* (if there is a complete circuit). This is called *electro-magnetic induction.*

To remember the direction of the induced current, we use *Fleming's* right *hand rule*:
Hold your *right* hand in a fist and then spread out the thumb, first finger and second finger so that they are at right-angles to each other.
—Point your **F**irst finger in the direction of the magnetic **F**ield (from N to S)
—Rotate your hand about that finger until your thu**M**b points in the direction of the **M**ovement of the wire.
—Then your se**C**ond finger points in the direction of the **C**urrent.

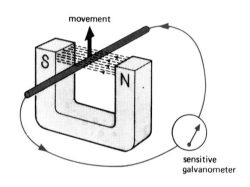

Check this rule with the direction of the induced current in experiment 39.1.

Experiment 39.2

Connect a coil with a large number of turns to a centre-zero galvanometer.

a) Then move a bar magnet quickly into the coil.
What happens?

b) Hold the magnet still inside the coil.
Does any current flow?

c) Then pull the bar magnet out of the coil.
Which way does the current flow?

Move the magnet in and out repeatedly.
What do we call this kind of current?

d) Move the magnet very slowly and then quickly.
Which method gives the greater current?

e) Move in a weak magnet and then a strong magnet at the same speed.
Which method gives the greater current?

f) Try the experiment with another coil having only a few turns.
Which coil gives the greater current?

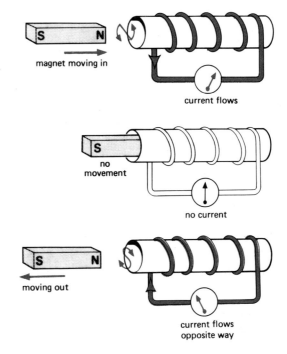

magnet moving in

current flows

no movement

no current

moving out

current flows opposite way

Faraday found that the strength of the induced current depends on
1) the speed of the movement
2) the strength of the magnet
3) the number of turns on the coil.

The *direction* of the current in the coil can be found by **Lenz's law**:

> **the direction of the induced current is such that it <u>opposes</u> the change producing it.**

This is shown in the diagrams above. When the N-pole of the bar magnet is moving *in*, the current flows so as to produce a N-pole at that end of the coil. This means that the current is opposing the movement of the magnet (by trying to repel it). When the N-pole moves *out*, the current flows the other way to produce a *S*-pole which again *opposes* the movement of the magnet (by trying to attract it).

Use Lenz's law to find the direction of the current in this diagram.

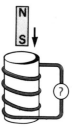

Current generators (dynamos)

Experiment 39.3
Place a magnet on a gramophone turn-table.
Connect a coil to a centre-zero galvanometer and
then hang the coil between the poles of the
magnet.

What happens to the pointer of the galvanometer
when the magnet rotates?
What happens if you speed up the turn-table?

A current that reverses to and fro like this is
called *alternating current* (a.c.).

Bicycle dynamo
A bicycle dynamo works in the same way. A
magnet rotates near a coil of wire so that the lines
of force are cut by the wire.
The coil is wound on a soft-iron core so that the
magnetic field is stronger.

Experiment 39.4
Connect a bicycle dynamo to a lamp and turn it
slowly and then quickly. What happens?
Turn the dynamo by hand and explain what you
feel.

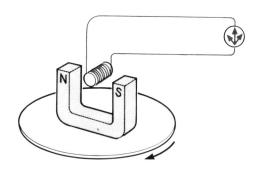

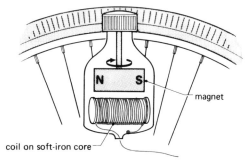

coil on soft-iron core

magnet

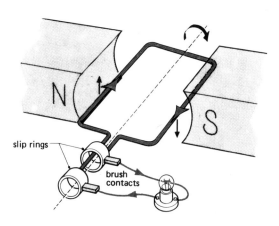

slip rings
brush
contacts

The simple a.c. generator
A generator can be built rather like a motor, with
a rotating coil and a fixed magnet. As the coil
rotates, it cuts lines of force and a current is
induced.

Use Fleming's *right* hand rule to find the direction of the current at the instant shown in the
diagram (if the coil is rotating clockwise). Do you
agree with the arrows on the diagram?

As the coil rotates, the current is conducted in and
out by way of *slip-rings* and carbon brushes.

Sketch a diagram of this generator when the coil
has turned through half a revolution and mark in
the direction of the current.
Which way does the current flow through the
lamp now?

Can you find three ways in which this generator
could be improved to increase the current?

330

Alternating current (a.c.)

Experiment 39.5
Connect an a.c. generator with slip-rings (or a bicycle dynamo) to an *oscilloscope*.
An oscilloscope is a kind of television set which will display the voltage output of the generator (see also page 342).

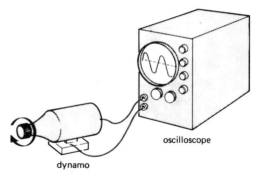

oscilloscope

dynamo

Rotate the generator steadily so that the oscilloscope draws a graph of the voltage against time. Sketch the shape of the graph.

Now connect the oscilloscope to a safe low voltage a.c. supply, so that it is really connected by way of the mains cable to the Electricity Board generator in the power station.

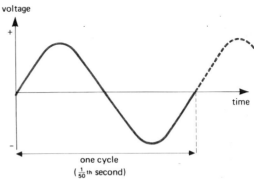

Look at the shape of the graph on the oscilloscope (this shape is called a sine curve).
Each complete cycle of the graph corresponds to one complete revolution of a simple a.c. generator. For the mains supply (in Britain), each cycle takes $\frac{1}{50}$ th second. That is, the frequency of the mains supply is 50 cycles per second (50 Hz).

Measuring a.c.
We have seen that a moving-iron meter will measure a.c. (see page 323). But should we measure the peak value or the average value or some other value?
The value we measure is the *effective value* (also called the 'r.m.s.' or 'root-mean-square' value). This is the value which is equivalent to a steady direct current and gives the same heating effect in an electric fire.

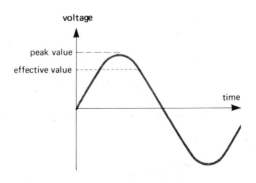

It is found that:

effective value = 0.7 x peak value	or	peak value = 1.4 x effective value

For example: the effective mains voltage in Britain is 240V.
∴ peak value = 1.4 x 240 = <u>340V</u>

This means that mains wiring must be insulated to at least 340V.

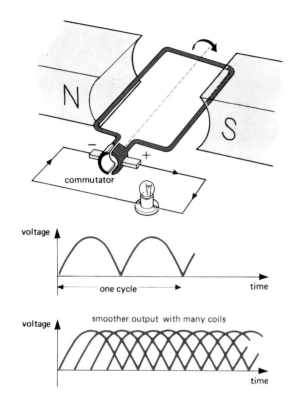

The simple d.c. dynamo

This is built in just the same way as a d.c. motor — it has a *commutator* so that the current in the brush contacts always flows the same way.

Use Fleming's right hand rule to find the direction of the current in the diagram.

Experiment 39.6
Connect a d.c. motor to a lamp and spin the motor rapidly. What happens?

Experiment 39.7
Connect an oscilloscope to a simple d.c. dynamo. Sketch the graph of the output voltage.

There is a smoother output if the dynamo has many coils and the commutator is split into many segments.

Commercial generators
The output of a dynamo can be increased in three ways, by:
1. Increasing the number of turns on the coil.
2. Increasing the magnetic field by
   a) winding the coil on a soft-iron core
   b) using stronger magnets or electromagnets.
3. Increasing the speed of rotation.

In large generators, it is usually the magnet that rotates, as in a bicycle dynamo.

In power stations, the generators are most often turned by steam power. In experiment 17.5 (page 131) you saw a model steam engine driving a dynamo to light a lamp. If you repeat this experiment you will find that the steam engine slows down when the lamp is switched on. How does Lenz's law help us to explain this?

In some countries there are hydro-electric power stations (see page 13).

A car speedometer

Experiment 39.8

Suspend a thick aluminium disc or dish over a strong magnet which can rotate on a gramophone turn-table. Use a straw as a pointer.

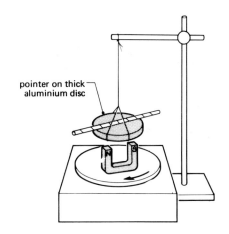

pointer on thick aluminium disc

What happens when the magnet rotates?
Does the pointer turn through a larger angle if you speed up the turn-table?

The movement of the magnet induces currents (called *eddy currents*) in the disc.
By Lenz's law, these currents try to oppose the relative movement between the two objects, so that the disc tries to follow the magnet.

How can this be converted into a practical design for a speedometer?

If you repeat the experiment with a *thin* disc, the eddy currents are much smaller and hardly turn the disc.

Two coils

Experiment 39.9   Mutual induction

Place two coils side by side as shown.
Connect one coil to a battery and a switch.
Connect the other coil to a centre-zero galvanometer.

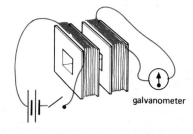

galvanometer

What happens as you close the switch?
Does anything happen if you keep the switch closed?
What happens as you open the switch?

A current is induced here (even though the coils are not moving) because when you close the switch, a magnetic field grows quickly and cuts the second coil to induce a current.
Similarly, when you switch off, the magnetic field collapses and cuts the second coil to induce a current (in the opposite direction).
With a steady current, the magnetic field is not moving and so no current is induced.

Now put a soft-iron core through both coils and repeat the experiment.
Are the deflections larger now? Why is this?

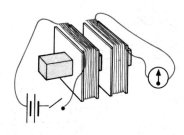

If the current is switched on and off repeatedly, what happens in the second coil?

If the first coil carried alternating current, what would you expect to find in the second coil? This is a simple *transformer*.

The transformer

Experiment 39.10
(! Danger – high voltages – demonstration only!)
Two coils are placed on a soft-iron core as shown.
One coil (the *primary* coil) is connected to an a.c.
voltmeter and a source of alternating current. The
secondary coil is connected to an a.c. voltmeter.

The two coils are connected to each other by a
magnetic field (as in the last experiment).
This magnetic field is growing and collapsing as
the alternating current rises and falls.
This means that there is an induced alternating
current in the secondary coil.

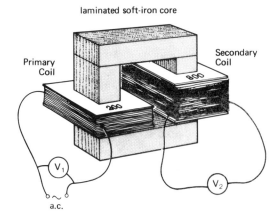
laminated soft-iron core

Primary Coil

Secondary Coil

V_1

V_2

a.c.

Use the voltmeters to measure V_2 the voltage
across the secondary coil and V_1 across the
primary coil. Calculate $V_2 \div V_1$

At the same time use the labels on the coils to find
N_2 the number of turns on the secondary coil and
similarly N_1 on the primary. Calculate $N_2 \div N_1$

What do you notice?

Repeat this for different voltages and different coils.

We find that:
$$\boxed{\frac{V_2}{V_1} = \frac{N_2}{N_1}}$$

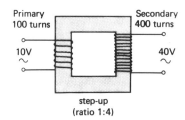

Primary 100 turns — Secondary 400 turns

10V ∼ — 40V ∼

step-up
(ratio 1:4)

If the secondary coil has more turns than the primary coil, then it is
a *step-up* transformer, because the secondary voltage is greater than
the primary voltage.
If the secondary has fewer turns than the primary, it is a *step-down*
transformer.

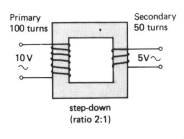

Primary 100 turns — Secondary 50 turns

10 V ∼ — 5V ∼

step-down
(ratio 2:1)

If the transformer was 100% efficient, the power in the secondary
would be equal to the power in the primary. This means that if the
voltage is stepped *up* by a certain ratio, the current in the secondary
is stepped *down* by the same ratio.

In practice, the efficiency is never 100% although it may be 98%.
One of the reasons for the loss of energy is that eddy currents are
induced in the soft-iron core. The core is laminated in thin strips to
reduce these eddy currents (see experiment 39.8).

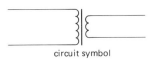
circuit symbol

Example.

A step-down transformer is required to transform 240V a.c. to 12V a.c. for a model railway. If the primary coil has 1000 turns, how many turns should the secondary have?

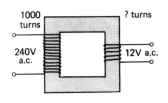

formula first:
$$\frac{V_2}{V_1} = \frac{N_2}{N_1}$$

then put in the numbers:
$$\frac{12}{240} = \frac{N_2}{1000}$$

$$N_2 = \frac{12}{240} \times 1000 = \frac{1}{20} \times 1000 = 50 \text{ turns}$$

Because the secondary voltage is $\frac{1}{20}$ th of the primary voltage, the secondary current will be 20 times the primary current.

Uses

1. Transformers are used to get low voltages from 240V a.c. mains, for door bells, accumulator chargers, hi-fi equipment, etc. etc.

2. *Experiment 39.11   Resistance welding*
   If the primary coil has 800 turns and the secondary only 5 turns, it is a step-down transformer with a ratio of 800:5, which is 160:1. In a perfect transformer, the current is stepped up in the same ratio, so that a current of 1A in the primary might give a secondary current of almost 160A.
   This large current heats the nail until it melts. The current can also be used to weld together two nails.

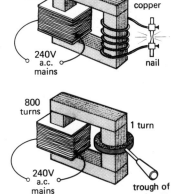

3. *Experiment 39.12   An induction furnace*
   If the secondary coil has just one turn, then the secondary current is even greater. This large current will quickly boil water or melt solder.

4. *The national grid system*
   To transfer energy from the power station to your home we could use *either* (a) a low voltage and a high current
   or (b) a high voltage and a low current.
   Method (a) is wasteful as the high current heats the power lines.
   Method (b) is used, with transformers at each end of the grid system. (This is the main reason for using a.c.).

A grid engineer called Sid,
Thought he'd change what he
usually did,
He very soon found,
With transformers changed round,
That he melted the national grid.

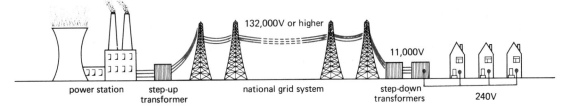

The induction coil

This works like a step-up transformer but with a *direct* current. We saw in experiment 39.9 that a steady direct current does not induce a current in the secondary coil, but if the d.c. is switched on and off then a current is induced.

The main use of an induction coil is to produce the sparks for the engine of a car (see page 72). A rotating cam switches on and off the primary circuit. This induces a large voltage across the many turns of the secondary coil. A rotating distributor applies this high voltage to the correct sparking plug.

Experiment 39.13   (!Danger!)
Look at a laboratory induction coil which uses a make-and-break vibrator, like an electric bell (page 316). Estimate the voltage if 30 000 V gives a 1 cm-spark between two spheres.

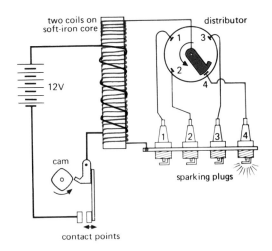

Summary

If there is relative movement between a magnetic field and a wire, then a current may be induced in the wire.

Fleming's *right* hand rule:
First finger for Field (N to S)
thuMb for Movement
seCond finger then gives direction of Current

Lenz's law: the direction of the induced current is such that it *opposes* the change producing it.

A simple a.c. generator has a rotating coil with slip-rings.

Effective value of a.c. = 0·7 x peak value.
In a simple d.c. generator, the coil is connected to a commutator.
Transformers change alternating voltages: $\dfrac{V_2}{V_1} = \dfrac{N_2}{N_1}$

If the voltage is stepped-down, the current is stepped-up in almost the same ratio.

Eddy currents are used in a speedometer but avoided in a transformer by using a laminated core. The grid system: electricity is distributed at high voltage and therefore low current.

An induction coil produces a high voltage by switching d.c. on and off.

Questions

1. Copy out and complete:
   a) When there is relative movement between a wire and a magnetic . . . . . then . . . . . may be . . . . . in the wire.
   b) Fleming's rule for the dynamo effect uses the . . . . . hand.
   c) Lenz's law: the . . . . . of the . . . . . current is such that it . . . . . the change producing it.

   d) The strength of the induced current depends on the . . . . . of the movement, the . . . . . of the magnet and the number of . . . . . on the coil.
   e) An a.c. generator is fitted with . . . . . whereas a d.c. generator is fitted with a . . . . .
   f) The . . . . . value of a.c. = 0.7 x . . . . . value.

2. Copy out and complete:
   a) A transformer will work only with . . . . . current.
   b) The formula for a transformer is . . . . .
   c) When the voltage is stepped-down, the current is stepped- . . . . . in the . . . . . ratio.
   d) The core of a transformer is . . . . . to reduce . . . . . currents.
   e) In the national grid system, the electricity is distributed at . . . . . voltage and . . . . . current.
   f) An induction coil works like a . . . . . but uses . . . . . current which is switched on and off.

3. A coil is connected to a galvanometer. When the N-pole of a magnet is pushed in the coil, the galvanometer is deflected to the right. What deflection, if any, is observed when a) the N-pole is removed, b) the S-pole is inserted, c) the magnet is at rest in the coil. State 3 ways of increasing the deflection on the galvanometer.

4. Professor Messer walks around muttering "When I go for a walk, my motor is *left* behind. A dynamo is the *right* equipment to make electricity." What do you suppose he is trying to remember?

5. a) Explain, with the aid of a diagram, the action of a simple a.c. generator.
   b) Sketch a graph to show how the current varies during one complete revolution of the coil.
   c) How is the generator modified to produce direct current?

6. Explain, with diagrams, the action of
   a) a bicycle dynamo b) an induction coil.

7. Complete the following table of transformers:

8. Draw a wave-form diagram for alternating current. Mark on the diagram one cycle and state the a.c. mains frequency in Britain. If the effective alternating current in a circuit is 10A, what is the peak value of the current?

9. Prof. Messer says that he is careful with the 240V a.c. mains because he does not want to be killed by a 340V shock. Explain this.

10. a) 240 000W might be transmitted through the national grid at (i) 240V (ii) 240 000V. Calculate the current in each case (hint: see page 306).
    b) Why would 240 000V be used, despite the dangers? c) Why is a.c. used?
    d) What are the energy changes in the process which begins with coal arriving at a power station and ends in a lamp in your home?

11. Explain what is wrong in each diagram:

WHEN I MOVE THE WIRE OUT, I THINK THE CURRENT SHOULD FLOW DOWN AB

I'M MOVING THE WIRE FROM N to S. WHY IS THERE NO CURRENT ?

WHY IS THE LAMP NOT GLOWING ?

What is the best way to revise? See page 390.

primary p.d.	secondary p.d.	primary turns	secondary turns	step-up or step-down
100V a.c.		10	100	
100V a.c.		100	10	
240V a.c.	12V a.c.	200		
11 000V a.c.	132 000V a.c.		12 000	

Further questions on page 369.

Multiple choice questions on electricity

Most examination papers include questions for which you are given 5 answers (labelled **A, B, C, D, E**) and you have to choose the *ONE* correct answer. Take care to read *all* the answers before you choose the one that you think is correct.

Example.
The kilogram is a unit of

   A   speed
   B   force
   C   mass
   D   length
   E   time

The correct answer is **C**.

Now write down the correct letter for each of the following questions (but do not mark these pages in any way).

1. Which of the following tests is the best for deciding whether a metal bar is magnetised?
   A   It is affected by a wire carrying a current.
   B   A compass needle is attracted by the bar.
   C   A compass needle is repelled by the bar.
   D   Iron filings stick to the bar.
   E   When freely suspended, the bar points toward the nearest magnet.

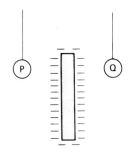

2. P and Q are metal balls hanging from nylon threads. When a negatively-charged rod is placed between them, as shown, P is repelled and Q is attracted by the rod.
Which of the following statements is correct?

	P	Q
A	positive	negative
B	positive	uncharged
C	negative	positive
D	uncharged	positive
E	positive	positive

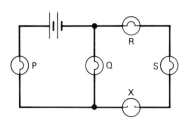

3. In the circuit shown, all the lamps were glowing until lamp X fused and went out.
Which lamp or lamps remain on?
   A   P only
   B   P and Q only
   C   R and S only
   D   P, Q, R and S
   E   None of them

4. P and Q are 12V lamps connected to a 12V battery.

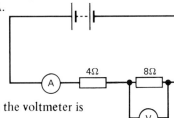

When the switch S is closed,

A P glows brighter and Q goes out.
B Q glows brighter and P goes out.
C Both lamps glow more brightly.
D Both lamps glow less brightly.
E Both lamps go out.

6. A 1500W electric fire is used on a 250V supply. What is the best value of fuse to be fitted in the plug?

A 1A
B 2A
C 5A
D 10A
E 13A

5. In the circuit shown, the reading on the ammeter is **2A**.

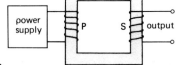

The reading on the voltmeter is

A 2V
B 4V
C 6V
D 8V
E 16V

7. Two 100W lamps and a 300W television set are switched on for 6 hours. What is the cost at 3p per kilowatt-hour?

A 1p
B 3p
C 4p
D 9p
E 36p

8. The diagram shows a wire carrying a current between the poles of a magnet. In which direction does the wire tend to move?

A Into the paper
B Out of the paper
C Towards the south pole of the magnet
D Towards the top of the page
E In some other direction

9. Which of the following will *NOT* increase the voltage produced by an a.c. generator?

A Increasing the speed of rotation
B Increasing the number of turns on the coil
C Increasing the strength of the magnetic field
D Using a soft-iron core inside the coil
E Increasing the size of the gap in which the coil rotates

10. The diagram shows a simple *step-up* transformer. Which of the following correctly describes each part of the transformer?

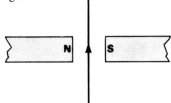

	Power supply	Core	Primary coil P	Secondary coil S
A	d.c.	steel	100 turns	10 turns
B	d.c.	iron	10 turns	100 turns
C	a.c.	iron	100 turns	10 turns
D	a.c.	iron	10 turns	100 turns
E	a.c.	steel	10 turns	100 turns

When you have finished:
a) check all your answers carefully
b) discuss your answers (and why you chose them) with your teacher.

More questions can be found in *Multiple Choice Physics for You.*

chapter 40

ELECTRON BEAMS

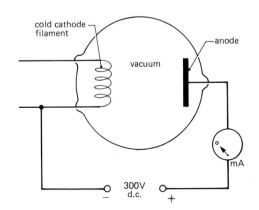

cold cathode
filament

anode

vacuum

mA

300V
d.c.
− +

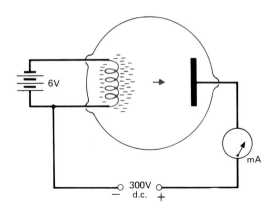

6V

mA

300V
d.c.
− +

Experiment 40.1   Diode valve
(! Danger − high voltages − demonstration only!)
The apparatus, called a *diode*, consists of a glass
bulb containing two metal electrodes in a vacuum.
One electrode, a metal plate, is called the anode.
The other electrode (the cathode) is a filament
which can be heated as in an electric lamp.

Connect a sensitive ammeter and a high voltage
battery (or power-pack) across the diode as
shown in the diagram.

Does any current flow?
Reverse the battery connections. Does any current
flow through a vacuum?

Now connect a 6V battery across the cathode
filament so that it is heated and glows.

Does a current flow through the ammeter now?
Reverse the connections to the high voltage
battery, as before. Does a current flow through
the ammeter now?

A current can flow only when the cathode is hot.
We believe that *electrons* are "boiled" off the
hot cathode because at this high temperature they
have a lot of energy.
This process is called *thermionic emission*.

The negative electrons form a *space charge* near
the cathode.
If the anode is *positive*, these negative electrons
are attracted across the gap and a current flows.
If the anode is *negative*, the negative electrons are
repelled and no current flows.

Because the current can flow only one way
through the apparatus, it is called a diode *valve*.

Experiment 40.2　Maltese cross tube

This glass bulb has a hot cathode with an anode close to it. The anode has a hole in it so that when it is positive and attracts electrons, some electrons pass through the hole and shoot across the vacuum. This arrangement is called an *electron gun*.

In the middle of the bulb is a second anode in the shape of a maltese cross. The end of the tube is a *fluorescent screen* as in a television set.

Switch on the cathode so that it glows white hot. Now switch on the 3000V supply to shoot electrons across the vacuum.

What do you see on the screen?

The streams of electrons leaving the cathode and shooting across the vacuum are called *cathode-rays*. The edges of the cathode ray shadow on the screen are sharp. What does this tell you about cathode-rays?

Bring a strong magnet near the tube.
What happens to the shadow?
What happens if you reverse the magnet?

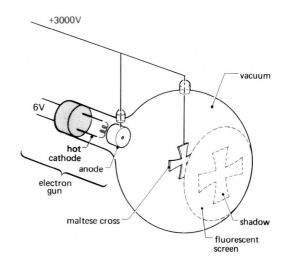

Experiment 40.3　Cathode-ray deflection tube

In this tube an electron gun sends a beam of electrons (cathode-rays) through the vacuum so that the beam cuts across a fluorescent screen. The screen is placed between two horizontal metal plates (labelled Y_1 and Y_2).

The electron beam can be deflected in two ways:
a) by *magnetic deflection* using bar magnets as in the last experiment. Alternatively electro-magnets (large coils) can be used.
b) By connecting plate Y_1 to the positive (+) terminal of the power supply and plate Y_2 to the negative (−) terminal, as shown. Why do the electrons deflect away from the negative plate and move nearer the positive plate? What happens if you reverse these connections? This *electrostatic deflection* is used in a cathode-ray oscilloscope.

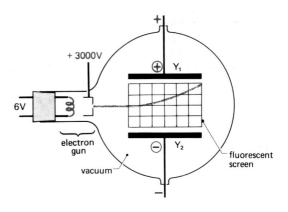

Mrs Messer: How do you make a Maltese cross?
Professor Messer: Pour water down his trousers.

The cathode ray oscilloscope (C.R.O.)

This contains a cathode ray tube very like the one in the last experiment: an electron gun sends electrons through the vacuum to a screen.

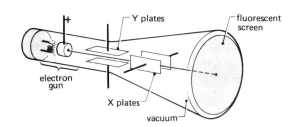

Two Y-plates can deflect the beam vertically (as in the last experiment). Two X-plates can deflect the beam horizontally.

The diagrams show how charging the plates positively and negatively can deflect the beam to any position on the screen.

How could the spot be deflected to the bottom left-hand corner?

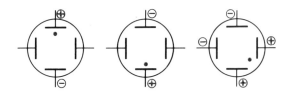

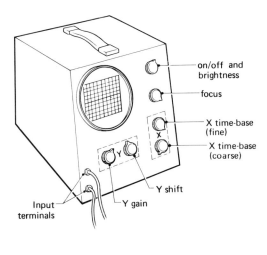

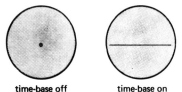

time-base off   time-base on

The controls
The complete oscilloscope has several controls to adjust, usually in the following order:

1. The *brightness* and *focus* controls.

2. With a microphone or other apparatus connected to the input terminals, the *Y-shift* control can be used to move the display vertically to the best position. The *Y-gain* control varies the amount of amplification in the vertical direction. A smaller input needs a larger amplification.

3. When we use a cathode ray oscilloscope (C.R.O.) for drawing graphs, we must use the *time-base* control. When this control is switched on, the spot moves at a constant speed across the screen and then jumps back very quickly to start again.
   The time-base controls are used to vary the speed of the spot. At high speeds it appears as a line because of the persistence of vision (see page 213).

When fast electrons are stopped suddenly at the screen, their kinetic energy is converted to heat energy and light energy. Some of this energy is in the form of invisible *X-rays* with a very short wavelength (see page 344).

Uses of a C.R.O.

1. Measuring voltages
The deflection of the spot depends on the *voltage* applied to the deflection plates. A C.R.O. is a good voltmeter, taking almost no current.

Experiment 40.4

With the time-base switched *off*, connect one 1.5V dry cell to the input terminals.
Measure the size of the deflection.

Now connect two dry cells in series to the input terminals.
What happens to the deflection?

What happens with three cells?
What happens if you reverse the battery?

What happens if you connect a low-voltage *alternating* current supply?

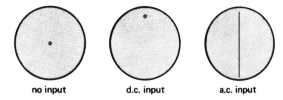

no input   d.c. input   a.c. input

2. Studying wave-forms
With an alternating input, the time-base should be used to move the spot horizontally at the same time, so that we can see the shape of the voltage wave-form.

A C.R.O. with a microphone can be used to display the wave-forms of sounds (see page 244).

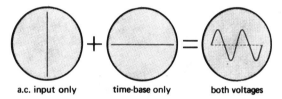

a.c. input only   time-base only   both voltages

3. Television
A television set is a C.R.O. with two time-bases — one to move the spot across the screen, the other to move it vertically (like the way your eyes move when you are reading).
The spot marks out 625 lines on the screen and it does this 25 times in each second.
The signal from the aerial varies the brightness of the spot so that a picture is built up of bright and dark spots (as in a newspaper photograph).

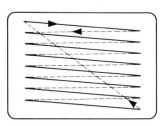

A VDU (Visual Display Unit) for a computer is built in the same way.

In a colour television set there are *three* electron guns (see the diagram on page 347). These build up a full colour picture using just the three primary colours: red, green and blue.

In a radar set, a C.R.O. is used to display a picture of the radio echoes received by a rotating aerial.

LONDON'S RIVER

The River Thames from Gallions Reach to Southend Pier
A composite picture from the five Radars of the Thames Navigation Service

X-rays

X-rays were discovered by Röntgen in 1895.

An X-ray tube is really a high voltage diode valve. A hot cathode emits electrons which are accelerated to high speed by a high voltage (perhaps 100 000 V).

When the fast moving electrons hit the tungsten target, some of their kinetic energy is converted to X-radiation (most of their energy is converted to heat which is conducted away to the cooling fins).

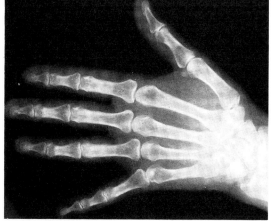

Controls for the X-ray tube
a) To give a more intense beam of X-rays, the cathode is made hotter, to give more electrons and so give more X-rays.
b) To give 'harder' X-rays (more penetrating, with shorter wavelength), the voltage across the tube is increased to give the electrons more kinetic energy. 'Softer' X-rays (less penetrating, longer wavelength) are given by lower voltages.
TV sets emit a small amount of soft X-radiation.

Properties of X-rays
1. !X-rays are dangerous! They can cause cancer.
2. X-rays pass through substances but are absorbed more by dense solids (e.g. bones or lead).
3. X-rays affect photographic film.
4. X-rays ionise gases, so that the gases become conducting (and so discharge electroscopes).
5. X-rays are not deflected by magnetic or electric fields.
6. X-rays can be diffracted (so they are *waves*).
7. X-rays are electromagnetic waves of very short wavelength (about 10^{-10} metre). See page 224.

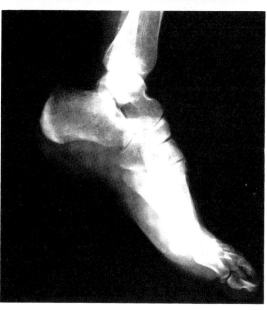

Uses of X-rays
1. In *medicine*: to inspect teeth, broken bones etc. With care, X-rays can be used to kill cancer cells.
2. In *industry*: to inspect metal castings and welded joints for hidden faults.
3. In *science*: photographs of X-rays diffracted by crystals give information about the arrangements of the atoms in different substances.

Summary

A hot cathode produces a space charge of negative electrons by thermionic emission. The electrons are attracted by a positive anode.

An electron beam ('cathode rays') can be deflected by a magnet or by charged plates (as in a C.R.O.).

When fast electrons are stopped by a target, X-rays may be produced.

Questions

1. Copy out and complete:
   a) When a wire is heated in a vacuum, it emits . . . . . This process is called . . . . . emission.
   b) A current flows in a diode valve only when (i) the cathode is . . . . . and (ii) the anode is . . . . . to attract the . . . . . electrons.
   c) A beam of electrons (". . . . . rays") can be deflected by (i) a . . . . . or (ii) . . . . . The . . . . . electrons are deflected away from a . . . . .-charged plate.
   d) In a cathode ray oscilloscope (C.R.O.) the vertical deflection of the spot is proportional to the . . . . . applied across the Y-plates. When studying waveforms, the spot is moved horizontally by a . . . . . circuit.
   e) When fast electrons are stopped, ... of very . . . . . wavelength may be produced.

2. Draw a simple labelled diagram of a cathode ray tube for a C.R.O. Name and explain the purpose of the controls on a C.R.O.

3. Professor Messer's full of woe —
   He can't control his C.R.O.

4. How are X-rays produced? List a) their properties b) their uses.

5. A time-base is adjusted so that it draws a horizontal line, as shown, 50 times per second. Draw diagrams of what is seen when the following are connected to the input terminals.
   a) a battery which makes the upper Y-plate positive b) an a.c. supply of 50 Hz
   c) an a.c. supply of 100 Hz d) an a.c. supply of 100 Hz in series with a diode.

6. When a 20V battery is connected to a C.R.O., the spot is deflected 1cm vertically.
   a) If another battery deflects the spot 3cm vertically, what is its voltage?
   b) The diagram shows what happens when a.c. is connected. What is the peak voltage? What is the r.m.s. (effective) voltage? (see page 331).

Further questions on page 370.

345

Physics at work: In Your Home

Records

Music is recorded on an LP record disc by making a narrow groove in the plastic (see also page 238). The vibrations of the sound are copied into the wobbles of the groove. The groove is an *analogue* of the sound (see page 358). A louder sound gives a larger wobble (with a large amplitude). A higher note gives a more rapid wobble (with a higher frequency).

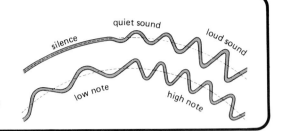

Magnetic pickup

The wobbles of the groove are converted to electricity by means of a needle, or *stylus*, attached to a small magnet. The magnet is near a small fixed coil of wire. As the needle wobbles with the groove, the magnet also wobbles. This makes the magnetic field in the coil vary and so the coil has a varying current in it (by *electromagnetic induction*, see page 328).

The varying current in the coil is passed to an amplifier (page 358) and then to a loudspeaker (p 319).

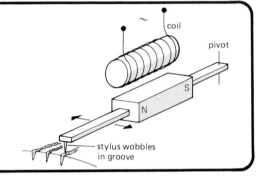

Stereo-records

Old records had only one soundtrack and were played back through only one loudspeaker.

Stereo records give a more realistic sound. They are recorded by two microphones (one to the left of the singer and one to the right) and played back through two loudspeakers. Both sounds are combined into one groove, with the left-hand side of the groove carrying the wobbles for the left-hand sound, and the right-hand side of the groove carrying the right-hand signal.

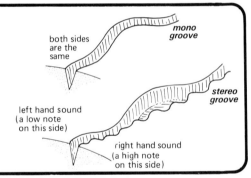

Compact discs

Compact discs are different: they are *digital* discs (see also page 358). Instead of the wobbling grooves there is a pattern of 'pits' cut into the shiny disc. There are billions of pits, with a varying length and spacing, depending on the sound.

The pattern of pits is read by using a weak laser beam (see page 203). The laser beam is focussed on one track and moves across the disc like a stylus. The pits and flats reflect the beam differently, producing a beam which is off or on. This shines on to a light-detector and de-coder which converts it to a varying electric current for an amplifier and loudspeaker.

Video discs work in a similar way and can store 50 000 pictures on one disc.

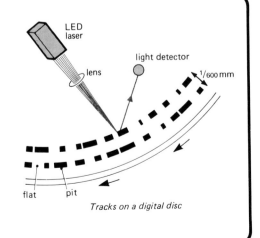

Tracks on a digital disc

Physics at work: In Your Home

Tape-recorder

A tape-recorder uses a 'tape head' to magnetise the iron oxide on the tape. The tape head is an electro-magnet with a very small gap (about $\frac{1}{1000}$ mm). The strong magnetic field near the gap magnetises the tape (with the same frequency as the sound). The electromagnet changes the direction of the tiny molecular magnets (see page 261).

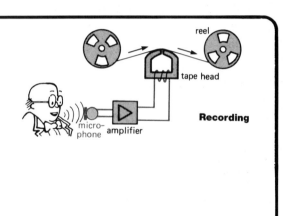

Recording

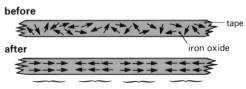

before

after

frequency of sections equals frequency of sound

On playback, the magnetised tape goes past a tape head again. The magnetic field of the tape induces a small current in the coil (by electromagnetic induction, see page 328). This current is amplified and passed to the loudspeaker.

A recording can be erased by a similar 'erase head'. This carries an alternating current at a very high frequency (80 kHz). It demagnetises the tape (so the molecular magnets point in many directions).

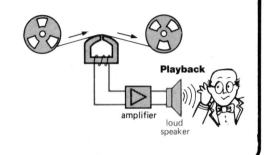

Playback

Television

A television tube is built like a C.R.O. (page 342). However, in a TV, the electron beam is deflected by currents flowing in the *deflection coils:*
The magnetic field of a coil bends the electron beam to different parts of the screen (see page 341). The beam moves to build up a new picture every $\frac{1}{25}$ second (see page 343).

In a colour TV set there are *three* electron guns. One gun varies in brightness according to the *red* signal from the TV station. The other two guns vary with the *green* and *blue* signals. Because these are the three primary colours of light (see page 228) a complete range of colours can be produced.
If you look closely at a TV screen you will see it is covered with groups of red, green and blue dots. These dots glow when the electrons hit them.

A *shadow mask* is fitted just behind the screen. This has a hole in it for each group of coloured dots. It ensures that the beam from the 'red' gun only hits the red dots; and similarly for the green and blue.

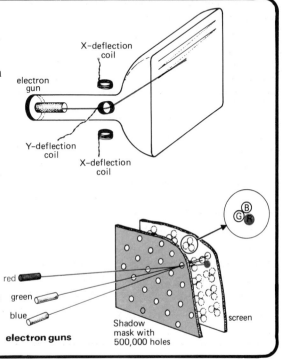

chapter 41

ELECTRONICS

Semi-conductors

Some substances (like germanium and silicon) are
neither good conductors nor insulators. These
substances are called *semi-conductors*.

Germanium or silicon semi-conductors are made by
adding to them very small amounts of other
substances (called 'impurities').
These *doped* semi-conductors are of two kinds:
n-type and p-type.
In *n*-type, the conduction is by *ne*gative electrons.
In *p*-type, the conduction is by *po*sitive charges,
called holes.

The semi-conductor diode (junction diode)

If a p-type material and an n-type material are
joined together, we get a very useful device,
called a p-n junction *diode*.

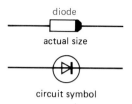

diode

actual size

circuit symbol

Experiment 41.1
Connect a battery, a lamp and a diode in series.
Connect the marked or narrow end of the diode (the
'cathode') nearest to the negative terminal of the
battery. What happens to the lamp?

Now turn the diode the opposite way round in the
circuit. Does the lamp still light?

A diode allows a current to flow *only one way*.
The arrow on the diode symbol shows the direction
in which conventional current can flow. The diode
is then said to be *forward-biassed*.

If the diode is *reverse-biassed*, the current is
nearly zero.

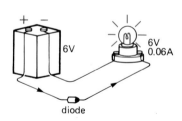

6V

6V
0.06A

diode

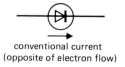

conventional current
(opposite of electron flow)

Experiment 41.2   I:V graph
Repeat experiment 35.23 (page 295) measuring the
current *I* and the voltage *V* for a junction diode.

As the graph shows, almost no current flows if the
voltage is applied in the reverse direction.

If this reverse voltage or the forward current is
increased too much, then the diode will be damaged.
See p 295 for graphs of other *non-ohmic* conductors.

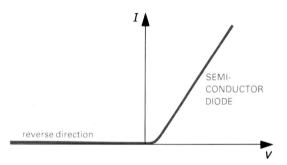

SEMI-
CONDUCTOR
DIODE

reverse direction

348

Uses of diodes

1 Protecting equipment

Diodes are used to protect radios and computers which would be damaged if the battery or power supply was connected the wrong way round. If the battery in the diagram was connected the other way round, no current would flow and no damage could be done.

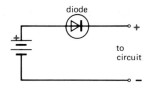

2 Switching in stand-by batteries

Here is a circuit that could be used on your bicycle:
Normally, if you are riding the bike, the dynamo will produce a higher voltage than the battery, and light the lamp using the *left*-hand circuit only.
But if you stop at traffic lights, the dynamo will stop, and then the *right*-hand circuit will light the lamp. What does each diode do?
A similar circuit could keep a computer running even if there was a mains power cut.

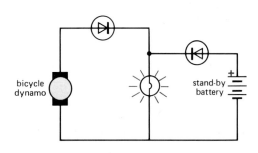

3 Rectification

Experiment 41.3   Half-wave rectifier

Connect a diode in series with a 1000Ω resistor and a low voltage a.c. supply.
Now use an oscilloscope (or C.R.O., see page 342) to look at the wave-form: first connect the C.R.O. across the a.c. supply (the 'input') and sketch the a.c. wave-form (see also page 343).
Then connect the C.R.O. across the load resistor as shown in the diagram. Sketch the new wave-form.
Why is only half the a.c. wave-form still present?
The diode conducts only for half the a.c. cycle (only when it is forward-biassed).
The alternating current has been *rectified* to an uneven direct current.

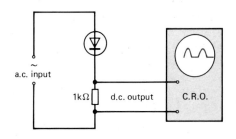

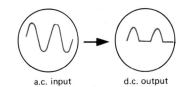

a.c. input   d.c. output

Experiment 41.4   Smoothing

Connect a 10μF capacitor (see page 278) in parallel with the resistor.
What happens to the wave-form?
The output is said to be *smoothed*.
The capacitor stores some energy when the voltage is high, and releases it when the voltage falls.

smoothed d.c. output

These *solid-state* diodes have many uses as rectifiers in radios, TV and computers.

The Light-Emitting Diode (L.E.D.)

Some special diodes called LEDs will give out light when they are passing a current.
The usual colour is red, but it is possible to get yellow or green LEDs.
A light-emitting diode is an example of a *transducer* – it changes energy from one form to another.
What are the two forms of energy in a LED?

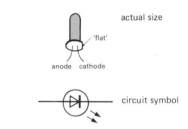
actual size
'flat'
anode   cathode

circuit symbol

Experiment 41.5
Connect a LED to a potential divider (see p 294) as shown. Make sure you have a protective resistor in the diode circuit.

What happens as you vary the voltage?
Connect a voltmeter across the LED. What voltage is needed before the LED will light?

Does the LED light if it is reversed in the circuit?

A LED will not light until the applied voltage is greater than about 2V. However if the full 6V was applied, the LED would be damaged. A LED must always have a *protective resistor* is series with it.

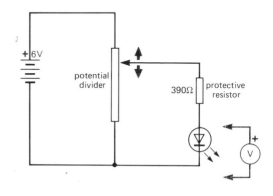

potential divider
390Ω protective resistor

Example
A conducting LED has a voltage drop across it of 2V and should pass a current of 10mA (= 0.01A). If it is connected to a 9V supply, what protective resistance is needed?
If there is 2 V across the LED, the other 7V must be across the protective resistor R.

formula first: $V = I \times R$ (see page 289)

then numbers: $7 = 0.01 \times R$

$$\therefore R = \frac{7}{0.01} = \underline{700\Omega}$$

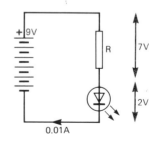

R   7V
2V
0.01A

Uses:
LEDs are often used as indicator lamps. LEDs are small, reliable, need only a small current and last much longer than filament lamps.
LEDs are sometimes used in *seven-segment displays* in digital clocks, cash registers and calculators.

Professor Messer was doing some sums on his calculator, sitting opposite Mrs Messer. Why did she laugh when he typed in 0.7734 and then 5508? How many other words can you make? Try: 0.37108, 5663. Or 35006, 5637. Are they 3781637?

seven-segment display

0 123456789

The Light-Dependent Resistor (L.D.R.)

When a semi-conductor has light shining on it, it can conduct
electricity more easily.
Its resistance depends on the brightness of the light, so it is
called a *light-dependent resistor* (or photo-conductive cell).

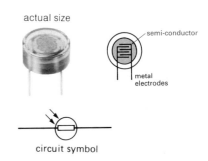

actual size

semi-conductor

metal
electrodes

circuit symbol

The LDR shown in the photo is made from a semi-conductor
called Cadmium Sulphide (CdS). In the dark its resistance is more
than 10MΩ. In sunlight its resistance is only about 100Ω.

Experiment 41.6   A light-meter

Connect an LDR in series with a cell, a milli-ammeter and a
variable resistor. Adjust the variable resistor to give a
deflection on the ammeter.
What happens to the current as you cover and uncover the LDR?

Uses:
This circuit could be used as a photographer's light-meter (if it
was calibrated). The same circuit can be used to alter the aperture
in automatic cameras. More uses are shown later (page 356).

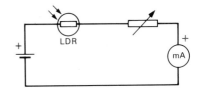

LDR

mA

The Thermistor

When a semi-conductor is heated, it can conduct electricity more
easily. This is because the rise in temperature makes more free
electrons available to carry the current. The higher the
temperature, the less the resistance.

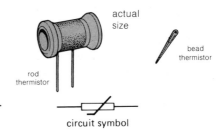

actual
size

bead
thermistor

rod
thermistor

circuit symbol

Experiment 41.7   An electronic thermometer

Repeat experiment 41.6 but with a thermistor in place of the LDR.
Heat the thermistor with your hand or a hair drier. What happens?

Uses:
This circuit could be used as an electronic thermometer if the
ammeter was calibrated in °C (see p 33). It would have a *non-linear
scale*, with *un*equal divisions.
Thermistors are used to protect the filaments of projector lamps
and TV tubes from a current surge as they are switched on. The cold
thermistor keeps the current low at first; as the filament heats
up, so does the thermistor and it gradually lets more current pass.
This extends the life of the filament.
More uses are shown later (page 355).

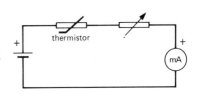

thermistor

mA

Experiment 41.8   I:V graph

Repeat experiment 35.23 (page 295) measuring the current I and the
voltage V for a thermistor.
The graph is shown on page 295.

Reed switch

A reed switch is often used in electronic circuits. It consists of a glass tube with two iron reeds sealed in it. If the iron is unmagnetized and there is a gap between the reeds, the switch is said to be *normally open* (NO) and no current can flow through it.

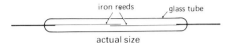

iron reeds   glass tube

actual size

If a magnet is brought near a reed switch it magnetizes each iron reed as shown (by 'magnetic induction', see page 262). The result is that the two reeds are attracted to each other and they bend to touch each other. Now that the switch is closed it can pass a current. If the magnet moves away, the switch will open again. Why?

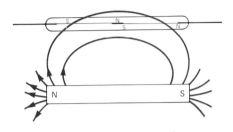

Experiment 41.9
Connect a reed switch in series with a cell and a lamp.
Use a magnifying glass to look closely at the reeds while you move a magnet nearby.
What happens to the lamp?

Normally-closed reed switch

If an extra non-magnetic contact is added to the reed switch as shown, it can be used as a *normally-closed* (NC) switch. The current flows from B to C *until* a magnet is brought near. What happens then?
How can it be used as a change-over (CO) switch?

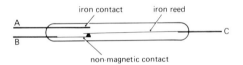

iron contact   iron reed

non-magnetic contact

Uses:
A normally-closed reed switch can be used in burglar alarms. The reed switch is fixed in the door frame and a magnet is fixed to the door. What happens if a burglar opens the door?
What are the disadvantages of this simple alarm?

NC reed switch

6V

bell

Reed relay

If a reed switch is put inside a coil of wire (an electromagnet, see page 315), then it can be operated by a current in the coil.
This is a *reed relay*.
A small current in the coil can switch on a larger current in the reed switch. (It is a simple kind of amplifier.)

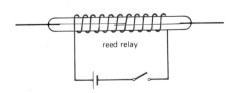

reed relay

Electromagnetic relay

The diagram shows another kind of relay, which can switch on a larger current than a reed relay.
If enough current flows in the coil it will attract the soft iron armature (see also p 316). The armature is pivoted and so pushes the springy contacts together so that a current can flow through them. What happens if the coil current is switched off?

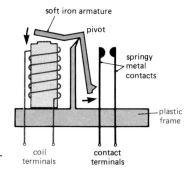

The diagram shows only one pair of normally-open contacts, but others can be added side by side so that several circuits can be switched. Normally-closed and change-over contacts can also be used.

Experiment 41.10   A burglar alarm
Connect a relay and an LDR to a battery and bell as shown:
Shine a light on the LDR. What happens to its resistance? (p 351).
What happens to the current in the LDR?
What happens to the relay? What happens to the bell?
If you put this circuit in a drawer (where it is dark), what would happen if a burglar opened the drawer?

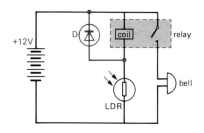

Using a relay
In most circuits it is important to include a *diode* as shown at **D** in the diagram. This is because the relay coil acts like an induction coil (see page 336) and can give a large reverse voltage. The diode protects the circuit by shorting out this large voltage.

Transducers

The LED, LDR, thermistor and reed switch are all examples of *transducers* – they convert energy from one form into another. The rest of this chapter is about electronic *systems*. All the systems have three main parts as the diagram shows:
For example, a record player has an input transducer (the pickup), a processor (the amplifier) and an output transducer (loudspeaker).

Here is a summary of the transducers you have met:

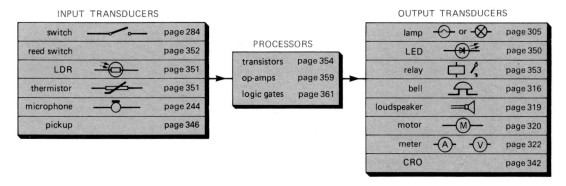

INPUT TRANSDUCERS			PROCESSORS		OUTPUT TRANSDUCERS		
switch		page 284			lamp	or	page 305
reed switch		page 352	transistors	page 354	LED		page 350
LDR		page 351	op-amps	page 359	relay		page 353
thermistor		page 351	logic gates	page 361	bell		page 316
microphone		page 244			loudspeaker		page 319
pickup		page 346			motor	M	page 320
					meter	A V	page 322
					CRO		page 342

The transistor

This is a device for amplifying small electric currents. It is made of three layers of n, p, n semi-conductor material. The three layers are called the *collector, base* and *emitter*.

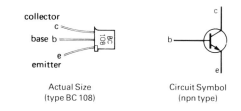

Actual Size
(type BC 108)

Circuit Symbol
(npn type)

The diagram shows one kind of transistor, but there are many different sizes and shapes.

Experiment 41.11   Current amplifier
To investigate a transistor, set up the circuit shown in the diagram:

Do you find that lamp L_2 lights but lamp L_1 does not?

Now unscrew lamp L_1 so that there is a break in the circuit to the base b.
Do you find that lamp L_2 (in the collector circuit) goes out?

Now replace lamp L_1. Does L_2 light again?

This shows that **a small current in the base circuit causes a larger current to flow in the collector circuit.**

In your circuit, a current of about 1mA in the base circuit has switched on a current of about 60mA in the collector circuit.
A transistor acts as a *current amplifier.*

A transistorized radio contains several transistors to amplify small currents until they are big enough to drive a loudspeaker.

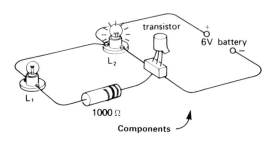

Components

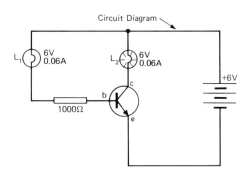

Circuit Diagram

Experiment 41.12   Moisture detector
Change your circuit so that instead of lamp L_1, there are two probes as shown. The probes are two pieces of copper wire placed close together but not touching.
What happens to lamp L_2 if you connect the probes together? Why is this?

What happens if you place a drop of water touching both probes? Why?

How could you place the probes in a water tank to light the lamp when the tank fills to a certain depth? How could you use this circuit to light a warning lamp when it is raining outside?

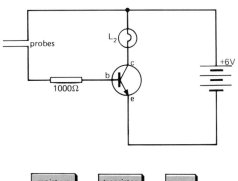

Experiment 41.13 Help for a blind person
Change your circuit so that it includes a relay to
switch on a buzzer.
(Remember to include a diode D to protect the
transistor – see experiment 41.10).
Place the probes so they are inside the top of a
cup or beaker.
What happens when you fill the beaker with water?

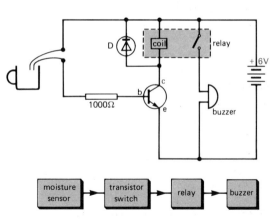

Uses:
How could this be used to help a blind person know
when to stop pouring from the tea-pot?
How could this circuit be used to warn when a
baby's nappy is wet?
How could it be used to switch on windscreen-
wipers automatically?

Experiment 41.14 A fire alarm (a heat-operated switch)
Connect a thermistor and a variable resistor to a transistor switch
as shown. Adjust the resistor until the lamp is just turned off.
What happens when you heat up the thermistor? (see page 351).

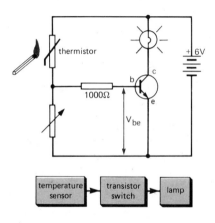

The thermistor and the variable resistor are the two parts of a
voltage divider (see page 294).
As the thermistor gets hotter and its resistance *de*creases, its
'share' of the 6V *de*creases and so the voltage across the variable
resistor *in*creases. This increases V_{be} (see the diagram) and
switches on the transistor.
If V_{be} is less than 0.6V the transistor switch is off.
If V_{be} rises above 0.6V the transistor starts to conduct.
If V_{be} is above 1.5V the transistor is fully on.

Why is the variable resistor called a 'sensitivity control'?

Uses:
Draw a diagram showing how you could add a relay so that a bell
would ring as a fire warning.
How could this be used to warn if a chip pan was getting too hot?

Experiment 41.15 A frost alarm
Invert the temperature sensor part of your circuit, as shown:

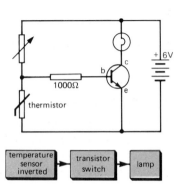

Does the lamp light now when the thermistor is hot or when it is
cold? Why?

Uses:
How could this be used as a frost alarm to warn a fruit farmer?
How could this be adapted to keep a tropical fish tank from
becoming too cold?

355

More uses of electronic circuits

Experiment 41.16   A light-sensor circuit
Connect an LDR to a transistor switch as shown:
Cover up the LDR so it is dark. Is the lamp off?
Shine light on the LDR. What happens?

Which component can you adjust to make your
circuit switch on:
– even in a dim light?
– only in a bright light?   Try it.

Use:
How could you use this as a burglar alarm for a
drawer? (It is usually dark inside a drawer.)

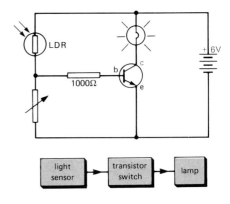

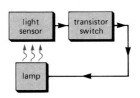

Experiment 41.17   An electronic candle
Place the LDR near the lamp and facing it, in a dark
room. Adjust it so the lamp is just off.
What happens when you bring a lighted match near
the lamp? Why does the 'candle' light?
What happens if you snuff out the 'candle' by
covering it with your fingers? Why?
This is an example of *positive feedback* from the
lamp to the LDR.

Experiment 41.18   A burglar alarm (a dark-operated switch)
Invert the light-sensor part of your circuit as shown:
Adjust the sensitivity control until the lamp is
off in daylight.
Now cover up the LDR. What happens?

What happens to the resistance of the LDR when it
is covered (see p 351)? Does this increase or
decrease the voltage V_{be}? Why does the lamp light?

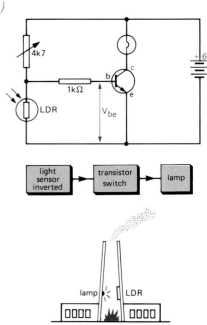

Uses:
How can this circuit be used to switch on
street-lamps automatically at dusk?
Draw a diagram to show how you could add a relay
to ring a bell if a burglar interrupts a beam of
light. What is the disadvantage of this circuit
(see experiment 41.31)?

It is important that factory chimneys do not
pollute the air with too much smoke. Invent a
device to warn a factory manager if the chimney is
producing too much smoke.

Experiment 41.19   A delay circuit
Connect a resistor and a capacitor (see page 278) in series with a
battery as shown:
Add a high resistance voltmeter and make sure the capacitor and
voltmeter are connected the correct way round (correct *polarity*).
Watch the voltmeter and close the switch. What happens?

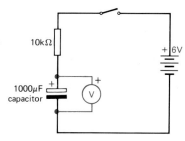

The battery is charging up the capacitor through the resistance.
Because it is a large capacitance and a large resistance it takes a
long time. The larger the resistance and the larger the
capacitance, the longer the *time constant* of the circuit.

Experiment 41.20   A time-operated switch on
Connect a capacitor and a resistor to a transistor switch as shown:
Press switch S_1 and wait. What happens?
Why is there a delay before the transistor is switched on?

Press switch S_2 to discharge the capacitor and so re-set the
circuit. How can you vary the time delay?   Try it.

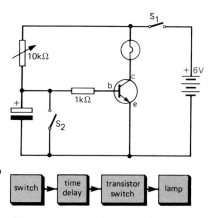

Uses:
How could this circuit be used to make an egg-timer for a deaf
person? Draw the circuit you would use if the person was blind.

Experiment 41.21   A time-operated switch off
How could you change your circuit so that it does the opposite? (So
that S_1 switches the lamp on and it goes *off* after a time delay.)
Try it.

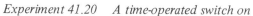

Use:
How could you use this circuit to switch your radio off after you
have gone to sleep?

Experiment 41.22   A sound-operated switch
Connect a microphone to a transistor switch as shown:
This circuit also contains a *thyristor* (or *silicon-controlled rectifier*,
or *SCR*).
This is a special diode with a third connection called the gate (g).
A thyristor does not conduct *until a current enters the gate*;
then it continues conducting *even if the gate current stops*.

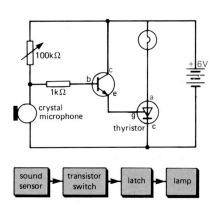

Tapping the microphone gives a voltage for a short time (see page
244). Why does the lamp light and stay lighted?
This is a 'latching' circuit (see also page 364). To turn off the
lamp you must disconnect the battery.

Uses:
How could this be used to sound an alarm if a baby cries?
How could this be used as a burglar alarm?

Amplifiers

In experiment 41.11 we saw that a small current in
the base of a transistor causes a larger current to
flow in the collector circuit
This is a *current amplifier*.

Experiment 41.23   Single-stage amplifier (an intercom)
The diagram shows a simple one-stage amplifier:
The microphone gives a varying *input* voltage
across a–b (see also page 244). This is amplified
to give a larger *output* voltage across A–B.
It is a *voltage amplifier*.
If the output is 100 times the input voltage, the
circuit has a *voltage gain* of 100.

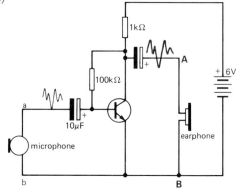

Use:
How could you use this as an intercom to connect
your bedroom with another room?

In practice, amplifiers for radios or record-
players usually have two or more stages to give
greater amplification. This is done by connecting
the output A-B of the first stage to the input a-b
of the second stage.

Analogue or digital?
The last circuit is an example of an *analogue* (or
linear) system: the voltage can vary smoothly and
have any voltage between the lowest and the
highest.
Another kind of system is *digital*. Digital systems
can only have certain definite values, usually
just *on* or *off*. A transistor switch (page 354) and
a logic gate (see page 360) are digital systems.

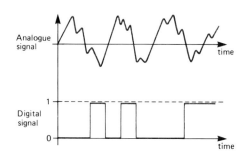

Integrated circuits (IC)
Both analogue and digital systems can be built
into an *integrated circuit* (IC). The photo shows a
full-size IC for an op-amp (see opposite page).
The actual circuit or 'chip' is very small but can
easily contain the 20 transistors, 11 resistors
and 1 capacitor needed to make the op-amp circuit.

Nowadays some ICs have more than a million
components on one chip. This is *micro*electronics.

actual size
of I.C.

□

actual
size
of
'chip'
inside

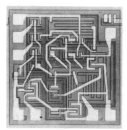

magnified photo
of the circuit

358

Operational amplifier (op-amp)

One very useful IC is called an *operational amplifier* or *'op-amp'* (see photo on opposite page).
It is an amplifier with a very high gain. It has 2 inputs and two main uses:

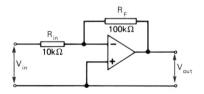

+9V supply

inverting input

non-inverting input

output

−9V supply

circuit symbol

1 Inverting amplifier

For this use, the op-amp is connected as shown in the diagram (the power supplies are not shown).
This circuit includes a *feedback resistor* R_F. This resistor feeds back some of the output into the *inverting* input. This is called *negative feedback*.
Although this reduces the gain of the amplifier, it also reduces distortion and makes sure the output is an accurate copy of the input.
The output is larger than the input and *inverted*.
The formula for the *voltage gain* (see opposite page) is:

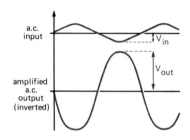

$$\text{Voltage gain} = \frac{V_{out}}{V_{in}} = \frac{R_F}{R_{in}}$$

With the resistance values shown in the diagram, the gain will be 10. By changing the resistors, an op-amp can give a gain up to 1000.

a.c. input

V_{in}

V_{out}

amplified a.c. output (inverted)

2 Comparator (to compare two voltages)

If the op-amp is used without any feedback, it has a very high gain but is unstable. Its output swings quickly between a high voltage and a low voltage. It is now a digital circuit.
The op-amp *compares* the 2 inputs and switches the output (to high or to low).
If V_1 is greater than V_2 then the output is low.
If V_1 is less than V_2 then the output is high.

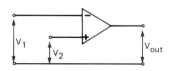

V_1

V_2

V_{out}

The op-amp is a very sensitive and rapid switch. It can be used to make the sensor circuits on pages 354–357 much more sensitive.
For example, here is a light-sensitive circuit from experiment 41.18:

A fixed potential divider (see p 294) provides a steady voltage V_1 to one input. The second potential divider includes an LDR, so the voltage V_2 varies with the light shining. If the LDR is dark, V_2 is larger than V_1 and the output is high, so it switches on the transistor and the lamp.
This circuit is very sensitive to changes in light. How could you use it to compare the whiteness of two sheets of paper?

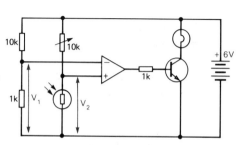

10k

10k

1k

+6V

1k

V_1

V_2

Logic circuits

Experiment 41.24   AND circuit

Connect the circuit shown in the diagram:
Press down the switches one at a time and then both together.
What do you have to do to get the lamp to light?

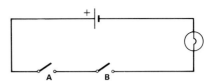

The lamp lights only if switch **A AND** switch **B** are pressed.
We can describe the way this circuit or 'gate' works in a diagram called a *Truth Table*:

Check each row of the Truth Table carefully.
This is called the *logic* of the circuit.

switch A	switch B	lamp
open	open	off·
closed	open	off
open	closed	off
closed	closed	on
inputs		output

This kind of logic is used in a washing machine where there is a main 'on' switch and another switch operated by the door of the machine. For safety, the washing machine will only work if the main switch is on **AND** the door is closed.

Truth Tables are usually shown with numbers rather than words. Because each switch has only 2 states (open and closed) we use only 2 numbers: **0** and **1**.
If the switch is open and so electricity cannot pass, it is given the value **0**.
If the switch is closed so electricity can go through, it is called **1**.
In the same way, a lamp which is off is **0**, and a lamp which is on is **1**.

switch A	switch B	lamp
0	0	0
1	0	0
0	1	0
1	1	1
inputs		output

Experiment 41.25   OR circuit

Here is a circuit with a different logic:

The lamp comes on if switch **A OR** switch **B** is pressed. This is called an *OR gate*.
Check its Truth Table carefully:

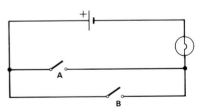

The doors of cars have switches connected with OR logic: the light inside the car comes on if either the driver's door **OR** the passenger's door is opened.

The circuits on this page only use mechanical switches. In practice it is much more useful to have logic circuits that are made of transistor switches, and use electrical signals for inputs and outputs. These logic gates can be made very small, inside integrated circuits. They are used in digital watches, robots and computers.

switch A	switch B	lamp
0	0	0
1	0	1
0	1	1
1	1	1
inputs		output

AND gate

A tiny integrated circuit can easily contain several AND gates,
built from transistor switches (see page 354).

Each transistor switch operates on the *voltage* applied to it. The
voltage can be 0 (called logical 0) or it can be higher (called
logical 1).

The circuit symbol for an AND gate with two inputs is shown here:
(Remember its shape: it looks like the D of AND)

The Truth Table is the same as the AND gate in experiment 41.24.
There is an output only if input **A AND** input B are at a logical 1.
Remember: 0 is zero volts and 1 is a higher voltage (usually
between 5V and 9V).

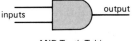

AND Truth Table

inputs		output
0	0	0
1	0	0
0	1	0
1	1	1

OR gate

The circuit symbol for an OR gate with two inputs is shown here:
Notice the different shape from an AND gate.

The Truth Table is the same as the OR gate in experiment 41.25.
There is an output if either input **A OR** input B are at a higher
voltage.

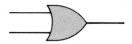

OR Truth Table

inputs		output
0	0	0
1	0	1
0	1	1
1	1	1

NOT gate

The third kind of logic gate is the NOT gate or inverter.
Its circuit symbol is shown here:

Its Truth Table is very simple: the output is always the *opposite*
of the input. It is an *inverter.*

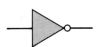

NOT Truth Table

input	output
0	1
1	0

Other gates

Sometimes these gates are combined.
For example: AND with NOT = NAND gate. OR with NOT = NOR gate.
Compare these symbols and Truth Tables with the AND and OR gates
What do you notice?

   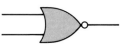

NAND Truth Table

inputs		output
0	0	1
1	0	1
0	1	1
1	1	0

NOR Truth Table

inputs		output
0	0	1
1	0	0
0	1	0
1	1	0

Logic gates
Have just 2 states:
Just off or on
With **0** or **1**

361

Using logic gates

Experiment 41.26   Wet AND light circuit
Connect up this system and look at it carefully:

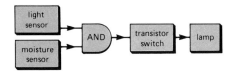

If the moisture detector is dry, then its voltage
input to the AND gate is a logical 0 (because
resistor R_1 keeps the voltage low).
If the LDR is dark, its resistance is very high
and so its voltage to the AND gate is also a
logical 0 (see also experiment 41.16).

From the Truth Table for an AND gate (see p 361) you can see that
there will be no output from the AND gate and the lamp will stay off.
In fact, to light the lamp you will have to change *both* inputs to a
logical 1. This happens when the moisture detector is wet **AND** the
LDR has light shining on it (see the red numbers on the diagram).

moisture sensor	light sensor	output
dry 0	dark 0	0
wet 1	dark 0	0
dry 0	light 1	0
wet 1	light 1	1

Uses:
You could use this circuit to light a lamp (or sound a bell) whenever
it rained **AND** it was day-time. This could tell you to bring in the
washing from the line, or bring in the baby from the pram (but it
wouldn't bother you if it rained at night).
If the moisture detector was replaced by a door bell switch, you
could ensure that your door bell didn't ring at night!

Experiment 41.27   Wet AND warm circuit
Replace the LDR circuit with a thermistor circuit:

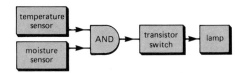

What do you have to do to get the lamp to light?

Uses:
How could you add a relay to use this circuit to
switch on a washing machine only when it was full
of water **AND** hot enough?
How could you fix this circuit to a pan to tell a
blind person if their hot milk is boiling over?

Experiment 41.28   Adding to a test button
It is often useful to add a test button so that you can
check if the circuit is working. Here is one way to do it:

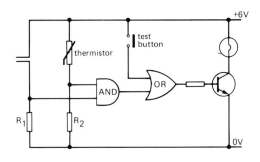

The lamp will light if the sensors are wet **AND**
warm, **OR** if the test switch is closed.

Experiment 41.29 Wet AND cold circuit

To invert the condition from 'warm' to 'cold' we can add a NOT gate to the last circuit:
(Another way would be to invert the temperature sensor circuit.)

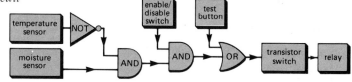

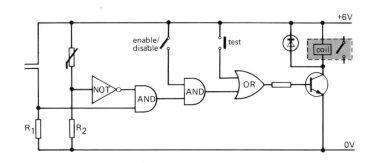

The circuit also has an extra AND gate and an 'enable/disable' switch. Can you see that the second AND gate can never operate unless the enable/disable switch provides a logical 1?

Uses:

If this relay operated a heating coil in an electric kettle or your hot water tank, it would be a safety device: the heater would not come on unless the water was cold **AND** there was enough water to cover the probes.

Experiment 41.30 A combination lock

Connect three push switches A, B and C to a system of gates as shown:
When each switch is pressed, it changes from a logical 0 to a logical 1.

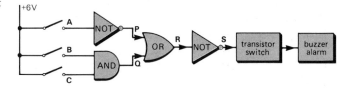

To analyse this circuit we need to draw up a Truth Table to cover all possibilities. The first three columns show all the eight possible combinations of A, B and C:
The next column shows the result at P (of NOT A) while column Q shows the result at Q (of B AND C). You should work through this table carefully, moving down each column in turn.

The circuit might be part of a combination lock on a safe. Looking at the final column you can see that some combinations would set off the alarm. What happens if you press switch A alone? Or A and B together? Or C, then A?
The only way to press all the switches safely, is to make sure you press both B and C *before* you press A. Any other combination of A, B and C will set off the alarm.

Suppose you replaced the OR gate with a NAND gate, and the AND gate with an OR gate. What is the right combination for this circuit?

A	B	C	P	Q	R	S	
0	0	0	1	0	1	0	
0	0	1	1	0	1	0	
0	1	0	1	0	1	0	
0	1	1	1	1	1	0	
1	0	0	0	0	0	1	=alarm
1	0	1	0	0	0	1	=alarm
1	1	0	0	0	0	1	=alarm
1	1	1	0	1	1	0	

The bistable

Some circuits can store information – they have a memory. The simplest memory circuit uses 2 NOR gates which are cross-connected.
This is called a *bistable* or *flip-flop* because it has *two* stable states (see the diagrams):

The circuit flips between state A and state B.
You should check the logic of each NOR gate (see page 361).
The circuit remembers or 'latches' the last thing that happened to it. It is also called a *latch* or a *bistable multi-vibrator* and is used in computer memories.

Experiment 41.31   A burglar alarm with a latch

If a burglar breaks a light beam, the alarm should continue to ring even after he has moved past the light beam. Try this system:

What happens if you briefly cover the sensor? How can you re-set the circuit?

How could you add a NOT gate (inverter) to your circuit to make it respond to the flash of a burglar's torch at night?

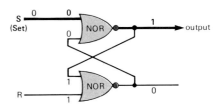

Stable state A
If the S input is briefly connected to logical 0, it sets the bistable to state A.

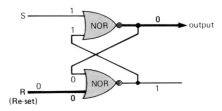

Stable state B
If the R input is briefly connected to logical 0, it re-sets the bistable to state B.

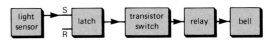

Summary

A diode allows a current to flow one way only. This can be used to rectify a.c. to d.c.
The resistance of an LDR decreases if it is illuminated. The resistance of a thermistor decreases when it is heated. An LDR, thermistor, and a reed switch are input transducers.
A relay and a LED are output transducers.
A transistor amplifies small currents. A transistor (or an op-amp) can be used as an electronic switch.
There are several kinds of logic gates: AND, OR, NOT, NAND, NOR.

Professor Messer's in one of his states. He can't work out his logic gates:

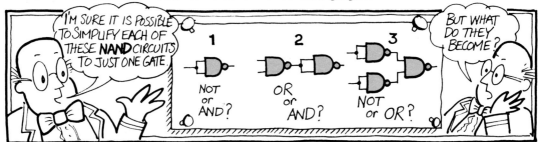

Questions

1. Copy out and complete:
   a) Germanium and silicon are called . . . . . .
      A diode will pass a current only when it is
      . . . . . -biassed.
   b) A diode is useful as a half-wave . . . . . to
      convert . . . . . current to . . . . . current. The
      output can be smoothed by a . . . . .
   c) A LED (. . . . .) is a transducer; it changes
      . . . . . energy to . . . . . energy.
   d) When an LDR (. . . . .) is illuminated, its
      resistance . . . . . When a thermistor is
      heated, its resistance . . . . .
   e) A . . . . . (NO) reed switch does not pass a
      current until a . . . . . is brought near it.
   f) A . . . . . current in the coil of a relay can switch
      on or off a . . . . . current through the contacts.
   g) A transistor is used to . . . . . small currents.
      A small current in the . . . . . circuit causes
      a . . . . . current to flow in the . . . . . circuit.

2. Draw the symbol and truth table for a) AND
   b) OR   c) NOT   d) NAND   e) NOR gates.

3. The diagram shows a diode circuit used to
   switch in a battery automatically if the main
   power supply fails.

   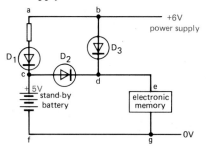

   a) Which route does the current flow when the
      power is on? Why?
   b) Which route to 'trickle charge' the battery?
   c) Which route when the power fails? Why?

4. Draw block 'systems' diagrams for the following
   applications:
   a) to light a lamp if it rains.
   b) to ring a bell if it rains.
   c) to switch on a fan motor if a greenhouse
      gets too hot.
   d) to sound an alarm if a thief opens a
      light-tight drawer.

5. Draw transistor circuits for the systems in
   question 4.

6. Draw block 'systems' diagrams:
   a) to switch on a water pump if a house plant
      is too dry.
   b) to switch on a heater if a fish tank gets
      too cold.
   c) an anti-burglar device to switch on your
      house lights at night even when you are away
      on holiday.

7. Draw transistor circuits for question 6.

8. Draw block 'systems' diagrams:
   a) to switch on a water pump if a ditch fills
      with water.
   b) to switch on a roadworks light at night.
   c) to switch on a fan in a storage heater if it
      heats up.
   d) to ring a bell if a baby cries.
   e) a fire alarm detecting the smoke
      ('smoke is more dangerous than fire')
   f) to switch on a pump to fill a water tank if
      the level is too low.
   g) to ring a bell when an oven is up to the
      correct temperature.
   h) for an oven timer.

9. Draw block 'systems' diagrams:
   a) to sound an alarm only if the baby cries *and*
      its nappy is wet.
   b) to open a greenhouse window if it is warm in
      the daytime.
   c) to light a lamp if it is a good day for
      gardening because it is warm, dry and light.
   d) to sound an alarm if the tropical fish tank
      is dark or cold or the water is low.
   e) to allow a drilling machine to run only if
      the operator presses a foot switch and has
      not got her hands breaking a beam of light
      which is shining near the drill.
   f) add a test button to all these circuits.
   g) to show a deaf person if the doorbell has
      rung since last they looked. (Add a reset.)

Further questions on page 370.

Further questions on page 370.

365

Electric power

1. a) Draw a circuit diagram to show an electrical ring main with three power sockets.
   b) (i) What is meant by a 'fuse'?
      (ii) Why should an electrical circuit contain a fuse?
      (iii) If a 240 V ring main has a 30 A fuse what is the maximum power which may be taken from the circuit?
      (iv) What is the maximum current which will flow through any part of the ring circuit?   (N16+)

2. A television set is rated at 800W and operates on 250V mains.
   a) Which fuse should you put in the plug? (2A, 5A or 13A)
   b) If the set was used for 10 hours how many kilowatt-hours of electricity would be used?
   c) If electricity costs 6p per kilowatt-hour, what is the cost of using the set for 10 hours?   (AL)

3. A 2-kilowatt electric fire is used for 10 hours each week and a 100-watt lamp is used for 10 hours each day. Find the total energy consumed each week and the total cost for each week if one kilowatt-hour of electricity costs 6p.   (JMB)

4. For a bath, 200 kg of water at a temperature of 45°C is required. The cold water supply is at 15°C.
   a) Calculate the quantity of heat required to warm the water for the bath (specific heat capacity of water = $4 \cdot 2 \times 10^3$ J/kg K).
   b) 1 'unit' of electrical energy is 1 kilowatt-hour. Calculate the number of joules which are equivalent to 1 kilowatt-hour.
   c) If the hot water for the bath were heated electrically, calculate the number of 'units' which would be required.
   d) A shower requires only 20 kg of water at the same temperature. Calculate the number of 'units' which would be required for a shower.   (AEB)

5.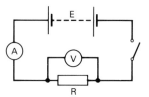

   The diagram shows an electrical circuit in which a current is passed through a resistor R. The supply voltage E can be altered, and the voltmeter has a high resistance. Readings of the current and potential difference are recorded for different supply voltages.

Current (A)	0·10	0·20	0·30	0·40	0·50	0·60	0·70
p. d. (V)	0·6	1·2	1·8	2·5	3·5	5·0	7·0

   a) Plot a graph of potential difference (along the y-axis) against current (along the x-axis).
   b) What is the resistance of R when the current is (i) 0·25 A, (ii) 0·60 A?
   c) Give one possible reason for the difference in these resistances.
   d) Calculate the power being transformed by resistor when the current is 0·60 A.
   e) The resistor is placed in an aluminium block of mass 0·5 kg and connected to a potential difference of 5V. If no heat is lost, calculate the rise in temperature of the aluminium block after 10 minutes. The specific heat capacity of aluminium is 900 J/kg °C.   (O&C)

6. A heating coil of resistance 9 ohm is connected in series with a moving-coil ammeter (of resistance 0·1 ohm) and a d.c. power supply (of e.m.f. 20V and internal resistance 0·9 ohm). The coil is placed in a vessel of negligible thermal capacity containing 0·20 kg of a liquid of specific heat capacity 4000 J/kg K (4000 J/kg °C)
   a) Draw a circuit diagram and label the components with their values.
   b) Calculate the current which flows in the circuit.
   c) Calculate the power which is developed in the coil.
   d) Calculate the power which is 'wasted' (i.e. in the meter and the battery).
   e) Calculate the rise in temperature of the liquid which takes place in 10 minutes. (N16+)

7.

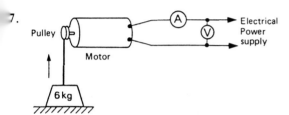

The diagram illustrates an arrangement for measuring the energy input and the energy output of an electric motor. It is found that the 6 kg mass, initially at rest, can be raised 3 m in 5 s and that the readings of the ammeter and voltmeter are 2 A and 24 V respectively. (The gravitational field strength is 10N/kg)

a) How many coulombs of electricity have passed through the motor? [2 marks]

b) What is the energy input from the supply? [3]

c) How much energy was used in lifting the 6 kg mass? [3]

d) It is found that the gain in energy of the mass is never equal to the energy put in by the electrical supply. Account for this difference in energy. [2]

(MEG)

Chemical effect

8. This question is about some of the many problems with which an electrician might be faced at work. Imagine that you are an electrician and that the following problems and questions are brought to you by some of your customers. In each case describe how you would answer them. Remember, your business depends on satisfied customers.

In each of the first four cases, explain a possible cause for your customer's trouble and how it might be solved.

Customer 1 'A lamp in my house has gone out.' [2 marks]

Customer 2 'The lights downstairs have all gone out.' [2]

Customer 3 'All the lights in the house have gone out.' [2]

Customer 4 'All the lights in my street have gone out.' [2]

The next customer would like a clear wiring diagram.

Customer 5 'I would like two lamps in the dining room both worked from the same switch.' Draw a suitable wiring diagram. [4]

The next customer is quite bewildered by the problem and you must give this customer a clear, full explanation of how you make the calculation.

Customer 6 'How much a week will it cost me to run my 1 kW electric fire and 2 kW electric radiator if they are both used for 4 hours each evening?' (1 kWh costs 5p) [4]

The last customer does not even know there is a problem!

Customer 7 'I should like you to come and fit a mains socket in the bathroom so that I can have the electric fire on when I have a bath.' Explain what the problem is. [3]

(NEA)

9. You are asked to copper plate a piece of brass with copper.

a) What electrolyte would you use?

b) What substance would you use for the anode?

c) Why would the brass have to be very clean?

d) What type of a supply of electricity would you use?

e) State **two** ways in which the mass of copper deposited could be halved.

f) Name **two** other metals that could be deposited by electrolysis. (SE)

Magnetic effect

10. Draw diagrams to show the pattern and direction of the magnetic field produced by

a) a solenoid carrying a current,

b) a bar magnet. (C)

11. Draw a labelled circuit diagram of the vibrator of an electric bell or buzzer. Explain how the vibrator works. (JMB)

12. Draw a labelled diagram showing the essential features of an electrical relay and describe how the relay works. Name one situation in which such a relay could be used and explain why an ordinary type of switch is not suitable. (JMB)

Magnetic effect (continued)

13. The diagram shows the construction of a
    simple moving-coil galvanometer.

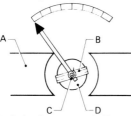

a) Label the four parts indicated.
b) The coil of such a galvanometer has a
   resistance of 10 Ω and gives a full-scale
   deflection for a current of 0·1 A. Explain
   why a resistance must be placed in parallel
   with the meter if larger currents than this
   are to be measured.
c) The galvanometer is to be used to
   measure currents up to 1 A which gives
   full-scale deflection. What current must
   pass through the parallel resistor when
   1 A is being measured?
d) What must be the value of the parallel
   resistor?   (S16+)

14. a) Draw a clearly labelled diagram to
       illustrate the mechanism of a moving coil
       galvanometer.
    b) When a direct current passes through a
       moving coil galvanometer, the pointer
       indicates a steady reading.
       (i) Explain what causes the pointer to
            deflect, and why it comes to rest.
       (ii) Explain why the reading will increase
            if the current is increased.
    c) State and explain briefly the step which is
       usually taken in the design of the galvano-
       meter to ensure that the scale is linear. (NI)

15. A radio enthusiast wishes to convert a
    milli-ammeter into a voltmeter capable of
    reading up to 20 V. Show how this can be
    done, if the milli-ammeter has an internal
    resistance of 6 ohm and gives a full-scale
    deflection with a current of 10 mA.
                                        (AEB)

16. a) Describe fully the structure and mode of
       action of an ammeter which can measure
       alternating currents.
    b) A moving coil meter, for measuring *direct*
       current, has a full scale deflection 15 mA
       and a resistance of 5 Ω. Explain how the
       meter could be converted
       (i) to act as an ammeter reading up to 5 A,
       (ii) to act as a voltmeter reading up to 5 V.
                                        (L)

17. A moving coil meter has resistance 5 Ω and
    gives full scale deflection for a current of 10
    mA. Show clearly how an appropriate
    resistor may be used to convert this meter
    into (a) an ammeter reading to 1·0 A, (b) a
    voltmeter reading to 20 V. Calculate the
    resistance value in each case.   (S)

18. a) Two parallel conductors carrying currents
       in opposite directions are set up close
       together. Draw separate diagrams of
       (i) the magnetic field due to one wire, and
       (ii) the resultant magnetic field due to the
       two wires, and explain the effect of the
       conductors on each other showing the
       current and the field directions. State a
       rule relating current and field directions
       with the force produced. Describe a prac-
       tical arrangement to demonstrate this effect.
    b) Define (i) the ampere, (ii) the coulomb.
                                        (JMB)

19. a) Name the unit of electric charge.
    b) How much charge has passed round a
       circuit in 20 minutes if a steady current
       of 0·5 A flows?
    c) If the total energy supplied by the source
       in this time is 3600 J, what is the e.m.f.
       of the source?   (N16+)

20. a) State the relationship between the
       ampere and the coulomb.
    b) (i) 600 C pass round a circuit in 5
            minutes. If the current is constant,
            what is its value?
       (ii) If the total resistance of the circuit is
            50 Ω calculate the electrical energy
            consumed in this time.   (N16+)

Electromagnetic induction

21. Draw and label a diagram of a simple rotary d.c. generator (dynamo). What modifications would be necessary to make the generator produce a.c. instead of d.c.?

Draw two sketch graphs of the voltage outputs, one clearly labelled d.c. and the other clearly labelled a.c., to illustrate **two** revolutions of the coil in each case.

On the a.c. graph, indicate the peak voltage and the r.m.s. voltage. What is the mathematical relationship between these two voltages?   (AEB)

22. Draw a labelled diagram of a simple apparatus that could be used to demonstrate electromagnetic induction.

a) Describe what you would do with the apparatus.

b) Describe the observations you would make.

c) The right-hand rule is used to help us in finding the direction of the induced current. Write down the quantities that relate to:

thumb, first finger, second finger.
        (S16+)

23. In the transmission of electrical energy over great distances in the grid system, low currents are employed at high voltage.

a) State and explain what advantage is gained by having low currents flowing in the transmission cables.

b) What disadvantage arises from employing very high transmission voltages?

c) Why is a.c. used in this system?   (N16+)

24. A transformer is used with a 240-V mains supply at a frequency of 50 Hz to operate a 12-V electric bell.

a) What type of transformer is this?

b) Calculate the turns ratio of this transformer.

c) State the frequency of the voltage in the secondary circuit.   (W)

25. a) Electricity is to be supplied from a power station to a factory by means of a transmission line of overall resistance 100 Ω. The power station produces 10^7 W. This power is delivered to the line at a voltage of 50 kV. Calculate

(i) the current that would flow in the transmission line,

(ii) the power converted to heat in the transmission line.

As an alternative, the voltage at which the power is delivered to the line can be raised, by means of a transformer, to 500 kV before it is transmitted along the line and then lowered, by another transformer, before being passed to the factory. Calculate

(iii) the current that would now flow in the line,

(iv) the power that would now be converted to heat in the line.

Which do you regard as the better voltage for the transmission?

b) Describe the essential features of a step-up transformer.

c) Transformers are not 100% efficient. Give **two** reasons why this is so.

        (O&C)

26.

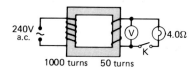

240V a.c.   1000 turns   50 turns   (V)   4.0Ω K

The circuit shows a step-down transformer used to light a lamp of resistance 4·0 Ω under operating conditions.

Calculate

a) the reading of the voltmeter, V, with K open,

b) the current in the secondary winding with K closed (the effective resistance of the secondary winding is 2·0 Ω),

c) the power dissipated in the lamp,

d) the power taken from the supply if the primary current is 150 mA, and

e) the efficiency of the transformer.

        (L)

Electronics

27. Draw a clearly labelled diagram to show the essential features of a cathode ray oscilloscope. An oscilloscope is arranged so that a spot is produced on the screen. Describe and explain the effect on the appearance and position of the spot in each of the following cases:
a) the size of the accelerating p.d. in the electron gun is increased;
b) a p.d. is applied to the horizontal pair of deflecting plates (the Y-plates) so that the uppermost plate is positively charged;
c) a horizontal magnetic field is applied in the direction shown in the diagram;
d) while the horizontal magnetic field of (c) is applied, the size of the accelerating p.d. in the electron gun is increased.

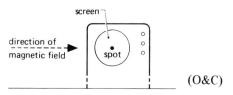

(O&C)

28. Draw a diagram of a simple half-wave rectifier. Include a step-down transformer to feed the rectifier from the mains a.c. supply. Label the diagram clearly and explain how it works.

29. The diagram shows a 'diode bridge' circuit which is a 'full-wave' rectifier used in power supplies. Alternating current is applied to it (from a transformer).

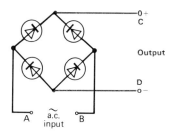

a) Copy the diagram and show the path of current if A is more positive than B.
b) On another diagram show the path if B is more positive than A.
c) Why does it give a d.c. output?
d) Draw the whole rectifier circuit by including a transformer and a smoothing capacitor.

30. a) Draw the symbol used to represent the following components in a circuit.
(i) n.p.n. transistor
(ii) electrolytic capacitor
(iii) thermistor
(iv) loudspeaker
(v) microphone
(vi) relay
b) What happens to the resistance of a thermistor as the temperature is increased?
(AL)

31.

The diagram above shows an incomplete electronic circuit.
a) Copy the diagram and make the following additions to it.
(i) In the circle draw the symbol for an npn transistor.
(ii) Add the symbol for a battery, correctly connected between B and D.
(iii) Draw the symbol for a thermistor between points A and X.
(iv) Draw the symbol for a variable resistor between points X and C.
(v) Draw the symbol for a lamp in the collector circuit.
(vi) Connect the base and the emitter of the transistor to the correct places in the circuit.
b) As the circuit is arranged, the lamp will not light. If there is a rise in temperature the lamp will light. Explain this.
c) A simple re-arrangement of this circuit would make the lamp light when the temperature falls. Explain what changes should be made, and how they will affect the working of the circuit.
d) In the original circuit, instead of a variable resistor, a fixed resistor could have been used between points X and C. Explain why it might have been considered desirable to use a variable resistor for these two circuits. (N16+)

32. a) When a transistor is connected into a circuit to determine the characteristic curve, the following results are obtained.

I_b (μA)	20	40	60	80	100
I_c (mA)	1·1	2·3	3·2	4·4	5·5

(i) Plot the graph of I_c against I_b (I_c on the y-axis). (ii) Draw the best straight line through the points. (iii) Calculate the gradient I_c/I_b (show clearly on your graph which part of the line you have used). (iv) What property of a transistor does this illustrate?

b) Draw a circuit which can be used to produce the results in (a). (EA)

33. This circuit is designed to switch on the beacon automatically when daylight fades.

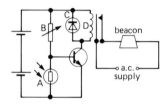

a) (i) What is the full name of component A?
(ii) What is the abbreviation of its name?
(iii) What changes the value of this component? (iv) Explain fully how the properties of the component change.
b) Name component B.
c) Component D is the coil of a reed switch:
(i) Why has this coil no metal core?
(ii) What is the function of the coil?
d) (i) Explain how the transistor circuit works, mentioning the function of each component.
(ii) Explain how the transistor circuit controls the operation of the beacon.
e) The beacon operates on a.c. Explain why this supply cannot be used for the transistor circuit. (N16+)

Further Reading

Inside the Chip – Davies (Usborne)
Robotics – Potter (Usborne)
Mastering Electronics – Watson (Macmillan)
Everyday Electronics Magazine (IPC)

34. Draw block 'systems' diagrams:
a) For a device to test the walls of buildings for dampness.
b) To tell an angler if his fishing line moves (hint: use a reed switch).
c) To photograph a bursting balloon by operating the camera flash from the sound.
d) For a safety device in an oil central-heating system so that no more oil will be pumped to the boiler if the pilot light is off.
e) For an alarm that rings in a neighbour's house if an old person living alone does not flush the toilet within 10 hours.

35. a) A market gardener wants a circuit that will operate a pump to water his tomatoes only when the soil is dry and it is night-time (so the sun won't blister the tomatoes):

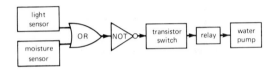

Analyse this circuit and explain what happens.
b) Modify the circuit for the owner of an apple orchard, who wants an alarm to sound whenever it is frosty at night (so he can light smoky fires to protect his crop).
c) How could you re-design these circuits using only AND and NOT gates?

36. The diagram shows a circuit for a combination lock. When a switch is pressed, it changes its input from 0 to 1.

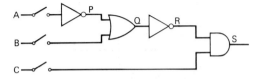

a) Draw up a truth table showing the state of P, Q, R, S for all combinations of A, B, C.
b) If S rises to logic 1, it opens the lock. What is the correct combination?

Have you started revision yet? See page 390

chapter 42

In 1896, Henri Becquerel discovered almost by accident that some substances (like uranium and radium) can blacken a photographic film even in the dark.

These *radioactive* substances are dangerous and the experiments in this chapter will be teacher demonstrations only.

Detecting radioactivity

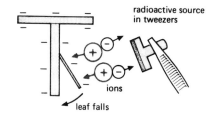

1. Photographic film
The photograph shows how some film has been blackened by radioactivity except in the shadow of a key.

2. The gold-leaf electroscope
Experiment 42.1
Charge a dry gold-leaf electroscope as in experiment 32.10 (page 270).

radioactive source in tweezers

ions

leaf falls

Does the leaf stay up for several minutes? What happens if a radioactive source (e.g. radium) is brought near the leaf, using tweezers?

The leaf falls because the air is *ionised* by the radioactivity (see also experiment 32.17 on page 272). A negative electroscope attracts the positive *ions* and so it is discharged. This *ionisation* is used in other detectors of radioactivity.

3. The cloud chamber
In a cloud in the sky, the water droplets tend to form on dust particles or on charged ions.
In a cloud chamber, a ray from a radioactive source causes a line of ions on which a thin cloud forms (rather like the cloud trail behind a high flying aircraft).

The photograph shows the tracks of 'alpha' particles in a cloud chamber.

4. The spark counter

Experiment 42.2

A spark counter consists of a fine wire stretched just below a piece of metal gauze.
A high voltage between the wire and the gauze is adjusted until it is almost, but not quite, sparking. Using tweezers, a radioactive source (radium) is brought close to the gauze.

What happens? Why?

Rays from the radioactive source ionise the air so that it is a better conductor and sparks are produced.

Do the sparks occur regularly or randomly?

Place a thick piece of paper between the source and the gauze. What happens? Why?

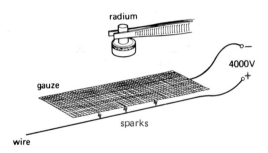

5. The Geiger-Müller Tube (G-M tube)

This is a metal tube with a thin wire down the centre. It contains a gas at low pressure. It works on the same principle as a spark counter, but the voltage is lower so that no spark is formed.
Instead a pulse of current is amplified and passed to either:

a *scaler*, which counts the pulses and shows the total number, *or*

a *ratemeter*, which shows the rate in "counts per second". In addition, a loudspeaker clicks each time a ray from the source passes through the tube.

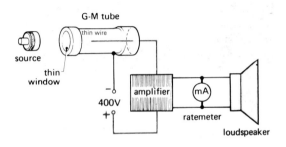

Experiment 42.3

A radium source is placed near a G-M tube connected to a ratemeter and loudspeaker.

What do you hear as the source moves closer?

Place a thick piece of paper between the source and the tube.
Does the paper stop all the rays?

Because paper stopped the sparks in the last experiment, but not the clicks in this experiment, it seems that there are at least two kinds of radiation from radium.
In fact there are three kinds, called *alpha, beta* and *gamma* rays.

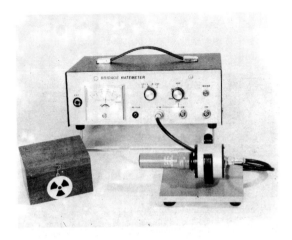

373

Alpha, beta and gamma rays

Experiment 42.4
Use a Geiger-Müller tube with a very thin window
and connect it to a ratemeter.
Switch on and notice the *background count* due
to natural radioactivity.

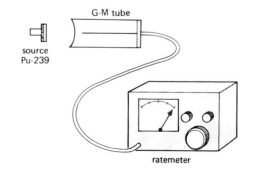

Using tweezers, a plutonium-239 source is placed
close to the window of the tube.

Note the count-rate on the ratemeter.

Place a piece of paper between the source and the
tube. What happens to the count-rate?

This radiation is called **alpha (α) radiation.**
It is easily stopped by paper or by your clothing.
It can be deflected by electric or magnetic fields.
It can be shown that α-**radiation is positively
charged particles moving at high speed.**
In fact an α-particle has the same positive charge
and mass as the nucleus of a helium atom (see page 370).

Experiment 42.5
Using tweezers, the source is replaced by a
strontium-90 source.
Note the count-rate.

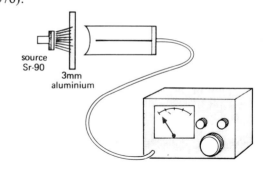

Does a sheet of paper affect the count-rate?
Does aluminium foil affect the count-rate?
What happens if an aluminium sheet, 3mm thick,
is placed between the source and the tube?

This radiation, from strontium-90, is called
beta (β) radiation. It is more penetrating than
α-particles, but cannot penetrate through 3mm
of aluminium.

Experiment 42.6
Place a powerful magnet between the source and
the tube.
What happens to the count-rate?
Move the G-M tube until the count-rate increases
again.

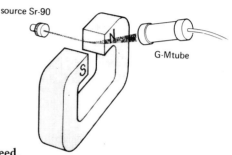

The direction of the deflection shows that β-rays
must be *negatively-charged particles* (using
Fleming's left hand rule, see page 318).
In fact, β-**particles are electrons travelling at very high speed.**

Experiment 42.7
Use the same apparatus as before, but with a cobalt-60 source.

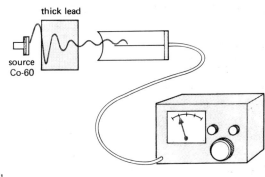

Note the count-rate.

Does a sheet of paper stop this radiation?
Does 3mm aluminium sheet stop this radiation?

This radiation from cobalt-60 is called **gamma (γ) radiation.**
It is not affected by a magnetic field and is very penetrating. Although it is reduced, it is not stopped even by thick pieces of lead.

If the distance between the source and the G-M tube is increased, it is found that at twice the distance the count-rate is only one quarter. At three times the distance, the intensity of γ-rays is only one ninth.
The intensity of light from a lamp changes in just the same way.
In fact, it can be shown that these penetrating and dangerous **gamma-rays are really invisible electro-magnetic waves of very short wavelength** (see page 224).

Repeat all these experiments with a radium source.
Do you find that it emits alpha- *and* beta- *and* gamma-rays?

Properties	α-particle	β-particle	γ-rays
Nature:	Positive particle (Helium nucleus) (about 7000 times the mass of an electron)	Negative e⁻ electron	Electromagnetic waves
Affected by electric and magnetic fields?	Yes	Yes, strongly.	No.
Penetration:	Absorbed by paper (or 6cm of air)	Absorbed by 3mm aluminium	Reduced but not stopped by lead.
Causes ionisation?	Strongly.	Weakly.	Very weakly.
Detectors:	Photographic film Cloud chamber Spark counter Gold-leaf electroscope Thin-window G-M tube	Photographic film Cloud chamber G-M tube	Photographic film Cloud chamber G-M tube

Atomic structure

Experiments show that α-particles can pass through thin gold foil. This suggests that a gold atom might be almost entirely empty space.

To find more about the structure of an atom, Ernest Rutherford suggested an experiment in which α-particles were fired at thin gold foil. Detectors were used to find how the α-particles were scattered by the gold atoms. It was found that some α-particles were scattered back towards the source — rather like firing a machine-gun at tissue paper and finding that some of the bullets bounce back!

In 1911, Rutherford showed that this could be explained if each atom has a tiny core or *nucleus* with a *positive charge*. A positive nucleus repels the positive α-particles so that they are scattered in different directions.

When Rutherford calculated the size of a nucleus, he found that it was very small even compared with a very small atom. If you imagine magnifying an atom until it is about the size of your School Hall, then the nucleus in the centre would be about the size of this full stop.

The simplest atom is the **hydrogen (H) atom**. The nucleus is a single positive charge called a *proton (p)*. There is a single negative *electron (e)* in an orbit round the nucleus. Because the charges are equal but opposite, the atom is neutral.

A slightly more complicated atom is the **helium (He) atom**. This has two protons in the nucleus and, to be neutral, two electrons outside. The helium nucleus also has two uncharged particles, called *neutrons (n)*.

A neutron and a proton are roughly equal in mass — each is about 1800 times the mass of an electron.

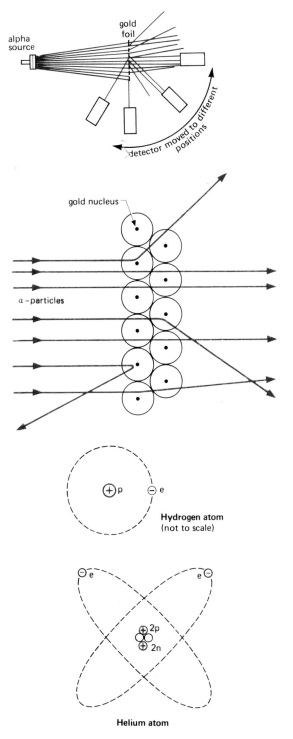

Hydrogen atom
(not to scale)

Helium atom

376

Atomic number and mass number

Different elements have a different number of protons — for example, a *lithium* atom has 3 protons, a *beryllium* atom has 4 protons, a *boron* atom has 5 protons, a *carbon* atom has 6 protons and so on.

The number of protons in an atom is called its *atomic number*.

What is the atomic number of lithium?
A list of some elements in order of atomic number is shown at the bottom of the page.

Another important number is the *mass number*: this is the total number of *nucleons* (= protons + neutrons).

For example, the mass number of lithium is 7 (= 3 protons + 4 neutrons).
A shorthand way of describing an atom is shown here:

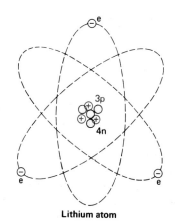

Lithium atom

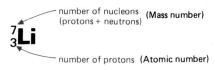

Isotopes

The oxygen that you are breathing has an atomic number of 8. That is, it has 8 protons in the nucleus (and 8 electrons to make a neutral atom).
Most of the oxygen that you are breathing has also 8 neutrons (so the mass number = 8 + 8 = 16). This is $^{16}_{8}O$.
Look at the diagram of this atom.

Some of the oxygen you are breathing has got 9 neutrons (but still 8 protons, making a mass number of 17). This is $^{17}_{8}O$.
Draw a diagram of this atom.

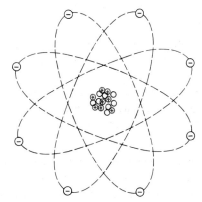

an atom of $^{16}_{8}O$

$^{16}_{8}O$ and $^{17}_{8}O$ are both atoms of oxygen.
They have the same chemical properties but have different masses. They are called *isotopes* of oxygen.
Oxygen has a third isotope, $^{18}_{8}O$. Draw a diagram of an atom of this isotope.

All elements have more than one isotope. Many of these isotopes are *unstable* — the nucleus breaks up into smaller parts.
When this happens α-particles, β-particles or γ-rays may be emitted from the nucleus. This is called *radioactive decay*.

Atomic Number	Element + symbol		Commonest Isotope
1	Hydrogen	H	$^{1}_{1}H$
2	Helium	He	$^{4}_{2}He$
3	Lithium	Li	$^{7}_{3}Li$
4	Beryllium	Be	$^{9}_{4}Be$
5	Boron	B	$^{11}_{5}B$
6	Carbon	C	$^{12}_{6}C$
7	Nitrogen	N	$^{14}_{7}N$
8	Oxygen	O	$^{16}_{8}O$
92	Uranium	U	$^{238}_{92}U$

Radioactive decay

Experiment 42.8

For this experiment you need a large number of wooden cubes or dice, each with one face painted red. Imagine each cube is an atom.

If you shake and throw all the dice-atoms, some will land with the marked face upwards.

Let us pretend that these are atoms which have disintegrated and fired out α-particles or β-particles, and γ-rays.

Remove these disintegrated 'atoms' and count how many atoms have survived.

Repeat this process with the surviving dice and continue until almost all of them have 'decayed'. Plot a graph of the *number of dice-atoms surviving* against the *number of throws*.

Real atoms behave in a similar way: each atom disintegrates in a random, unpredictable way. A large number of atoms gives a smooth *radioactive decay curve*.

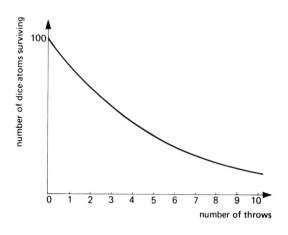

Half-life

The time taken for *half* the atoms to decay is called the *half-life* of the substance.

The graph shows a decay curve for a radio-active substance. On the graph you can see that after 1 half-life, half the atoms have disintegrated and half have survived.

After a further half-life, the activity has halved again so that only $\frac{1}{4}$ of the atoms survives. After 3 half-lives it has halved again and only $\frac{1}{8}$ th survives.
What fraction survives after 4 half-lives?
What fraction survives after 5 half-lives?

The half-life of different substances varies widely – from fractions of a second up to millions of years.

The half-life of radium is 16 centuries.
This means that in a radium source in school (mass of radium only about five millionths of a gram), the atoms are disintegrating at the rate of 700 million in a physics lesson and yet it will do this for 16 centuries before half of the atoms have decayed!

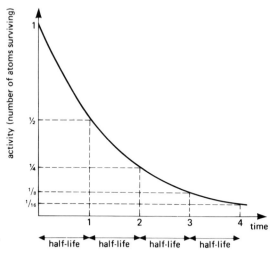

By mistake (in his lunch) Freddy Furze,
Ate radium and died with a curse,
But the point about Fred,
Is that now he is dead,
His half-life is sixteen hundred years!

Alpha-decay

An alpha-particle is a helium nucleus (see page 370). It is $^4_2$He.
It has 4 nucleons: 2 protons + 2 neutrons.

Radium-226 ($^{226}_{88}$Ra) decays by α-emission. When it loses the
α-particle, its mass number (226) must decrease by 4 (to 222).
Also, its atomic number (88) must decrease by 2, to become 86. It
has changed to a *different element*. It is now *radon* (Rn), a radioactive gas.
The nuclear equation is: $^{226}_{88}$Ra $\longrightarrow$ $^{222}_{86}$Rn + $^4_2$He ↗

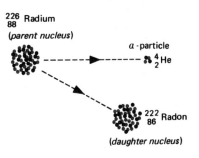

226 Radium
88
(parent nucleus)

α-particle
$^4_2$He

222 Radon
86
(daughter nucleus)

Notice that the top numbers balance on each side of the equation.
Similarly with the bottom numbers.

Radon-222 is itself radioactive and decays by α-emission. What does
it become? Write down the equation (element 84 is polonium, Po).

Beta-decay

The $^{218}_{84}$Po from the last equation can decay by β-emission. This is
the emission of an electron *from the nucleus*. But there are no
electrons in the nucleus!
What happens is this: one of the neutrons changes into a proton
(which stays in the nucleus) *and* an electron (which is emitted as a
β-particle).
This means that the atomic number *increases* by one, while the total
mass number stays the same.
The polonium changes into another element, called astatine (At).

The nuclear equation is: $^{218}_{84}$Po $\longrightarrow$ $^{218}_{85}$At + $^0_{-1}$e ↗

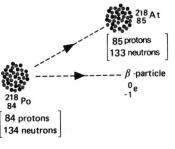

218 At
85
[85 protons
133 neutrons]

β-particle
$^0_{-1}$e

218 Po
84
[84 protons
134 neutrons]

Notice again that the top numbers balance and so do the bottom numbers.

Gamma-emission

When an α-particle or a β-particle is emitted, the nucleus is usually
left in an 'excited' state. It loses its surplus energy by emitting a γ-ray.

'Splitting the atom' (transmutation)

In one of Rutherford's experiments, he fired α-particles at the
nuclei of nitrogen atoms.
He found that some of the nitrogen atoms turned into oxygen atoms!
The nuclear equation is:

$^{14}_7$N + $^4_2$He $\longrightarrow$ $^{17}_8$O + $^1_1$H

nitrogen   alpha-   oxygen   hydrogen
nucleus   particle   nucleus   nucleus
                                (proton)

Notice again how the numbers balance.

The photograph shows how this appears in a cloud chamber.

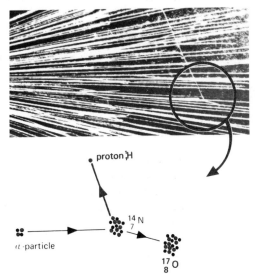

proton $^1_1$H

14 N
7

17 O
8

α-particle

Uses

Radioactive isotopes can be made by bombarding substances with α, β and γ-rays. These radioactive isotopes can then be used, with care, in many different ways:

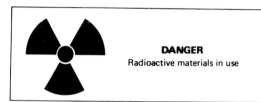

1. Radioactive tracers.
Radioactive fertiliser can be fed to plants and then traced through the plant using a Geiger-Müller counter. This is used to develop better fertilisers and insecticides.

Similarly, in hospitals, an isotope with a short half-life may be injected into a patient's body. A G-M counter is used to trace the movement of the chemical through the body, to check that certain glands are functioning properly. Tumours can be located because they take up more radioactive isotope than other parts of the body.

If radioactive pistons are fitted to a car then a G-M counter can be used to test the oil for traces of wear from the piston.

Leaks from a pipeline carrying oil or gas can be traced by injecting a radioactive isotope into the pipeline. This saves digging up the whole pipeline.

2. Sterilising
γ-rays can be used to kill bacteria in foods or in hospital blankets without making them radioactive. In the same way, hospital equipment can be sterilised after it has been sealed in a plastic bag. Cancer cells in a patient's body may be killed by careful use of γ-radiation.

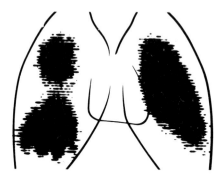

This is a scan of a pair of lungs using radio-isotopes. The light area in the middle of the left lung shows a cancer growth.

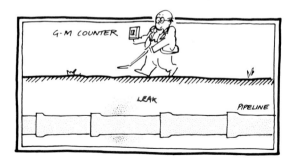

3. Thickness control
In paper mills the thickness of paper can be controlled by measuring how much β-radiation passes through the paper to a G-M counter. The counter controls the pressure of the rollers to give the correct thickness.

4. Checking welds.
If a γ-source is placed on one side of the welded metal and photographic film on the other side, weak points will show up when the film is developed.

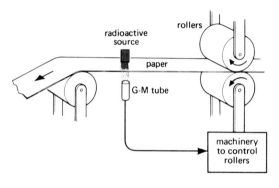

Nuclear energy

If an atom of uranium-235 ($^{235}_{92}$ U) is bombarded with slow moving neutrons, it breaks up into two smaller atoms and three fast neutrons.
This process is called *nuclear fission.*

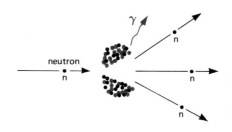

Mass and energy
The mass of the pieces after fission is found to be *less* than the mass of the pieces before the collision.
The mass that is lost has been converted to energy.
Albert Einstein showed how to calculate this:

$$\text{Energy released } (E) = \text{mass lost } (m) \times \text{ velocity of light squared } (c^2) \quad or \quad E = mc^2$$

The fission of 1 kg of uranium-235 produces more energy than 2 million kg of coal.

When a uranium-235 atom splits, three neutrons are emitted and these may hit three other uranium atoms so that they split and emit more neutrons which in turn hit other atoms and so on. This is called a *chain reaction.*

If a piece of uranium is above a certain *critical size* (about the size of a tennis ball), then this chain reaction takes place very quickly, as in an atomic bomb.

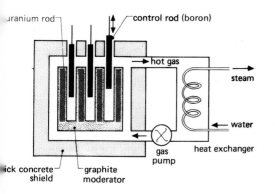

Gas-cooled Reactor

Mrs Messer: "What do nuclear scientists have for dinner?"
Professor Messer: "Fission chips".

Nuclear power station
In a *nuclear reactor*, the chain reaction is controlled by movable control rods made of boron. These absorb neutrons to slow down or stop the chain reaction.
Graphite is included as a *moderator* to slow down the neutrons so that the reaction works more efficiently.
The heat produced is used to make steam to drive generators in a power station.

Nuclear fusion
In a hydrogen bomb and in the Sun, hydrogen nuclei fuse together and release nuclear energy. This process has not yet been successfully controlled for power stations (see page 11).

381

Summary

The properties and detectors of alpha-particles, beta-particles and gamma-rays are shown in the table on page 375.

An atom has a small, heavy, central nucleus containing protons (+) and neutrons (no charge). The number of protons is called the atomic number. The number of nucleons (= protons + neutrons) is called the mass number.

In a neutral atom, the number of electrons in orbit = the number of protons.

An isotope has the same atomic number but a different mass number.

The half-life of an isotope is the time taken for half the atoms to decay.

Questions

1. Copy out and complete:
   a) An alpha . . . . . is . . . . .-charged and has the same charge and mass as the nucleus of a . . . . . atom.
   b) Beta . . . . . are . . . . .-charged electrons travelling at very . . . . . speed.
   c) Gamma rays are . . . . . rays of very short . . . . .
   d) Beta . . . . . are more penetrating than . . . . . but less penetrating than . . . . .
   e) . . . . . and . . . . . can be deflected by electric and magnetic fields, but . . . . . cannot.
   f) Alpha, beta and gamma-rays can be detected by . . . . . film, a . . . . . chamber, or a . . . . . tube. A gold-leaf electroscope or a spark counter can be used to detect . . . . .

2. Copy out and complete:
   a) An atom has a small heavy . . . . . containing . . . . .(-charged) and . . . . .(. . . . .-charged).
   b) A . . . . . and a . . . . . are each about 1800 times the mass of an electron.
   c) The atomic number is the number of . . . . .
   d) The mass number is the number of . . . . .
   e) An isotope has the same number of . . . . . but a different number of . . . . .
   f) The half-life of an element is the . . . . . taken for . . . . . of the atoms to decay. After . . . . . half-lives, only one-eighth of the element remains.

3. Complete the table:

	mass	charge
proton	1 unit	+ 1 unit
neutron		
electron		
α-particle		
β-particle		
γ-ray		

4. Explain, with diagrams, *four* uses of radioactive substances.

5. a) Most of the carbon in your body is $^{12}_{6}C$. Draw a diagram of this atom.
   b) Another isotope is $^{14}_{6}C$ with a *half-life* of 6000 years. What is its *atomic number* and *mass number*? Explain the words in italics and draw a diagram of this atom. What fraction of $^{14}_{6}C$ remains after 24 000 yrs?

6. Strontium-90 has a half-life of 28 years. It is a β-emitter and may be absorbed into human bone. How much time must pass before its activity falls to $\frac{1}{32}$ of its original value?

7. Explain why workers in a nuclear power station wear badges containing photographic film.

8. Professor Messer fell asleep and dreamed that the nuclei of atoms could not be changed in any way. Why was his dream a nightmare?

Further Questions on Radioactivity

1. Of the three kinds of radiation, α-particles, β-particles and γ-rays, emitted by a radioactive substance, state which one
   a) travels at the greatest speed,
   b) carries negative charge,
   c) forms dense straight tracks in a cloud chamber,
   d) is not deflected by a magnetic field,
   e) must be emitted when $^{232}_{90}$Th decays to $^{228}_{88}$Ra,
   f) is similar in nature to X-rays,
   g) is similar in nature to cathode rays. (O)

2. a) $^{20}_{10}$Ne and $^{22}_{10}$Ne are isotopes of neon. The number 10 is known as the . . . . . The numbers 20 and 22 are known as the . . . . . Each $^{22}_{10}$ Ne atom contains . . . . . neutrons and . . . . . protons.
   b) The half-life of a certain radio-active substance is 2·5 years. Of an original sample of mass 0·8 g, the mass remaining unchanged after ten years would be . . . . . g (AEB)

3. A ratemeter (counter) was used at intervals of 10 minutes to measure the activity of a radio-active source and the following results were obtained:

Time (min.)	0	10	20	30	40	50	60	70
Count rate (counts per second)	650	520	416	333	300	213	170	136

   a) On a graph, plot these results.
   b) Which count rate appears to have been misread and should therefore be ignored?
   c) Draw a smooth curve through the other points.
   d) What count rate would you expect after 25 minutes?
   e) At what time was the count rate 500 counts per second?
   f) At what time was the count rate 250 counts per second?
   g) What is the half-life of the source? (AEB)

> **Further reading**
> *Finding out about Nuclear Energy* (CRAC)
> Physics Plus: *Using Radioactive Tracers* (CRAC)

4. The equation below represents part of the reaction in a nuclear reactor.
   $$^{235}_{92}U + ^{1}_{0}n \longrightarrow ^{236}_{92}U$$
   a) Explain the significance of the numbers 235 and 92. [2 marks]
   b) $^{236}_{92}U$ atoms are unstable and disintegrate spontaneously into fragments approximately equal in size, together with two or three fast-moving neutrons and a large amount of energy.
   What is this process called and what is the source of the energy? [2]
   c) $^{235}_{92}U$ is much more likely to absorb slow-moving (thermal) neutrons than fast-moving neutrons. Describe how neutrons may be slowed down in the reactor core. [4]
   d) Explain what is meant by a chain reaction. [4]
   e) How is the rate of energy production in the reactor core controlled? [4]
   f) The energy is produced in the form of heat in the reactor core. How is the heat removed from the core and how is it converted into electricity? [2]
   g) When the fuel rods are withdrawn from the reactor core they are *radioactive*, containing *isotopes* with long *half-lives*. Explain the terms in italics. [8]
   h) Outline three precautions which must be taken to ensure safe operation of the reactor. [6] (NEA)

5. A lithium *nucleus* is denoted by the symbol $^{7}_{3}$Li. Draw a model of a neutral lithium atom, clearly labelling each part.
   Suppose a beta-particle is now emitted. Draw another model of the atom to show the result of such an emission. (L)

6. Explain the significance of the numbers in the symbol $^{60}_{28}$Ni. Comment briefly on the fact that it is also possible to have $^{58}_{28}$Ni. Part of the uranium decay series is written $^{234}_{90}$Th$\rightarrow ^{234}_{91}$Pa$\rightarrow ^{234}_{92}$U$\rightarrow ^{230}_{90}$Th. State what particles are emitted at each stage of this decay. (L)

Multiple choice questions on electronics and nuclear physics

Most examination papers include questions for which you are given
5 answers (labelled **A, B, C, D, E**) and you have to choose the *ONE*
correct answer.

Although the answers are given, the questions are not always as easy
as they look. Take care to read *all* the answers before you choose the
one that you think is correct.

Example.
The newton is a unit of

A	speed	**C**	mass	**E**	time
B	force	**D**	length		

The correct answer is **B**.

Now write down the correct letter for each of the following questions
(but do not mark these pages in any way).

1. Chris and Melanie are trying to use an oscilloscope to display the
waveform of an a.c. supply but the traces they get are as shown.
Which controls should they adjust?

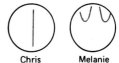

Chris   Melanie

	Chris	Melanie
A	Y-shift	X-timebase
B	X-timebase	Y-shift
C	X-shift	Y-shift
D	X-timebase	Y-gain
E	Y-gain	X-shift

Questions 2 and 3
With no input, a horizontal line was obtained on an oscilloscope.
The diagrams show some traces that were obtained with different
inputs.

2. Which trace would you get if the Y-plates were connected across
the secondary of a transformer?

3. Which trace would you get if the Y-plates were connected to the
output of a single diode rectifier?

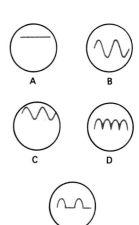

4. Which one of the following statements about alpha-particles is
correct?
   - **A**   They carry a negative charge.
   - **B**   They are emitted by all radioactive substances.
   - **C**   They are very penetrating.
   - **D**   They are deflected by magnetic fields.
   - **E**   They are identical with the nuclei of hydrogen atoms.

5. The diagram shows a beam of beta-particles passing between a positively-charged plate and a negatively-charged plate.
The beta-particles are

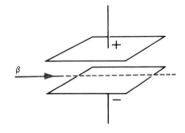

    A   deflected toward the positive plate
    B   deflected toward the negative plate
    C   deflected into the paper
    D   deflected out of the paper
    E   not deflected

6. A radioactive source is tested by adding sheets of different material between the source and a G-M counter, the whole experiment being performed in a vacuum. The results are as follows:

sheet added	effect on G-M counter
paper 3mm aluminium sheet thick lead sheet	large fall in the count rate no significant change count rate decreases, but not to zero

This means that the source is emitting

    A α-particles only
    B β-particles only
    C α-particles and γ-rays
    D β-particles and γ-rays
    E α-particles, β-particles and γ-rays

7. Which one of the following statements about a neutral atom is NOT correct?

    A   The number of protons equals the number of electrons.
    B   The nucleus is positively-charged.
    C   Most of the mass is in the nucleus.
    D   All atoms of oxygen have the same number of protons.
    E   All isotopes of an element have the same number of neutrons.

Questions 8 and 9
One of the isotopes of uranium is $^{238}_{92}$ U.

8. The atomic number of uranium is
    A 92   **B**   146   **C** 238   **D** 330   **E** variable

9. The number of neutrons in this isotope is
    A 92   **B**   146   **C** 238   **D** 330   **E** variable

10. The radioactive isotope $^{25}_{11}$ Na has a half-life of 1 minute.
What fraction of $^{25}_{11}$ Na remains after 3 minutes?

    A $\frac{1}{3}$
    B $\frac{1}{4}$
    C $\frac{1}{6}$
    D $\frac{1}{8}$
    E none

When you have finished.
a) check all your answers carefully
b) discuss your answers (and why you chose them) with your teacher.

More questions can be found in *Multiple Choice Physics for You.*

Doing your practical work

The GCSE exams have at least 20% of the marks awarded for practical work. Often these marks will be awarded by your teacher while watching you do experiments in the laboratory. You may feel self-conscious when your teacher watches you doing practical work but it is important to overcome this feeling and get on with the work.

Your marks will be awarded for particular skills. The details vary from one Exam Board to another but there are 6 basic skills. Your teacher will be looking to see how well you can do these 6 skills:

1) Follow instructions accurately and carefully.
2) Select and use the most suitable equipment.
3) Make observations accurately.
4) Design and carry out an investigation.
5) Record results in tables, graphs and charts.
6) Draw a conclusion from the results.

When marking you on these skills, your teacher will usually have a detailed checklist of what to look for. For example, on skill 3, 'Making Observations', the checklist might be as in this table:

There will be a similar checklist for each of the other skills. Ask your teacher for more details.

Here are some more details on each of these 6 skills:

1 Following instructions
This sounds simple but it can be an easy way to lose marks! If you are given written instructions you should read them at least *twice*. Concentrate on each sentence and each word — sometimes even the commas are important. Follow the instructions step by step, doing exactly what they say, and tick off each step when you have done it.

2 Selecting and using the most suitable apparatus
It is important to choose the most suitable equipment — this usually means the most accurate equipment.
For example, to measure the volume of 100 cm³ of water you should choose a measuring cylinder (see page 93) not a beaker with a 100 cm³ mark on it. (Why?) To measure a smaller volume of water you should choose a narrower measuring cylinder. (Why?)

The table shows a list of equipment you should be familiar with:

Can you use each of them accurately?

You might also be expected to know about a joulemeter (page 44), ticker-timer (p 149), C.R.O. (p 342), magnetic compass (p 264), microscope (p 215), G-M tube and ratemeter (p 373).

'Practical skills are important'

Checklist for Skill 3: MAKING OBSERVATIONS

Candidate's skill level	Marks awarded
The candidate reads the scales, but the observations are generally inaccurate. Makes frequent errors in estimation.	1
Some observations are inaccurate. Makes regular errors in reading scales. He/she has difficulty in handling units and forgets to check that the instrument reads zero before using it. Can make rough estimates with help.	2
Reads scales accurately and usually allows for zero-errors. He/she repeats readings only when instructed to. General observations tend to lack detail.	3
Uses instruments correctly but he/she sometimes needs prompting to repeat results, especially unexpected ones. General observations are careful but lack fine detail.	4
Uses instruments correctly and checks results. Handles zero-errors and units easily. Makes detailed and systematic observations and checks unexpected ones.	5

	page
metre rule	6
measuring cylinder	93
spring balance	86
balance (lever or electric)	89
stopclock, stopwatch	7
ammeter	286
voltmeter	288
thermometer	33

3 Making observations accurately

The checklist (opposite page) shows how this skill might be marked. You can see that accuracy is important as well as care in checking results. Remember to correct for *zero-errors* (for example, you should check that your spring balance or ammeter reads zero before you start to use it).

When reading a scale, make sure you look at right angles to it, so that you read the correct number. When using a measuring cylinder remember to read the *bottom* of the meniscus (see page 19).

In some experiments you might be given 'extra' equipment which may not have an obvious use. What do you think each of the following might be useful for: plumb line, spirit level, set square, magnifying glass, blotting paper or cloth.

4 Designing and carrying out an investigation

As well as following instructions to do an experiment, sometimes you may be asked to *design* an experiment to solve a problem. You may find it helpful to remember the seven steps in an investigation by remembering **PRISERC**:

The most important thing here is to devise a *fair test*. For example, look at experiment 8.10 on page 55. In this experiment it is important that we use two cans of the *same* size, and *same* shape, with the *same* amount of cold water, at the *same* temperature, placed the *same* distance from the *same* fire, with the temperatures taken at the *same* time. This is called 'controlling the variables' so that only *one* variable changes (the surface of the cans).

Skills 5 and 6 are discussed on the next page.

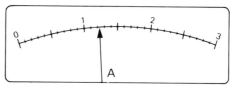

Reading scales

When reading scales, make sure you work out first what each small division stands for. It can help if you make a note of this before you start the experiment. For example, you could put a note next to your ammeter saying "0–3A, each mark = 0.1A".

Reading scales can be tricky. Manufacturers usually put as many marks on the scales as the accuracy of it will allow. This means that you should only read to the nearest scale division (ie. the nearest mark). For example, from the diagram above, you would write down 1.2A.

However in some papers the examiners expect you to 'interpolate' and try to judge the fractions of a division. In this case you might write down 1.23A. Your teacher will advise you which to do.

PRISERC: 7 steps of an investigation

P	PROBLEM	What is the problem? Make sure you understand it clearly.
R	RESEARCH	Do you know enough to tackle the problem? What do you need to look up?
I	IDEAS	What ideas can you suggest to tackle the investigation? Is each one a fair test?
S	SELECT	Select one idea and select the best apparatus for it.
E	EXPERIMENT	Do the experiment accurately and carefully.
R	RESULTS	Record your results, using a table, graph or chart.
C	CONCLUSION	Write down your conclusion: what have you discovered?

Professor Messer's not too bright. It's up to you to put him right:

5 Recording results

The results of your experiment often have to be recorded in a table. Make sure you label each column with the *quantity* being measured and its *unit*.
The example shows a table for the 'Hooke's Law experiment' (see page 86).

Sometimes you may use your results to draw a bar-chart or a pie-chart (see example on page 11).
More often you will have to draw a line graph of your results. You can do this in 4 steps as shown here:

Extension of a spring

Load (in N)	Extension (in mm)
0	0
10	3
20	7
30	10

4 steps in drawing a graph:

1) Choose simple scales.
   For example, 1 large square = 1 newton, (or 2N, 5N, 10N).
   Never choose an awkward scale like 1 square = 3N or 7N!
2) Plot the points and mark them neatly. Recheck each one.
3) If the points look as though they form a straight line, draw the *best* straight line through them with a ruler (and pencil). Check that it looks the *best* line.
4) If a point is clearly off the line you should always use your apparatus to repeat the measurements and check it.

Finding the gradient

This needs care, but can be done in 3 steps.
a) Draw a large right-angled triangle as shown in red:
b) Find the value of the two sides 'Y' and 'X', *in the units of the graph*. That is, you must find the size of Y and X using the scales on the graph (not just measuring with a ruler). For example, in the diagram, Y = 20mm and X = 60N. Can you see why?
c) Calculate the gradient (or 'slope') of the graph by dividing Y/X. Keep the units with the numbers, so that you find the unit of the gradient.
For example, from the diagram:

$$\text{Gradient} = \frac{Y}{X} = \frac{20 \text{ mm}}{60 \text{ N}} = 0.33 \text{ mm/N}$$

A calculator gave the result as 0.33333333. Why is it better to write it as 0.33?

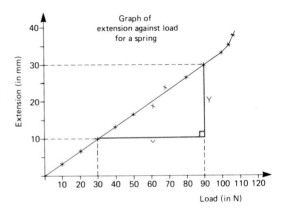

6 Drawing a conclusion

Every experiment has a 'conclusion'. This is a summary of what you found out (or sometimes what you didn't find!) Always look at your results or graph or chart and think hard to decide what you have discovered, or what reasonable or 'valid' deduction can be made from your results.

For example, from this Hooke's Law experiment (see also page 86) you might conclude:
a) The graph is a straight line through the origin (see also page 403). Therefore the extension is *proportional* to the load (up to the elastic limit).
b) The elastic limit is 100N.
c) Below this value, the spring extends by 0.33mm per newton.

On the opposite page there are some practical questions for you to try.

388

Questions

1. In a school Science Club there are 18 pupils: 6 pupils are aged 14 years, 5 pupils aged 13, 4 aged 12, 2 aged 15 and 1 aged 16.
   Plot a bar chart to show this information.

2. A waste bin is found to contain 50% paper, 25% glass, 20% plastic, 5% metal.
   Plot a pie chart to show this information.

3. The table shows the results of weighing a number of 20p coins:

No. of coins	6	12	17	22	25	28	31	34	36	39
Mass (g)	29	59	83	108	127	137	152	167	177	187

a) Plot a graph, discarding any doubtful results.
b) Does it go through the origin?
c) What is the mass of 20 coins?
d) How many coins should make a mass of 127g?
e) What is the mass of 1 coin?
f) What is the gradient of the graph?

4. The table shows the results of measuring the thickness of a number of sheets of paper:

No. of sheets	15	22	25	35	46	65	75	78	85
Thickness (mm)	1.7	2.4	2.7	4.4	5.1	7.2	8.1	8.6	9.4

a) Plot a graph, discarding any doubtful results.
b) Does it go through the origin?
c) What is the correct thickness of 35 sheets?
d) How many sheets would make 6.6mm?
e) What is the thickness of 1 sheet?
f) What is the gradient of the graph?

Other useful questions on drawing graphs, and analysing results and making conclusions:
Mechanics p 91 (Q6), p 158 (Q4), p 168 (Q2)
p 171 (Q24, 25), p 172 (Q26, 28, 29), p 393 (Q4).
Heat p 82 (Q31), p 80 (Q19).
Electricity p 302 (Q23, 26), p 366 (Q5), p 371 (Q32).
Radioactivity p 383 (Q3)

For each of the following **questions 5–17** use the ideas of PRISERC to plan an investigation. In each case you should:

a) Think of more than one method if possible and select the best one.
b) Ensure you make it a fair test.
c) Sketch the apparatus you would use and add notes to explain the method.
d) Draw a table with headings and, if possible, say what graph you would draw.

After you have done this individually, you should discuss the investigation with a group of two or three others in your class.
Talk through all the details until you are all sure you have thought of everything. Use steps (c) and (d) to keep a record of the final plan.

5. Which make of paper towel is best for soaking up water?
6. Does a candle heat faster than a spirit burner?
7. Do the sleeves of different anoraks insulate equally well? See page 59.
8. When stretched, do copper wire and nylon fishing line behave the same way? See p 86.
9. Who has the strongest hair in your class?
10. Which bounces better: a tennis or a golf ball?
11. Do pendulums with heavy bobs swing slower than pendulums with light bobs?
12. Which make of paper tissue is the strongest?
13. Does red paint absorb infra-red rays better than blue paint? See page 55.
14. How would you test the upper frequency limit of hearing for all members of your class? See p 240.
15. How could you test the eyesight of your class?
16. Are the poles of a bar magnet of equal strength?
17. How would you put copper, iron, tin, constantan in order, by how well they conduct electricity?
18. He's got his answers wrong again.
    Take a look — can *you* explain?

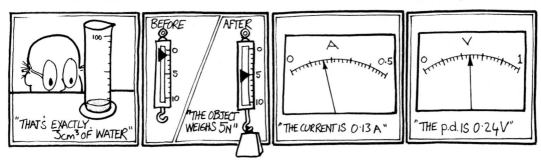

Revision techniques

Why should you revise?
You cannot expect to remember all the Physics that you have studied unless you revise. It is important to fix it all in your memory so that you can recall it in an examination.

Where should you revise?
In a quiet room (perhaps a bedroom), with a table and a clock. The room should be comfortably warm and brightly lighted. A reading lamp on the table helps you to concentrate on your work and reduces eye-strain.

When should you revise?
Start your revision early each evening, before your brain gets tired.

How should you revise?
If you sit down to revise without thinking of a definite finishing time, you will find that your learning efficiency falls lower and lower and lower.

If you sit down to revise, saying to yourself that you will definitely stop after 2 hours, then your learning efficiency falls at the beginning but *rises towards the end* as your brain realises it is coming to the end of the session (see the graph).

We can use this U-shaped curve to help us work more efficiently by splitting a 2 hour session into 4 shorter sessions, each of about 25 minutes with a short, *planned* break between them. The breaks *must* be planned beforehand so that the graph rises near the end of each short session.

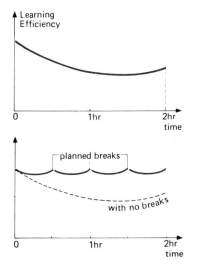

For example, if you start your revision at 6·00 p.m., you should look at your clock or watch and say to yourself, "I will work until 6·25 p.m. and then stop — not earlier and not later". At 6·25 p.m. you should leave the table for a relaxation break of 10 minutes (or less), returning by 6·35 p.m. when you should say to yourself, "I will work until 7·00 p.m. and then stop — not earlier and not later"

Continuing in this way is more efficient and causes less strain on you.

How often should you revise?

The diagram shows a graph of the amount of information that your memory can recall at different times after you have finished a revision session.

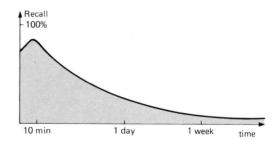

Surprisingly, the graph rises at the beginning. This is because your brain is still sorting out the information that you have been learning.
The graph soon falls rapidly so that after 1 day you may remember only about a quarter of what you had learned.

There are two ways of improving your recall and raising this graph.
● 1. If you briefly *revise the same work again after 10 minutes* (at the high point of the graph) then the graph falls much more slowly.
This fits in with your 10 minute break between revision sessions. Using the example on the opposite page, when you return to your table at 6.35 p.m., the first thing you should do is review, briefly, the work you learned before 6.25 p.m.

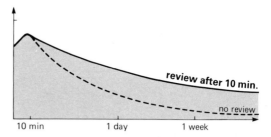

The graph can be lifted again by briefly reviewing the work *after 1 day* and then again *after 1 week*. That is, on Tuesday night you should look through the work you learned on Monday night and the work you learned on the previous Tuesday night, so that it is fixed quite firmly in your long-term memory.

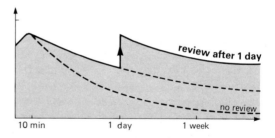

● 2. Another method of improving your memory is by taking care to try to *understand* all parts of your work. This makes all the graphs higher.
If you learn your work in a parrot-fashion (as you have to do with telephone numbers), all these graphs will be lower. On the occasions when you have to learn facts by heart, try to picture them as exaggerated, colourful images in your mind.

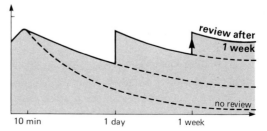

Remember: **the most important points about revision are that it must occur often and be repeated at the right intervals.**

Suggestions for a revision programme

1. Beginning with Chapter 3 (on 'Molecules'), read the summary at the end of the chapter to gain some idea of the contents.
   Then read through the chapter looking at particular points in more detail, before reading the summary again.

2. Covering up the summary, check yourself against the fill-in-the-missing-word sentences at the end of the chapter.

   Remember to *re-read* the summary and to *review* each chapter after the correct revision intervals of 10 minutes, then 1 day, then 1 week (as explained on the previous page).
   Continue in this way with all the other chapters in the book.

3. While reading through the summaries and chapters like this, it is useful to collect together on a single sheet of paper all the formulae that you need to know. See also the checklists on pages 394 and 395.

4. While you are going right through the book, doing this for every chapter:
   a) Attempt the questions on the 'Further Questions' pages. Your teacher will be able to tell you which are the most important ones for your syllabus.
   b) If there are multiple-choice questions in your examination, do the multiple-choice tests on Heat (page 76), Mechanics (166), Light (230), Electricity (338) and Nuclear Physics (384).
   c) There are further questions on each of the topics in a book called *Multiple Choice Physics for You*.

5. On the opposite page are some more questions, which contain work from many different parts of the syllabus. Sometimes a hint is given by showing the page number.

6. A few weeks before the examination, ask your teacher for copies of the examination papers from previous years. These 'past papers' will help you to see:
   — the particular style and timing of your examination
   — which experiments you may be asked to do or to describe
   — which topics and questions are asked most often and which suit you personally.

When doing these past papers, try to get used to doing the questions *in the specified time*.

It may be possible for your teacher to read out to you the reports of examiners who have marked these papers in previous years.

Professor Messer:
   If runners are defeated and cowboys are deranged, are examiners detested?

Revision questions

1. Professor Messer says that his vacuum cleaner is powered from the Sun. Do you agree? Explain in detail.

2. Give the *units* and, where possible, the *definitions* of the following quantities: mass, weight, density, force, work, power, energy, pressure, speed, acceleration, velocity, frequency, wavelength, specific heat capacity, specific latent heat, potential difference, current, resistance, kinetic energy.

3. An electric motor is used to raise a load of 1500N through a vertical height of 5m in 10 seconds.
   a) What is the average power needed to do this (p. 132)? The motor takes a current of 4A at a potential difference of 250V.   b) What is the power taken by the motor? (p. 306).
   c) What fraction of this power is used usefully to lift the load?   d) What happens to the remainder?

4. The table shows the speed, at different times, of a car which starts from rest, moves with constant acceleration, then travels at constant speed and finally slows down with constant deceleration. Its mass is 1000 kg.

Time t in seconds	0	2	4	5	8	10	11	12
Speed in m/s	0	4	8	10	10	10	5	0

   a) Plot a graph from the table   b) What is its maximum speed?   c) For how long does it travel?   d) What is its initial acceleration? (p. 148)   e) What is its final deceleration?
   f) How far does it travel between time $t = 6$ seconds and time $t = 9$ seconds?   g) How far does it travel during the first 5 seconds? After 5s what is   h) its K.E. and   i) its momentum? What is the resultant force while it is   j) accelerating and   k) decelerating? (p. 157)

5. The Physics of a motor-car. Using diagrams, explain the following:
   a) the 4-stroke cycle (p. 72)   b) the cooling system (p. 52)   c) the ignition system (p. 336)   d) lead-acid battery (p. 280)   e) dynamo for charging the battery (p. 332)   f) starter motor (p. 320)   g) relay for the starter motor (p. 317)   h) dipping headlamps (p. 193)   i) hydraulic brake system (p. 99)   j) gears (p. 142)   k) stability of racing cars (p. 124)   l) speedometer (p. 333)   m) pressurised cooling system (p. 65)   n) antifreeze (p. 63)   o) salting roads in winter (p. 63)   p) seat belts (p. 89)   q) driving mirror (p. 192)   r) reflectors (p. 198)

6. An electric water-heater is rated at 250V, 1250W and is run for 8 hours.
   a) What fuse should you fit?
   b) What is the resistance of the element?
   c) What is the cost at 6p per kWh? (p. 307)

7. A full water tank measures 2 m x 2 m x 50 cm.
   a) If the density of water is 1000 kg/m^3, what is the mass of water? (p. 92)
   A heating coil takes 7A at a p.d. of 400V.
   b) How much heat is produced in 50 minutes? (p. 306)
   c) If all this heat is given to the water (specific heat capacity = 4200 J/kg°C), what is its rise in temperature? (p. 45)
   d) What is the resistance of the heating coil?

8. A heating coil of resistance 25 Ω takes a current of 2A.
   a) What is the p.d. across the coil?
   b) How much heat is produced in 5 minutes?
   c) If all the heat is given to 0·1 kg of ice at 0°C and is just sufficient to melt the ice, what is the specific latent heat of ice? (p. 62).

9. An object 2 cm high is placed
   a) 18 cm   b) 6 cm
   from a converging lens of focal length 10 cm. Find the position and nature of the image in each case (p. 206).

✓ Revision checklist: formulas

Whenever you have a problem to solve in the Physics examination (and perhaps later in your job), it is essential to be able to write down the formula *quickly* and *accurately*.

The list below shows all the formulas you might need. **Your teacher will be able to tell you which of them you should tick and then learn.**

For each formula, the page number is given. When you look at each page, make sure that you read the whole page so that you understand the background of the formula.

Try to learn each formula by understanding it, not just learning it by heart. Many of these pages also have a worked example, so that you can see how the formula should be used.

It is a good idea to collect all the formulas on to one sheet of paper which you can use for regular revision right up to the day of the exam.

Revision checklist: definitions and laws

In some examination papers you may be asked to *define* a certain quantity, or to *state* a law or a principle.

As before, your teacher will be able to tell you which of the following list you should tick and then learn thoroughly.

Heat
○ Linear expansivity   23
○ Upper and lower fixed points   33
○ Gas Laws   37, 39, 40, 41
○ Specific heat capacity   45
○ Heat capacity   46
○ Specific latent heat of fusion   61
○ Specific latent heat of vaporisation   64

Mechanics
○ Force, Weight   85
○ Hooke's Law   86
○ Newton's First Law   89
○ Newton's Second Law   157
○ Newton's Third Law   114
○ Density   92
○ Relative density   95
○ Pressure   97
○ Archimedes' principle   108
○ Principle of flotation   109
○ Scalar and vector quantities   116
○ Parallelogram of forces   117
○ Moment of a force   120
○ Principle of moments   121
○ Centre of gravity   122
○ Work and energy, Joule   127
○ Principle of conservation of energy   129
○ Power, Watt   132, 306
○ M.A., V.R., efficiency   138, 139
○ Average speed, velocity   148
○ Acceleration   148
○ Momentum   162
○ Principle of conservation of momentum   163

Waves & Sound
○ Wavelength   175
○ Frequency   175
○ Amplitude   175
○ Transverse and longitudinal waves   174
○ Resonance   243
○ Pitch, loudness, quality   244, 245

Light
○ Law of reflection   185
○ Real image   192, 205
○ Virtual image   187, 205
○ Principal focus and focal length
    of a curved mirror   189
○ Refractive index and Snell's Law   195, 196
○ Critical angle   197
○ Principal focus and focal length of a lens   204
○ Magnification   205
○ Spectrum and dispersion   222
○ Primary colours of light   228

Electricity
○ Law of magnetism   259
○ Magnetic field and line of flux   264, 315
○ Angles of declination and dip   265
○ Neutral point   264
○ Law of electrostatics   267
○ Potential difference, volt   276, 297
○ Capacitance, farad   276
○ Series and parallel circuits   286, 287, 292, 293
○ Ohm's law and resistance   289
○ Micro-, Milli-, Kilo-, Mega-   288
○ E.m.f., internal resistance   297
○ Resistivity   296
○ Kilowatt-hour   307
○ Anode, cathode and electrolysis   311, 312
○ Maxwell's screw rule   314
○ Rule for polarity of a coil   315
○ Ampere, coulomb   325
○ Fleming's left hand rule   318
○ Fleming's right hand rule   328
○ Lenz's law   329
○ Truth tables   361

Nuclear Physics
○ Alpha, beta and gamma rays   374, 375, 379
○ Atomic number, mass number   377
○ Isotope   377
○ Half-life   378

Examination technique

In the weeks before the examinations:

Attempt as many 'past papers' as you can so that you get used to the style of the questions and the timing of them.

Note which topics occur most often and revise them thoroughly, using the techniques explained on previous pages.

Read through your list of essential formulae as often as possible if you need to memorise them (see page 394).

Just before the examinations:

Collect together the equipment you will need:
- Two pens, in case one dries up.
- At least one sharpened pencil for drawing diagrams
- A rubber and a ruler for diagrams.
  Diagrams usually look best if they are drawn in pencil and labelled in ink.
  Coloured pencils are usually not necessary but may sometimes make part of your diagram clearer (e.g. when drawing a spectrum).
- A watch for pacing yourself during the examination. The clock in the examination room may be difficult to see.
- For some examinations you may need special instruments e.g. compasses, a protractor or an electronic calculator.

Use your lists of essential formulas and definitions for last minute revision.

It will help if you have previously collected all the information about the length and style of the examination papers (for all your subjects) as shown below:

Date, time and room	Subject and paper number	Length (hours)	Type of paper: – multiple choice? – single word answers? – longer answers – essays? – practical?	Sections?	Details of choice (if any)	Approximate time per question. (minutes)
7th May 9.30 Hall	Physics 1	2	Multiple choice + long answers	2	A – all B – 3 out of 5	A – 2 min. B – 20 min.

In the examination room:

Read the introduction to the examination paper. It will often contain a phrase such as: "Carelessness and untidy work *will* be penalised". This means that you *must* take care to write and draw neatly or else you will lose marks.

Use the information on the front of the paper to calculate the length of time to be spent on each part of the paper. Then use your watch to pace yourself correctly.

Suppose for example that one section or a long question is supposed to occupy you for 20 minutes. If you finish in 15 minutes, this means that you know the work particularly well or, more likely, that you have not written in sufficient detail. On the other hand, if you take 25 minutes or longer, it means that you may not have enough time to finish all the questions and so you will be at a disadvantage compared with other candidates who do finish.

If there is a choice of questions, use your experience with past papers to choose a topic that suits you well.

When attempting a calculation, it usually helps to draw a diagram and label it with all the available information. In this way you can see more easily what you know and which formulae are likely to be useful. (See also page 403.)

If you cannot do part of a question, leave a gap so that you may return to it later.

If you finish with time to spare, look over your answers and add to them or re-attempt them.

NATIONAL EXAMINING BOARD

Physics

23rd May
9.30 a.m.

Careless and untidy work will be penalised.

Section A — 1 hour
(30 compulsory questions)

Section B — 1 hour
Answer 3 questions (out of 5)

In what ways is your examination paper *different* from this one?

For multiple-choice questions:

— Read very carefully the coding instructions for A, B, C, D, E.
— Mark the answer sheet *exactly* as you are instructed.
— Take care to mark your answer (A, B, C, D, E) opposite the *correct* question number.
— Even if the answer looks obvious, you should look at all the alternatives before making a decision.
— If you do not know the correct answer and have to guess, then you can improve your chances by first eliminating as many wrong answers as possible.
— Make sure you give an answer to every question.

Careers

Have you thought what you want to do when you leave school?

After English and Mathematics, Physics is the most important qualification for a great many careers. For example, unless you gain an adequate qualification in physics you will find it almost impossible to enter the Engineering industry at professional or technician levels.

Engineering covers a very wide range of interesting and rewarding careers for men and women, as shown in the list below:

Aeronautical engineering
Automobile engineering
Civil engineering
Communications engineering
Electrical engineering
Electronic engineering
Heating & ventilation engineering
Hydraulic engineering
Industrial engineering
Instrument & control engineering
Lighting engineering
Machine tool engineering
Marine engineering
Mechanical engineering
Mining engineering
Municipal engineering
Naval engineering
Nuclear engineering
Production engineering
Radio & TV engineering
Railway engineering
Refrigeration engineering
Structural engineering
Telecommunications engineering

Physics is needed or has a direct bearing on many other rewarding careers.

In the following list,   * = physics is essential
+ = physics is an advantage.

Air traffic control +
Architecture +
Army * or +
Astronomy *
Audiology *
Automobile industry:
    craft +
    technician *
Broadcasting technician *
Civil service scientific officer +
Computer programming +
Dairy technician +
Dental technician +
Dentistry +
Draughtsmanship +
Electronics *
Ergonomics *
Geophysics *
Laboratory technician *
Medical physics *
Medicine +
Merchant Navy:
    deck, navigating or
    engineering officer *
Metallurgy *
Meteorologist *
Ophthalmic optics *
Patent agent *
Pharmacy +
Physiotherapy +
Quantity surveyor +
Radio and TV repair *
Radiography *
Royal Air Force * or +
Royal Navy:
    officer *
    artificer +
    radio operator *
    mechanic *
Telecommunications (telephone, radio,
    satellite) *

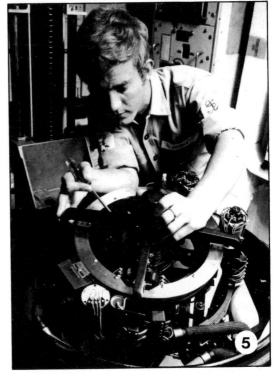

Some careers using Physics:

1. An engineer adjusting a rocket payload.

2. Checking the magnetic tape for a computer.

3. Repairing an alternator from an oil tanker.

4. Testing electronic components.

5. Electrical mechanic working on a gyro.

Of course, for many of the careers listed or illustrated on these pages, you will need further study in the Sixth Form or in a College of Further Education. You should consult your careers teacher for more detailed information.

Also useful are the following books published by the Careers Research and Advisory Centre (CRAC), Bateman Street, Cambridge:

Your choice at 13+   Your choice at 15+

Your choice at 17+   Connections: Physics

Dates of some inventions and discoveries mentioned in this book

Invention or discovery	Inventor or discoverer	Date	Page number in this book
Wheel and axle	—	before 2000 B.C.	141
Pulley	—	about 700 B.C.	144
Screw	—	about 400 B.C.	143
Pinhole camera	—	about 1500 A.D.	182
Compound microscope	Janssen	1590	217
Air thermometer	Galileo	1597	29
Refracting telescope	Lippershey	1608	219
Laws of refraction	Snell	1621	195
Mercury barometer	Torricelli	1644	101
Vacuum pump	Guericke	1654	100
Liquid-in-glass thermometer	—	1655	33
Boyle's Law	Boyle	1662	37
Spectrum	Newton	1666	222
Reflecting telescope	Newton	1669	219
Pressure cooker	Papin	1680	66
Law of gravity	Newton	1685	85
Laws of motion	Newton	1687	89, 114
Steam engine	Newcomen	1705	71
Piano	Cristofori	1709	247
Celsius scale	Celsius	1742	33
Capacitor	Kleist	1745	277
Lightning conductor	Franklin	1752	273
Latent heat	Black	1760	61
High pressure steam engine	Watt	1765	71
Diving suit	Périer	1772	98
Submarine	Bushnell	1776	107
Hot air balloon	Montgolfier	1783	53
Hydrogen balloon	Charles	1783	109
Parachute	Blanchard	1785	154
Gold-leaf electroscope	Bennet	1787	269
Hydraulic press	Bramah	1795	99
Aneroid barometer	Cante	1799	102

Something to do

1. Play the Inventions Game — someone names an invention and you have to answer: 5 marks for the inventor's name, 5 marks for the right date (2 marks if you are close — say within 20 years).
2. Plot a large labelled timechart of the inventions since 1500.

Simple cell	Volta	1800	279
Double-slit interference	Young	1801	234
Steam locomotive	Trevithick	1804	70
Diffraction grating	Fraunhöfer	1819	236
Magnetic effect	Oersted	1819	314
Electric motor	Faraday	1821	320
Photography	Niépce	1826	210
Ohm's law	Ohm	1826	289
Brownian movement	Brown	1827	16
Transformer	Faraday	1831	334
Dynamo	Faraday	1831	330
Laws of electrolysis	Faraday	1834	312
Semi-conductors	Rosenschöld	1834	348
Refrigerator	Perkins	1834	67
Cranked bicycle	Macmillan	1839	142
Pneumatic tyre	Thomson	1845	99
Cathode rays	(several)	1859	341
Telephone	Bell	1876	317
Four cycle petrol engine	Otto	1876	72
Gramophone	Edison	1877	238
Microphone	Hughes	1878	319
Electric lamp	Edison	1879	305
Electric iron	Seeley	1882	304
Motor car	Benz	1885	73
Ciné film	Friese-Green	1889	213
Diesel engine	Diesel	1892	73
X-rays	Röntgen	1895	344
Radioactivity	Becquerel	1896	372
Electron	Thomson	1897	268, 340, 376
Radium	Curies	1898	373, 378
Tape recorder	Poulsen	1900	347
Vacuum cleaner	Booth	1901	106
Vacuum flask	Burger	1904	57
Diode valve	Fleming	1904	340
$E = mc^2$	Einstein	1905	381
Idea of nucleus	Rutherford	1911	376
Cloud chamber	Wilson	1911	372
G-M tube	Geiger & Müller	1928	373
Jet engine	Whittle	1930	74
Electrostatic generator	Van de Graaff	1931	272
Radio telescope	Jansky	1931	219
Neutron	Chadwick	1932	376
Cat's eye reflectors	Shaw	1934	198
Nuclear reactor	Fermi	1942	381
Transistor	Shockley, Bardeen & Brattain	1947	354
Hovercraft	Cockerell	1954	112
Integrated circuit	Kilby	1958	358
LASER	Maiman	1960	203
Optical fibre	Kao, Newnes & Beales	1966	202

Check your maths

Directly proportional

Look at the table. Do you see how Y and X are connected?
If X doubles, so does Y. If X halves, so does Y, etc.

We say: Y is *directly proportional* to X
In symbols: $Y \propto X$

For an example, see Hooke's Law (page 86) or Ohm's Law (page 289).

Y	X
3	1
6	2
9	3
12	4

Making an equation

We can always change a proportionality into an equation by putting
in a constant, k:

$$Y = kX \quad or \quad \frac{Y}{X} = k$$

In the example in the table, $k = 3$.

Inversely proportional

Look at the table. Do you see how P and V are connected?
If V doubles, P halves. If V halves, P doubles, etc.

We say: P is *inversely proportional* to V
In symbols: $P \propto \dfrac{1}{V} \quad or \quad \dfrac{1}{P} \propto V$

We make an equation in the same way as before:

$$P = k \times \frac{1}{V} \quad or \quad P = \frac{k}{V} \quad or \quad P \times V = k$$

In the example in the table, $k = 12$.

P	V
12	1
6	2
4	3
3	4

Changing the subject of an equation

As long as you do the same thing to *both* sides of an equation, it will still balance:

Example 1

From page 92: $D = \dfrac{M}{V}$

a) To change the subject to M:

Multiply both sides by V,
then cancel: $D \times V = \dfrac{M}{\cancel{V}} \times \cancel{V}$

$$\therefore M = D \times V$$

b) To change the subject to V:

First multiply by V as
before, to get: $D \times V = M$

Then divide by D: $\dfrac{\cancel{D} \times V}{\cancel{D}} = \dfrac{M}{D}$

$$\therefore V = \frac{M}{D}$$

Example 2

From page 153: $a = \dfrac{v - u}{t}$

To change the subject to v:

First multiply by t: $a \times t = \dfrac{v - u}{\cancel{t}} \times \cancel{t}$

$$\therefore at = v - u$$
$$at + u = v - \cancel{u} + \cancel{u}$$
$$\therefore v = u + at$$

Often it is easier if you put in your numbers
before changing the equation.

Graphs

a) *Directly proportional*

For an equation like $Y = kX$ (see the opposite page), the graph is a *straight* line, *through the origin*.

The gradient (or slope) $= k$ (See page 388).
For an example, see Hooke's Law (page 86).

b) *Linear but not directly proportional*

Consider an equation like $Y = kX + c$ or like $v = at + u$ (see example 2 opposite and page 153).

For these equations the graph is a straight line but *not* through the origin.

c) *Inversely proportional*

A graph of P against V (see table opposite) would give a curve.

The equation is $P = k \times \dfrac{1}{V}$ so to get a straight line, we must plot P against $\dfrac{1}{V}$

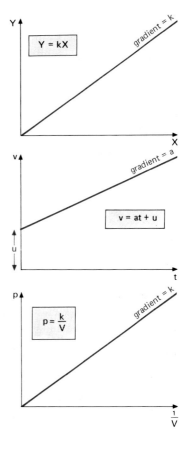

Dividing by fractions

Remember that dividing by $\frac{1}{2}$ is the same as multiplying by 2. Dividing by $\frac{1}{3}$ is the same as multiplying by 3, etc.

Many people make mistakes with the parallel-resistor formula on page 293. Read the example on that page carefully.

Large and small numbers – indices

The small *index* tells us how many decimal places to move.
The + or − sign tells us which direction to move.
That is, for 10^n, move the decimal point n places to the right.

Examples:

$2 \times 10^6 = 2\,000\,000$

$2{\cdot}1 \times 10^6 = 2\,100\,000$

$0{\cdot}21 \times 10^7 = 2\,100\,000$

For 10^{-n}, move the decimal point n places to the left.

Examples:

$4 \times 10^{-1} = 0{\cdot}4$

$4 \times 10^{-3} = 0{\cdot}004$

$4{\cdot}1 \times 10^{-3} = 0{\cdot}0041$

7 steps in solving a physics problem

1. Draw a diagram showing all the *information* given in the question.
2. Decide which *formula* to use. (If you are not sure, write down the possibilities and then decide.)
3. Decide the *units* for the formula. If necessary, change your information to the correct units (e.g. mm to metres).
4. *Substitute* the numbers in the formula.
5. *Calculate* the answer. Check that your answer is a sensible size.
6. Decide the *units* for your answer.
7. Read the question to see if you have *another part* to do.

Answers

HEAT

Chapter 4 (Expansion) page 31
13. 0·022 m   14. 0·198 km (198 m)
15. 0·000 017 per °C

Chapter 6 (Gas Laws) page 42
3. 4 cm^3 4. 900 K (627°C)
5. 3 atm   6. 100 cm^3
7. 400 mm Hg   8. 273 cm^3
9. 800 K (527°C)
10. 980 K (707°C)

Chapter 7 (Measuring Heat) page 47
2. Jack 4 MJ Wife 18 MJ   3. a) 42 000 J
b) 8400 J   c) 880 J   d) 3800 J
5. 500 J/kg °C   6. 4400 J/kg °C
8. 400 J/kg °C   9. 439 J/kg K

Chapter 9 (Changing State) page 69
5. a) 680 000 J   b) 170 000 J   c) 6 900 000 J
d) 230 000 J   6. 2 400 000 J/kg
7. 6 120 000 J (6·12 MJ)
8. 336 000 J/kg

Further Questions on Heat

Molecules (page 78)
4. 5 x 10^{-9} m

Gas laws (page 80)
14. b) 1·05 P ($\frac{21}{20}P$)
15. b) (ii) 50°C   (iii) 5·90 k Pa
16. 1·6 litres
17. 146 mm, 193 mm, –312°C
18. 142 000 N/m^2

Specific heat capacity (page 81)
21. d) (i) 21 000 J (21 kJ)   (ii) 875 J/kg °C
23. b) 4°C   c) 3600 J   d) 900 J/kg °C
24. (i) 720 J   (ii) 630 J

Latent heat (page 82/83)
30. a) 72 000 J   b) 2 400 000 J/kg
31. 480°C

MECHANICS

Chapter 11 (Pushes and Pulls) page 91
2. 40 N (39·2 N)   3. a) 600 N (588 N)
b) approx 100 N   c) 60 kg   5. a) 10 cm
b) 15 cm   6. a) 7 cm   b) 10·2 cm   c) 125 N

Chapter 12 (Density) page 96
1. 5000 kg/m^3 (5 g/cm^3)   2. 8000 kg/m^3
3. 18 000 kg   4. 2000 kg/m^3 32 000 kg   0·5 m^3
500 kg/m^3 a) B   b) C   c) A and C   d) D
5. 40 000 kg   6. 5 m^3 7. 130 kg
8. 3000 kg/m^3 (3 g/cm^3)   10. a) 19   b) 13·6
c) 0·92   11. 0·8

Chapter 13 (Pressure) page 106
2. 50 N/m^2 3. 1 000 000 N
6. a) 1 000 000 N/m^2 (100 N/cm^2)   b) 2000 N

Chapter 14 (Floating and Sinking) page 111
7. a) 10 N   b) 10 N   c) 1 kg   d) 0·001 m^3
e) 0·001 m^3 f) 4 kg   g) 4000 kg/m^3
8. 3000 kg/m^3

Chapter 15 (More about Forces) page 119
6. a) 50 N at 37° to 40 N force
b) 61 N at 25° to 40 N force
c) 36 N at 46° to 40 N force
8. a) 14 kg   b) 10 N
9. a) 150 N at N 53°E
   b) 150 N at S 53°W
10. 103 m/s at N 76°E
11. 94 N   34 N

Chapter 16 (Turning Forces) page 126
3. 40 Nm   4. a) 5 N   b) 4 N   5. 2 m

Chapter 17 (Work, Energy and Power) page 134

2. 10 J   3. 20 000 J   13. 200 W   15. 100 W
16. 500 W   17. a) 7500 W   b) 7·5 kW
c) 10 h.p.   18. a) 100 J   b) 10 000 J

Chapter 18 (Machines) page 147

2. 300 N   4. 3, once   5. a) 1·5   b) 2
c) 600 J   d) 800 J   e) 75%   8. (i) a) 2
b) 3   c) 20 J   d) 30 J   e) $66\frac{2}{3}$%   (ii) a) 3
b) 4   c) 600 J   d) 800 J   e) 75%

Chapter 19 (Velocity and Acceleration) page 160

2. a) 20 m/s   b) 2 m/s^2
5. a) 40 m/s   b) 200 m
6. a) 1 m/s^2 b) 10 sec
7. 20 m   8. 7·7 sec   9. a) 5 m   b) 2 sec
10. a) 20 m/s   b) 2 sec   c) 10 m
11. 8·45 m   13. 5 m/s^2 3 m/s^2
14. a) 3·0 x 10^7 N   b) 0·3 x 10^7 N   c) 1 m/s^2
d) zero
15. a) 30 m/s^2 b) 60 m   c) 90 000 N (90 kN)
18. a) 150 000 J (150 kJ)   b) zero   c) 150 kJ
d) 77·5 m/s^2
16. a) 1000 N   b) 170 N

Chapter 20 (Momentum) page 165

3. a) 2 m/s   b) 64 J   16 J
4. 2 m/s   5. 40 m/s   6. 0·1 m/s
7. a) 32 500 000 kg m/s
b) 32 500 000 N (3·25 x 10^7 N)

Further Questions on Mechanics

Hooke's Law (page 168)

2. a) 20 N   b) 12·5 N
4. a) 4 N   b) 0·5 N   c) 6 N   d) 2 N

Density (page 168)

7. c) 1000 kg/m^3

Pressure (page 169)

9. a) 0·015 m^3 b) (i) 37·5 kg   (ii) 375 N
c) 1250 N/m^2 (Pa)
d) 15 000 N/m^2 (15 kPa)
10. 9·0 x 10^7 N

Floating and sinking (page 169)

11. a) 3 N   b) 0·3 kg
c) 0·0003 m^3 (3 x 10^{-4} m^3)
12. d) 6700 kg/m^3
(i) 0·8 m^3 (ii) 640 kg   (iii) 6400 N
(iv) 0·64 m^3 (v) 1600 N   (vi) 400 N
13. (i) 0·81   (ii) 39·0 cm^3

Vectors (page 170)

14. b) 720 N   c) 56°
15. b) (i) 4 km/h   (ii) 8 km/h   (iii) 1 km
(iv) N 70·5°E   (v) 5·66 km/h
(vi) 0·53 h (32 min)

Moments (page 170)

18. b) 105 g   c) 1·25 N
19. e) (i) 600 N   (ii) 150 N

Energy and power (page 171)

21. a) 14 400 J   b) 600 W
22. 1250 W   375 J per second
23. a) 1080 J   b) 2520 J   c) 630 N

Machines (page 171)

24. a) 71·4%   b) 150 N
25. i) 1·0, 50, 1·5, 75, 1·6, 80   iii) 90

Velocity and acceleration (page 172)

26. b) (iv) 112·5 m   225 m   487·5 m
27. 7·5 km/s^2 1065 km
28. a) 0·2 sec   b) 7·5 cm/s   c) 15 cm/s
h) 0·8 sec
30. a) 200 m   b) 11 sec   c) 80 m
d) 15 sec (10 sec)   e) 8 m/s   g) zero

Force, acceleration and momentum (page 173)

31. a) 3 m/s^2 b) 178·5 m   c) 2·1 m/s^2
d) 4·2 N
32. b) (i) 2 m/s^2 (ii) 10 N   (iii) 20 kg m/s
(iv) 40 J   (v) 160 J
33. a) 800 m/s^2
b) (i) 0·067 sec
(ii) 0·027 m
34. a) 8 kg m/s   b) $2\frac{2}{3}$ m/s   c) 16 J   d) $10\frac{2}{3}$ J

WAVES

Chapter 21 (Waves) page 177
2. 340 m/s   3. 300 000 000 m/s (3×10^8 m/s)
4. a) 2 cm   b) 3 Hz

LIGHT

Chapter 23 (Reflection) page 188
1. a) $60°$ b) $60°$ c) $60°$
3. 2 m behind mirror; 5 cm wide

Chapter 24 (Curved Mirrors) page 193
6. 12 cm from mirror; 2 cm high; real, inverted, diminished   7. a) 24 cm from mirror; 8 cm high; real, inverted, magnified   b) 10 cm behind mirror; 4 cm high; virtual, erect, magnified   c) 5 cm behind mirror; 1 cm high; virtual, erect, diminished

Chapter 25 (Refraction) page 201
9. a) $65°$, $30°$ b) 1·8   10. 4 m

Chapter 26 (Lenses) page 209
6. 7·5 cm from lens; 2 cm high; real, inverted, diminished   7. a) 15 cm from lens; 4 cm high; real, inverted, magnified   b) 6 cm from lens; 3 cm high; virtual, erect, magnified   c) 5 cm from lens; 1 cm high; virtual, erect, diminished

Chapter 27 (Optical Instruments) page 221
3. 10 cm; 20 cm

Chapter 29 (More on Waves) page 237
4. 6×10^{-4} mm (6×10^{-7} m) (600 nm)
5. 4×10^{-4} mm (4×10^{-7} m) (400 nm)

SOUND

Chapter 30 (Sound) page 249
2. 6800 m (6·8 km)   3. 3·4 m
4. a) 20 000 Hz 20 Hz   b) 1·7 cm 17 m
6. 680 m   7. 2250 m   9. 360 m/s
15. a) 100 Hz   b) 400 Hz
16. a) 1·0 m   b) 340 m/s   c) 0·01 m (1 cm)

Further Questions on Waves, Light, Sound

Water waves (page 252)   4. a) 10 cm/s

Curved mirrors (page 253)
15. a) 9·0 cm   b) 2·0 cm
16. (i)   virtual, erect, same size, 15 cm behind mirror
    (ii) virtual, erect, magnified, 30 cm behind mirror
    (iii) virtual, erect, diminished, 10 cm behind mirror
17. b) 30 cm   18. 0·004 m (4 mm)

Refraction (page 254)
19. a) 2×10^8 m/s   b) 4 cm   20. $42°$ 1·49
21. a) $53·1°$ c) $38·7°$

Lenses (page 254)
24. a) 20 cm   b) 80 m
25. 3
26. 8·0 cm, real, inverted, same size, 1·5 cm
    9·3 cm, real, inverted, magnified, 2·0 cm
    4·0 cm, virtual, erect, magnified, 3·0 cm
27. b) 1·1 cm, 3·75 cm, virtual, erect, diminished

Optical instruments (page 255)
28. a) 50 mm   b) 52 mm   c) 1·3 m
31. a) 60 mm   b) 120 mm   c) 3

Sound (page 257)
44. a) (iii) 2 m   46. 11 m   47. b) (ii) 0·4 m
48. a) 150 m   b) 75 m

ELECTRICITY

Chapter 32 (Electrostatics) page 274
6. 5 km (3 miles)

Chapter 33 (More on Electrostatics) page 278
2. a) 10^{-5} F   b) 10^{-4} C (100μC)   c) 20V

Chapter 35 (Circuits) page 299
7. a) 1A, 0·5A, 1A
8. a) 10V   b) 3A   c) 10Ω
9. a) 2A   b) 2A   c) 6V   d) 14V
10. a) 5Ω b) 2A   c) 4V   d) 6V
11. a) 2Ω b) 12V   c) 12V   d) 4A   e) 2A
12. 12V   13. $1·0\Omega$ 14. 2A, 8V, 4Ω

Further Questions on Electricity (1)

Electrostatics (page 301)
15. b) (iii) $800\mu C$

Circuits (page 302)
16. a) 25Ω b) 4Ω 17. 1·2A   5·0A
18. a) 2Ω b) 2A   c) 10V   d) $1\frac{1}{3}$A
19. a) $1·25\Omega$ b) 5V   c) 0·5A   d) 4A   e) $0·75\Omega$
22. 1·95V   23. a) (iii) 2A   b) (iii) $3·2\Omega$
24. a) 20Ω b) $1\frac{2}{3}\Omega$ c) 6 m
26. c) 0.06(4)   d) 0·05

Chapter 36 (Heating Effect) page 310
2. 2A   3. 4A, 5A; 0·83A, 3A; 480W, 3A; 4A, 5A; 2400W, 13A   5. a) 0·25A   b) 960Ω 6. 96p
7. 80p

Chapter 37 (Chemical Effect) page 313
3. 6 g

Chapter 38 (Magnetic Effect) page 327
13. a) 101 mA   b) 10V
14. a) 0·1V   b) $0·101\Omega$ c) 9990Ω
17. a) 40C   b) 80 J
18. a) 30 J   b) 15 W   c) 20 W   d) $\frac{3}{4}$ (75%)

Chapter 39 (Electromagnetic Induction) page 337
7. 1000V a.c.; 10V a.c.; 10; 1000   8. 50 Hz, 14A
10. a) 1000A, 1A

Chapter 40 (Electron Beams) page 345
6. a) 60V   b) 100V, 70V

Further Questions on Electricity (2)

Electric power (page 366)
1. b) (iii) 7200 W
2. a) 5A   b) 8 kWh   c) 48p   3. 27 kWh 162p
4. a) 25·2 MJ ($2·52 \times 10^7$ J)
    b) 3·6 MJ   c) 7   d) 0·7
5. b) (i) 6Ω (ii) $8·3\Omega$ d) 3·0 W
    e) $4°C$
6. b) 2A   c) 36 W   d) 4 W   e) $27°C$

Magnetic effect (page 368)
13. c) 0·9A   d) $1·11\Omega$ 15. 1994Ω in series
16. b) (i) $0·0150\Omega$ in parallel   (ii) 328Ω in series
17. a) $0·0505\Omega$ in parallel
    b) 1995Ω in series
19. b) 600C   c) 6V
20. b) (i) 2A   (ii) 60 000 J (60 kJ)

Electromagnetic induction (page 369)
24. b) 20:1   c) 50 Hz
25. a) (i) 200A   (ii) 4 MW (4×10^6 W)
    (iii) 20A   (iv) 40 kJ (4×10^4 J)
26. a) 12V   b) 2A   c) 16 W
    d) 36 W   e) $66\frac{2}{3}$%

RADIOACTIVITY

Chapter 42 (Radioactivity) page 382
5. b) 1/16   6. 140 years

Further Questions on Radioactivity (page 383)
3. d) 370   e) 12 min   f) 43 min   g) 31 min

Graphs (page 389)
3. c) 98g   d) 26   e) 4·9g
    f) 4·9 g/sheet or 0·20 sheets/g
4. c) 3·9 mm   d) 60   e) 0·11 mm
    f) 0·11 mm/sheet or 9·0 sheets/mm

REVISION QUESTIONS (page 393)
3. a) 750 W   b) 1000W   c) $\frac{3}{4}$ 4. b) 10 m/s
c) 12 s   d) 2 m/s^2 e) 5 m/s^2 f) 30 m   g) 25 m
h) 50 kJ   i) 10 000 kg m/s   j) 2 kN   k) 5 kN
6. a) 5A(10A)   b) 50Ω c) 60p
7. a) 2000 kg   b) 8 400 000 J   c) $1°C$ d) 57Ω
8. a) 50V   b) 30 000 J   c) 300 000 J/kg
9. a) 22·5 cm from the lens; 2·5 cm high; real, inverted, magnified
b) 15 cm from the lens; 5 cm high; virtual, erect, magnified

Dotty Definitions

kilogram	— dangerous record player	principle	— royal tug of war
joule	— fight to the death	inclined plane	— there's writing on the walls of the aircraft
unit	— you do it with wool		
centigrade	— scale for perfumes	velocity	— but we didn't lose the coffee
change of state	— emigration	hertz	— painful
insulate	— getting home after time	amplitude	— well and truly eaten
crankshaft	— lunatic coalmine	dispersion	— or that person
newton	— up-to-date weight	circuit	— for making a teacher?
watt	— question	anode, cathode	— female debtors
power	— hit the lady	armature	— has strong teeth
M.A.	— mother	a.c./d.c.	— I view the ocean
maximum	— large mother	dynamo	— eat briefly

Something to do

For each of these terms give the correct definition or write as much as you can to describe them.

e.g. kilogram: unit of mass; 1 kg = 1000 gram; here on Earth 1 kg weighs 9·8 newton.

Wanted

A reward is offered for information leading to the arrest of Eddy Current, charged with assault and battery on a teenage coil named Milli Amp. He is also wanted in connection with the parallel theft of valuable joules from a bank volt.

Milli Amp tried to run but met series resistance and was overpowered. Later she was found by her friend Dinah Mo. After the couple had had a torque for a moment, Dinah said, "She almost diode but conducted herself well". Milli said, "Anode I would survive but it still hertz. It's enough to make a maltese cross".

Police say that the unrectified criminal escaped from a simple cell where he had been clamped in ions. First he had fused the electrolytes and then squeezed through a grid system. He was almost run to earth in a magnetic field by a line of force, but he has been missing since Faraday.

Watt seems most likely is that he stole an a.c. motor. He may decide to switch it for a megacycle and return ohm by a short circuit.

How many puns can you find? (I can find over 30) Write a sentence to explain the correct meaning of each one.

Index

413

Illustration acknowledgements

11, 12, 13, 258, 399 Central Electricity Generating Board
12 Daily Telegraph Colour Library   13 Vision
International/Louis Salou Explorer   18, 30 Bruce
Coleman   18, 251, 399 Format Photographers   18
Natural History Photographic Agency (NHPA)   19
Marley Waterproofing Products   23 French Tourist
Authority   24, 70 British Railways Board   24 G.
Maunsell & Partners   25 W.D. Hodgson   50 Poultry
World, Electrolux Ltd.   51 Fibreglass Limited   53
Camera Press (OBS) London J. Brown   54 Copper
Development Association, Honda (U.K.) Ltd.   55
Keystone Press Agency, Spanish National Tourist Office
58, 59, 251 Science Photo Library   59 Associated Press
59 British Aerospace   59 J. Allan Cash   70 N.A.S.A.,
Colin Taylor Productions, Air India   73 British
Leyland Ltd.   84 Ministry of Defence   89 Britax
(Wingard) Limited   95, 399 A Shell Photograph   102,
207, 211, 323, 324, 372, Keith Johnson   109
Skyfotos Ltd. for The Pacific Steam Navigation Co.
112 British Rail Hovercraft Ltd.   136 Mike Abrahams/
Network   138 Radio Times Hulton Picture Library
144 Thos. & Jas. Harrison Ltd. – The Harrison Line
147 Leo Mason   154, 165 C.O.I. Photograph   179
Imperial War Museum   181 Taken by W. Shackleton
on Baden-Powell's Expedition (R.A.S. 64)   199 P.W.
Allen & Co., British Airways   200 Tele Focus   211
British Tourist Authority, Colin Taylor Productions,
Greater London Council   213 British Films Limited
216 Rank Audio-Visual   219 Crown Copyright reserved
232, 233 W. Llowarch   234 D.G.A. Dyson   243
Memorex (U.K.) Ltd.   244 Paxman Musical Instruments
Ltd.   250 St. Bartholomew's Hospital   250 Womald
International Sensory Aids   251 Barnaby's Picture
Library   251 National Gallery   258 Electrolux Ltd.,
Rank Radio International, Industrial Magnets Ltd.,
Redditch   262 Institute of Ophthalmology 272, 278, 294,
351, 358 R.S. Components Ltd.   280 Chloride
Automotive Batteries Ltd.   282 H. Tinsley & Co.   307
The Electricity Council   313 W. Canning Ltd.   321
Stanley Power Tools Ltd.   332 G.E.C. Turbine
Generators Ltd.   343 Decca Radar Ltd.   358 Fairchild
Camera and Instrument Corporation   372, 379 Science
Museum, London   373 Cavallerie Ltd. and Bridage
Scientific Instruments   399 Deritend Electrical Services
Ltd., International Computers Ltd., M.O.D.

Acknowledgement is also made to the following
Examination Boards for permission to reprint questions
from their examination papers:

GCE Boards

AEB	Associated Examining Board
C	Cambridge Local Examinations Syndicate
JMB	Joint Matriculation Board
L	University of London
NI	Northern Ireland Examinations Board
O	Oxford Local Examinations Delegacy
O&C	Oxford and Cambridge Examination Board
S	Southern Universities Joint Board
W	Welsh Joint Education Committee
WAEC	West African Examinations Council

CSE Boards

AL	Associated Lancashire Schools Examining Board
EA	East Anglian Examinations Board
EM	East Midlands Regional Examinations Board
Middx	Middlesex Regional Examinations Board
NW	North West Regional Examining Board
SE	South-East Regional Examining Board
SR	Southern Regional Examining Board
W	Welsh Joint Education Committee

WM	West Midlands Examinations Board
WY	West Yorkshire and Lindsey Examining Board
N16+	Joint 16+ Consortium of the Associated Lancashire Schools Examining Board, the Joint Matriculation Board, the West Yorkshire and Lindsey Examining Board, the Yorkshire Regional Examinations Board (and earlier the North West Regional Examining Board)
S16+	Joint 16+ Consortium of the Oxford Local Examinations and the Southern Regional Examinations Board

GCSE Examining Groups

LEAG	London and East Anglian Group
MEG	Midland Examining Group
NEA	Northern Examining Association
SEG	Southern Examining Group
WJEC	Welsh Joint Education Committee

Other books by Keith Johnson

Physics for You Book 1

The first book of the course which covers the core syllabus in a lively and readable way. Book 1 covers Heat, Mechanics, Waves and Sound. It also contains advice on revision techniques and careers using physics.

Physics for You Book 2

The second book of the course. It covers Light, Electricity and Magnetism, and Nuclear Physics. It also contains suggestions for a revision programme, a revision checklist and advice on examination technique.

Multiple Choice Physics for You

This book provides carefully constructed and tested questions which can be used in a number of ways to supplement the course.

* It contains 300 questions grouped into 30 key topics which occur in all physics courses.

* The questions are for use during a course and may be used for classwork or homework, testing or revision, or for practice for examinations.

* Diagrams have been widely used to help the student to understand the questions and appreciate the principles involved.

* Simple numbers have been used in calculations to keep the arithmetic straightforward and to highlight the principles involved.

Timetabling

A complete and practical guide for those responsible for, or interested in, timetabling in schools and colleges. For those with access to a microcomputer the book contains a complete listing of a suite of computer programs which will automatically handle all the time-consuming analysis.

The following computer programs are available on disc.

OPT 1—6 (sorting option choices)
TT 1—6 (pre-timetable checks)
TT 7—9 (scheduling the timetable)
TT 10, 11 (printing out the timetable)
EXAM 1, 2 (analysing exam results)

Have you any comments to make?
Have you got any good cartoons or rhymes?
If you have, why not write to:
Keith Johnson
c/o Pat Rowlinson
Hutchinson Education
62—65 Chandos Place
London WC2N 4NW